Construction Technology

)8

)9

Construction Technology: an illustrated introduction

Eric Fleming

Former Lecturer
Construction Economics and Building Construction
Department of Building Engineering and Surveying
Heriot-Watt University

Blackwell
Publishing

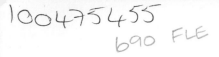

Editorial offices:
Blackwell Publishing Ltd, 9600 Garsington Road, Oxford OX4 2DQ, UK
   Tel: +44 (0)1865 776868
Blackwell Publishing Inc., 350 Main Street, Malden, MA 02148-5020, USA
   Tel: +1 781 388 8250
Blackwell Publishing Asia Pty Ltd, 550 Swanston Street, Carlton, Victoria 3053, Australia
   Tel: +61 (0)3 8359 1011

First published 2005 by Blackwell Publishing Ltd

Library of Congress Cataloging-in-Publication Data

Fleming, Eric.
   Construction technology / Eric Fleming.
      p.   cm.
   Includes index.
   ISBN 1-4051-0210-1 (pbk. : alk. paper)
   1. Building.   I. Title.

   TH146 .F58   2004
   690–dc22   2004008229

ISBN 1-4051-0210-1

A catalogue record for this title is available from the British Library

Set in 10/12 pt Palatino
by TechBooks
Printed and bound in India
by Replika Press Pvt. Ltd., Kundli

The publisher's policy is to use permanent paper from mills that operate a sustainable forestry policy, and which has been manufactured from pulp processed using acid-free and elementary chlorine-free practices. Furthermore, the publisher ensures that the text paper and cover board used have met acceptable environmental accreditation standards.

For further information on Blackwell Publishing, visit our website:
www.thatconstructionsite.com

Contents

Introduction   xi
Acknowledgements and Dedication   xiii
Abbreviations   xiv

1   Masonry Construction in Bricks and Blocks   **1**

Bricks and blocks standards and dimensions   2
Bricks   2
Terminology   2
Brick sizes   2
    Nominal sizing   3
Durability of bricks   3
Mortar joints   3
Coordinating sizes   3
Types of brick by shape   4
Kinds of brick by function   4
Brick materials   5
Testing of bricks   5
The bonding of bricks to form walls   5
Convention on thicknesses of walls   8
Types of bond   9
    Vertical alignment   14
    Honeycomb brickwork   16
    Quoins – an alternative definition 16
    Half brick thick walls   16
    Frog up or frog down   17
'Tipping'   17
Common and facing brickwork   18
Facing brickwork   18
Pointing and jointing   19
General principles of bonding   21
Blocks   22
Block materials   22
Concrete blocks   22
    Dense and lightweight concretes   23
    Autoclaved aerated concrete   23
Dimensions of standard metric block   23
Whys and wherefores of mortar   25
    Cement   25
    Lime   26
    Sand   27
    Water   27
Which mortar mix?   27
    'Fat' mixes   28
    General rules for selection of mortar 29

Mortar additives 30
    Mixing in additives 30
    Mixing mortar 31
Good or bad weather 32

2 Substructures 34

Excavation generally 34
Topsoil 35
Subsoils 36
    General categorisation of subsoils and their loadbearing capacities 37
Foundations 37
    The principal considerations 38
    Simple foundation calculations 39
    The mass of buildings 39
    Mass, load and bearing capacity 40
    Foundation width and thickness 41
    Reinforced concrete foundations 44
    Failure of wide, thin, strip foundations 44
    Trench fill foundations 45
    Critical levels and depths 46
    Level 46
    Finished ground level 47
    Bearing strata 48
    Depths and levels 48
    Step in foundation 49
Setting out 49
    The site plan 49
    Where do we put the building? 49
    Equipment required for basic setting out 49
    Setting out procedure 50
Excavation 53
    Marking out the excavation 53
    Excavation for and placing concrete foundations – and not
        wasting money doing it 53
Building masonry walls from foundation up to DPC level 57
Ground floor construction 59
    Detail drawings 59
Wall–floor interfaces generally 62
    Precautions 62
Solid concrete floors 62
    Single and double layer concrete floors with hollow masonry wall 62
Hung floors 64
    Hung timber floors 64
    Hung timber floor alternatives 66
    Hung concrete floors 67
Blockwork substructure 71

3 Walls and Partitions 73

General 73
    Requirements 74
Walls – environmental control 75
    Heat loss and thermal capacity 75

Resistance to weather – precipitation   75
Air infiltration   77
Noise control   79
Fire   79
Dimensional stability   79
Walls of brick and blockwork   81
Insulation of external walls   84
Timber frame construction   88
Traditional timber frame   88
Modern timber frame construction   91
Loadbearing and non-loadbearing internal partitions   96
Expansion joints   99

4   Timber Upper Floors   103

Upper floor joists   103
Linear and point loadings on upper floors   112
Openings in upper floors   113
For pipes   113
For flues   114
For stairs   116
Alternative materials for joisting   118
Sound proofing   120
Modern sound and fire proofing   121
Support of masonry walls   123
Floor finishes   124
Ceiling finishes   124

5   Openings in Masonry Walls   126

For small pipes and cables   126
For larger pipes and ventilators   127
Large openings in masonry walls   127
Alternative sill arrangements   136
Threshold arrangements   137
Partitions of masonry   139
Openings in timber frame walls   141

6   Roof Structure   148

Roof classifications   148
Prefabrication   149
Trussed roofs   150
The trussed rafter   153
Verges meet eaves   159
Roof bracing   160
Flat roofs in timber   162
Insulation, vapour control layers and voids and ventilation   164
Traditional roofs   167
Roof insulation   169

7   Roof Coverings   171

Tile and slate materials   171
Slates   175

Plain tiles   175
Interlocking tiles   176
Timber shingles   176
Bituminous shingles   176
Pantiles   177
Spanish and Roman tiles   177
Edges and abutments   178

8   Doors   **182**

Functions of doors and windows – obvious and not so obvious   182
Types of door   184
   Ledged and braced doors   185
   Bound lining doors   185
   Flush panel doors   187
   Panelled doors   188
   Pressed panel doors   189
   15 pane doors   190
Hanging a door   190
Fire resistant doors   193
   Smoke seals   195
   Glazing   196
Ironmongery   196

9   Windows   **204**

Timber casement windows   205
   Depth and height of glazing rebates   206
   Timber for casement windows   206
   Draught stripping materials   206
   Hanging the casements   207
   Joining the frame and casement members   209
Timber sash and case windows   211
   The case   212
   The sashes and case together   214
Vertical sliding sash windows   214
Glazing   218
   For ordinary glazing work   218

10   Stairs   **221**

Landings   222
Steps   222
Balustrades   223
Measurements   224
Joining steps to stringer   225
Winders   227

11   Mutual Walls   **228**

Transmission of sound   228
Calculation of surface density   228
Wall types   229
Fire resistance   231

12   Plumbing and Heating   233

    Pipework   233
    Pipe fittings – couplings and connections   234
        Range of fittings   239
    Valves and cocks   241
    Services generally   243
    Hot and cold water services   243
    Soil and ventilation stacks   246
    Overflows   246
    Water supply from the main   246
    Equipment   247
        Cold water storage cisterns   248
        Hot water storage cylinders   248
        Feed and expansion tanks   251
    Central heating   252
        Piping for central heating systems   253
        Emitters   255
    Appliances   255
    Waste disposal piping and systems   259
    Insulation   262
    Corrosion   263
    Air locking and water hammer   263
    First fixings   264

13   Electrical Work   266

    Power generation   266
    Wiring installation types   267
    Sub-mains and consumer control units   268
    Sub-circuits   270
    Work stages   272
        Electrician's roughing   272
        Earth bonding   273
        Final fix   275
    Testing and certification   275
    More on protective devices   275
    Wiring diagrams   276
    Accessories   277

Appendices:

A   Maps and Plans   279

B   Levelling Using the Dumpy Level   285

C   Timber, Stress Grading, Jointing, Floor Boarding   291

D   Plain and Reinforced In-situ Concrete   316

E   Mortar and Fine Concrete Screeds laid over Concrete Sub-floors or Structures   322

F   Shoring, Strutting and Waling   325

G   Nails, Screws, Bolts and Proprietary Fixings   328

H   Gypsum Wall Board   341

I   DPCs, DPMs, Ventilation of Ground Floor Voids, Weeps   344

J   Drawing Symbols and Conventions   353

K   Conservation of Energy   355

L   Short Précis of Selected British Standards   356

Index   380

Introduction

One of the many reasons for writing this book was the need to introduce students to a level of detail which they would gain only with practical experience on site or in workshops. The accusation that the text includes too much 'trade' material could be levelled, but bearing in mind that many of the students who might use this text will be potential builders, quantity surveyors and building surveyors, then the inclusion of the trade material is very necessary. One of the primary functions of certainly the builders and quantity surveyors is the need to be able to assess the cost of any building operation. Unless they understand the processes to be gone through it is impossible for these professionals to give an accurate cost. They don't have to be able to physically do the work but they must know exactly what is involved. So this text is for the 'early learner' who has no background in the construction industry. It is *not* intended to be an all embracing text; the physical size of the book could not allow that. So the author has been quite selective in what has been included, the reasoning behind the selection being the need to introduce the early learner to sufficient information to allow a general appreciation of the more common techniques used in domestic construction today.

Emphasis has been given to technical terms and terminology by having them printed in bold on at least the first occasion they are used. Where these terms are generally confined to one part of the UK, some alternative forms are given as well. References to Building Regulations should be understood to mean *all* the Regulations which are used in England, Wales and Scotland at the time of writing. References to particular Regulations will have the suffixes (England and Wales) or (Scotland) appended. Where the reference is to earlier editions of any particular Regulations, the date will be given, e.g. (1981).

A word about the drawings scattered through the text. None is to scale although, in the majority of instances, all component parts and components shown in any one drawing are in the correct proportion, with the exception of thin layers or membranes such as damp proof courses, felts, etc. which are exaggerated in thickness, following the convention in architectural drawing practice. Appendix J shows some of the conventional symbols used. The reader should get to know these; they are common currency when drawn information has to be read.

For the student who has recently left school there may be confusion, for the teaching of the use of centimetres in schools does not match up with the agreement by the construction industry to use *only* SI (Système International) units where only the millimetre, metre and kilometre are used to measure length. On architectural drawings dimensions are given only in millimetres and levels in metres to two places of decimals. Students will be expected to produce drawings in this manner during their courses. Following the convention on drawings etc., no mention of the unit of measurement will be made in the text when these are in millimetres. Any dimension given simply as a number must be assumed to be in millimetres. Any other measurements will have the unit of measurement following the number, e.g. 14.30 m meaning metres; 10 600 kN meaning kilonewtons and so on.

There are already hundreds of books on building construction or on just one aspect of it, be it a trade, material or technique(s). There must be many more technical papers and leaflets and books produced by various organisations with an interest in the industry.

They include the Building Research Establishment (BRE), Construction Industry Research and Information Association (CIRIA), Timber Research and Development Association (TRADA), the British Standards Institution, all the trade and manufacturing associations – the list is endless, but those mentioned are reckoned to be the experts. So why has this author chosen not to quote them at every opportunity? Well, I have quoted bits of the British Standards where they were appropriate, but so much of the rest of the material is on a higher plane as to confuse the early learner in the art of construction. There is enough in here to get someone started on domestic construction as it is today. Get that correct and then go on to read the more esoteric material, especially when so much is about what has gone wrong in the past and how it was put right.

A couple of areas which are sorely neglected by too many students are:

□ Manufacturer's literature – now widely available on the Internet
□ Using their own eyes.

On the first point above, there was a time not so long ago when manufacturers tended to have their literature about a product prepared by graphic artists who knew diddley-squat about building and so perpetrated some real howlers and horrors, so much so that many lecturers had to tell students to ignore that source of information until they could sort out the good from the ugly. There were notable exceptions and many will remember the competition to get hands on a copy of British Gypsum's White Book or the reception given to Redland's award winning catalogue on roofing materials – goodness, was it that long ago? Nowadays catalogues have to be considered as a serious source of information and they come out faster than any other form of information and so become almost the only way to keep up to date.

On the second point above, what better way to see how a wash hand basin is installed than to get underneath it with a good torch and have a good look. Look into the attic with that torch, probe into all the corners and see how the roof is put together. Look at the doors and windows and how they interface with the walls and floors and the ceilings.

Experience in teaching the subject to school leavers has brought one difficulty to the fore which many students have – the inability to visualise. Test this for yourself – describe something to a friend and ask them to draw it as you speak. I'm sure you'll get some funny results and some funny comments. It is a daunting task to be faced with technical construction drawings, especially detail drawings, and be expected to 'see' what is going on in terms of bricks, concrete in holes in the ground, joists and plasterboard, especially as you don't know what these are in their raw state. Hence the inclusion in this book of photographs of bits and pieces and of construction. Fewer and fewer students get the opportunity to see a building site, mainly due to the safety aspects of a site visit and ever increasing insurance premiums. And yet seeing for themselves is what so many desperately require.

When starting this book a year or two back, the idea was to include a detail drawing alongside a photograph of what it looked like on site, hoping that this would in some small way make up for lack of on-site experience. While there are a lot of photographs, the result is not as good as had been hoped. The author could easily spend another year just getting the photography up to scratch and would certainly do things differently. For this text I was unable to find herringbone strutting anywhere close to me so I made a mock-up of a pair of joists and put in timber and steel strutting. It makes the point adequately when viewed alongside the details. So many other photographs could have been of that type had I realised the value of mock-ups earlier.

If you think the book lacks something or has too much of one thing, or is a bit of a curate's egg or whatever, please write to me care of the publishers. If there is ever another edition it would be good – indeed vital – to have constructive feedback.

Eric Fleming

Acknowledgements and Dedication

I must formally thank Mitek Industries of Dudley and Eaton MEM of Oldham for giving me permission to reproduce images and providing the images on disc to include in this book. Also Simpson Strong Tie and their branch at Stepps near Glasgow who very kindly supplied me with samples. I must thank John Fleming & Co Ltd, timber and builders merchants of Elgin, Keith Builders Merchants and Mackenzie and Cruickshank, hardware retailers, both of Forres, who all allowed me to take photographs of materials, components and ironmongery.

I must also thank the many family, colleagues and friends from the half century I spent in the construction industry who have contributed to the information on which I have drawn so freely in writing this text.

Finally, I would like to dedicate this book to Myra, who has given me great encouragement with the writing and who has not complained when the book came between me and the renovation work we are attempting on our battered Georgian home.

F. W. 'Eric' Fleming FRICS
Forres
Scotland
March 2004

Abbreviations

aac	autoclaved aerated concrete
ABS	acrylonitrile butadiene styrene
ach	air changes per hour
bj	black japanned
BM	benchmark
BMA	bronze metal antique
BOE	brick on edge
BRE	Building Research Establishment
BS	British Standard
BSI	British Standards Institution
CAAD	computer aided architectural design
CAD	computer aided design
CCU	consumer's control unit
CH	central heating
CIRIA	Construction Industry Research and Information Association
cs	centres
csk	countersunk
CW	cold water
DLO	direct labour organisation
DPC	damp proof course
DPM	damp proof membrane
ELCB	earth leakage circuit breaker
EPDM	electronic position and distance measurement
EVA	ethyl vinyl acetate
FFL	finished floor level
FGL	finished ground level
FS	full sheet
galv.	(hot dipped) galvanised
HBC	high breaking capacity
H&C	hot and cold
HRC	high rupturing capacity
HW	hot water
IEE	Institution of Electrical Engineers
LH	left-hand
LPG	liquefied petroleum gas
MC	moisture content
MCB	miniature circuit breaker
MR	moisture resistant
m.s.	mild steel
m&t	mortice and tenon
OPC	ordinary Portland cement
OS	Ordnance Survey
OSB	oriented strand board
PCC	pre-cast concrete
PFA	pulverised fuel ash
PS	pressed steel
PTFE	polytetrafluorethylene
PVA	polyvinyl acetate
RCCB	residual current circuit breaker
RH	right-hand
rh	round head
RSJ	rolled steel joist
RWP	rainwater pipe
SAA	satin anodised aluminium
SLC	safe loadbearing capacity
SS	stainless steel
SSHA	Scottish Special Housing Association
SVP	soil and ventilation pipe
S/w	softwood
SWVP	soil, waste and ventilation pipe
t&g	tongue and groove
TC	tungsten carbide
TRADA	Timber Research and Development Association
TRV	thermostatic radiator valve
UB	universal beam
UV	ultraviolet
VCL	vapour control layer
WHB	wash-hand basin
WBP	water and boil proof
zp	zinc plated

1 Masonry Construction in Bricks and Blocks

Bricks and blocks standards and dimensions	2	Facing brickwork	18
Bricks	2	Pointing and jointing	19
Terminology	2	General principles of bonding	21
Brick sizes	2	Blocks	22
Nominal sizing	3	Block materials	22
Durability of bricks	3	Concrete blocks	22
Mortar joints	3	Dense and lightweight concretes	23
Coordinating sizes	3	Autoclaved aerated concrete	23
Types of brick by shape	4	Dimensions of standard metric block	23
Kinds of brick by function	4	Whys and wherefores of mortar	25
Brick materials	5	Cement	25
Testing of bricks	5	Lime	26
The bonding of bricks to form walls	5	Sand	27
Convention on thicknesses of walls	8	Water	27
Types of bond	9	Which mortar mix?	27
Vertical alignment	14	'Fat' mixes	28
Honeycomb brickwork	16	General rules for selection of mortar	29
Quoins – an alternative definition	16	Mortar additives	30
Half brick thick walls	16	Mixing in additives	30
Frog up or frog down	17	Mixing mortar	31
'Tipping'	17	Good or bad weather	32
Common and facing brickwork	18		

Why should we be starting a book on building construction with a discussion of bricks and blocks? Quite simply because bricks are one of the major construction materials instantly associated with construction in the mind of the novice or lay person, but more importantly because the sizes chosen for the manufacture of bricks and blocks affect practically everything in a building except the thickness of the coats of paint or the coats of plaster. This will be discussed in more detail as we proceed.

Bricks and blocks are entirely 'man-made' masonry units. A variety of materials are quarried, mined or salvaged from manufacturing processes and made into bricks or blocks.

Stone is quarried and shaped but occurs naturally and was often used as it was found below cliffs or outcrops or on beaches, or from the general stones on or in the ground.

Artificial stone and reconstructed stone are 'man-made'. Artificial stone is made by mixing particles of stone with a cement binder, water and occasionally a colouring material and then casting it into shapes. The idea is to create a 'look' of a particular kind of stone, even though none of that stone is used in the production. Reconstructed stone follows the same idea but generally omits the colouring agents since the stone particles used are the stone which is required at the end of the casting process. This is sometimes cheaper than

Fig. 1.1(a) Solid and perforated bricks.

Fig. 1.1(b) Single shallow frogged bricks.

the original stone and can sometimes be the only way to produce any quantity of something closely resembling the original stone where quarries are run down or closed.

We will consider only bricks and blocks. In Figure 1.1(a) there are two solid bricks on the left and two perforated bricks on the right. In Figure 1.1(b) there are two single shallow frogged bricks. There is obviously need for explanation so we will start by looking at materials, sizes and shapes and so on:

- Bricks and blocks can be made from a variety of **materials** other than fired clay or brick earth, e.g. calcium silicate and concretes.
- Bricks and blocks can be obtained in a variety of **sizes** and **types** and **kinds**.
- Bricks and blocks can be made in a variety of **shapes** other than the standard rectilinear shape discussed in this text but special shapes are the subject of British Standard 4729, *Dimensions of bricks of special shapes and sizes*.

- Bricks and blocks can be **cut** into different shapes and these we will discuss later in the chapter.

Bricks and blocks standards and dimensions

- Bricks and blocks of fired clay are the subject of British Standard BS 3921 (see précis in Appendix L).
- Brick is defined as a unit having all dimensions less than 337.5 × 225 × 112.5.
- Block is defined as a unit having one or more dimensions greater than those of the largest possible brick.

BRICKS

Terminology

The surfaces of a brick have names:

- Top and bottom surfaces are **beds**
- Ends are **headers** or **header faces**
- Sides are **stretchers** or **stretcher faces**.

Bricks and blocks are made using **mortar**; they are *not* made in cement. Cement, usually a dry powder, may or may not be an ingredient in a mortar depending on the type of wall, its situation, etc. Mortars are mixed with water into a plastic mass just stiff enough to support any masonry unit pressed into them. This is an important and fundamental issue which is discussed in detail a little later in the chapter.

Brick sizes

Bricks are made in many sizes; however, we will use only one size in this text – the standard metric brick. A standard metric brick has **coordinating dimensions** of **225 × 112.5 × 75 mm** and **working dimensions** of **215 × 102.5 × 65**.

Why two sizes? The coordinating dimensions are a measure of the physical space taken up by a brick together with the mortar required on one bed, one header face and one stretcher face. The working dimensions are the sizes to which manufacturers will try to make the bricks. Methods of manufacture for many units and components are such that the final piece is not quite the size expected but it can fall within defined limits. This can be due to things like shrinkage or distortion when drying out, firing, etc.

The difference between the working and co-ordinating dimensions of a brick is 10 mm and this difference is taken up with the layer of mortar into which the bricks are pressed when laying. The working dimensions are also known as the **nominal size** of a brick.

Nominal sizing

The term nominal sizing is used to describe a size which is subject to slight variation during the manufacture of a component or unit. The variation – larger or smaller – allowed is generally given in a British Standard. The differences – plus and/or minus – can be different.

The slight variation in size of individual bricks is allowed for by pressing the brick into the mortar layer a greater or lesser amount but always using up to the coordinating dimension or space of $225 \times 112.5 \times 75$.

Durability of bricks

Durability of bricks is very important when building in situations where freezing would be a problem and where the soluble salt content of the bricks would cause problems with the mortar – see sulphate attack later in the text. British Standard 3921 gives classifications of durability in terms of frost resistance and salt content, and an extract from the précis of BS 3921 given in Appendix L of the book is given here:

Durability of brickwork depends on two factors which arise from the use of any particular brick: resistance to frost and the soluble salts content. Frost resistance falls into three classes: Frost resistant (F), Moderately

Frost Resistant (M) and Not Frost Resistant (O). Soluble salts content is classed as either Low (L) or Normal (N). So, one could have a brick which is frost resistant with normal soluble salt content and this would be classified as FN. Similarly a brick which had no frost resistance and had low soluble salt content would be classed as OL.

Mortar joints

Mortar placed horizontally below or on top of a brick is called a **bed**. Mortar placed vertically between bricks is called a **perpend**.

Coordinating sizes

☐ The coordinating sizes allow the bricks to be built together in a number of different ways, illustrated in Figure 1.2. It is important to build brickwork to the correct coordinating size for the particular working size of brick specified.

Other components such as **cills**[1], **lintels**[2], door and window frames, etc. are manufactured to fit into openings whose size is calculated on the basis of whole or cut bricks displaced. This is illustrated in Figure 1.3. If non-metric sizes of brick are to be used then the components built into the brickwork should coordinate with that size.

☐ The height of the **lintel** is shown as three courses plus the joints between them measuring $3 \times 65 + 2 \times 10 = 215$ mm.
☐ The width of the window opening must be a multiple of half a brick plus the perpends, e.g. $8 \times 102.5 + 9 \times 10 = 880$ mm.
☐ The length of the lintel has to be the width of the opening plus the pieces which are built into the wall – the **rests**.

[1] **Cill**: Alternative spelling, sill, is a unit or construction at the bottom of a window opening in a wall designed to deflect water running off a window away from the face of the wall below.
[2] **Lintel**: A unit or construction over an opening in a wall designed to carry the loadings of the wall over the opening.

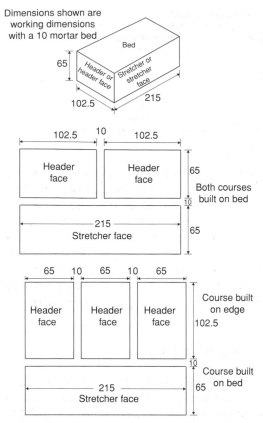

Fig. 1.2 Coordinating and working dimensions of the standard metric brick.

☐ Wall rests vary according to the load but assume in this case they are half the length of a brick each, less the mortar required in the perpends between the lintel ends and the adjacent brickwork.

☐ The length of the lintel is therefore $880 + 2 \times 102.5 = 1085$

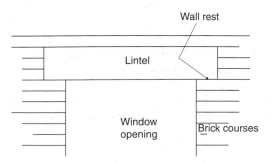

Fig. 1.3 A lintel over a window opening in a brick wall.

Types of brick by shape

Bricks can just be rectilinear pieces of material, and these are described as **solid**, but they might instead have a depression in one or both beds called a **frog**. Frogs can be quite shallow or quite deep but they will not exceed 20% of the volume in total.

Instead of being solid or having a frog(s), a brick might be:

☐ Cellular – having cavities or depressions exceeding 20% of the volume in total, *or*
☐ Perforated – holes not exceeding 20% of the volume in total; minimum 30% solid across width of brick.

All of these types by shape are illustrated in Figure 1.4.

Kinds of brick by function

Bricks can be manufactured to fulfil different functions, i.e. strength, resistance to water

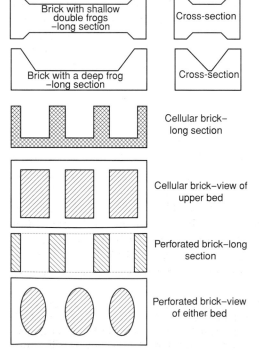

Fig. 1.4 Types of brick by shape.

Table 1.1 Brick types by end use – compressive strength and water absorption.

Type by end use	Compressive strength N/mm²	Water absorption %
Engineering A	≤ 70	≤ 4.5
Engineering B	≤ 50	≤ 7.0
Damp proof course 1	≤ 5	≤ 4.5
Damp proof course 2	≤ 5	≤ 7.0
All other	≤ 5	No limits

absorption, decoration or for no particular function other than to build a wall and be covered over with plaster or render.

The British Standard recognises five types classified by function or end use, as shown in Table 1.1.

The vast majority of bricks used are of the 'All other' category, the compressive strength being perfectly adequate for all but the most severe loadings. The 'No limits' category for water absorption must of course be tempered with any requirement to resist weather penetration of a wall. It would be foolish to build an external wall of facing brick if these bricks were very absorbent. On the other hand a water absorption less than required by damp proof course (DPC)[3] or engineering bricks would be an unnecessary expense and might even be counter productive in the long term. We will look at the effect of water absorption on wall faces in Chapter 3.

Brick materials

As well as fired clay or brick earth, bricks are also manufactured from **calcium silicate** (BS 187 and 6649) and **concrete** (BS 6073). Standard metric-sized bricks are manufactured in both kinds. Shapes of concrete bricks differ from those in clay and calcium silicate. Different strengths apply to all kinds of bricks. In parts of the UK a naturally occurring mixture

[3] The damp proof course is most commonly known by the initials DPC. It is a layer impervious to water built into walls and floors to prevent moisture in the ground rising into the structure of a building. It will be discussed in greater detail in Chapter 2 and Appendix I.

of clay and coal has been quarried or mined. This clay can be formed into bricks and fired in traditional brick kilns but uses less fuel because of the entrained coal particles. In Scotland the resulting bricks are known as **composition bricks**. They display a rather burnt look on the outer faces and when cut the core is frequently quite black. They are only used for common brickwork (see later is this chapter).

Testing of bricks

All brick kinds and types are subject to test in order to comply with the appropriate British Standard. Tests include dimensions, soluble salt content, efflorescence, compressive strength and water absorption. Please refer to the précis of BS 3921 in Appendix L where the tests are listed but not discussed in detail. The reasons for these tests will be discussed in Chapter 3.

The bonding of bricks to form walls

Bonding of bricks refers to the practice of laying the bricks in layers or **courses** and in any of a number of patterns or **bonds** to form a wall of a homogeneous construction, i.e. the individual bricks overlap each other in adjacent layers, the pattern alternating in adjacent layers or after a number of similar layers. The patterns in these layers are formed with whole and cut bricks as well as with bricks manufactured to a 'special shape' other than the standard rectilinear one.

We must first examine why we need to cut bricks in specific ways. The most simple cut

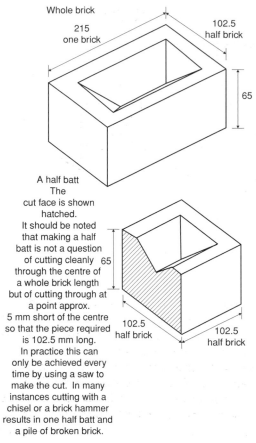

Fig. 1.5 Whole brick and half batt dimensions.

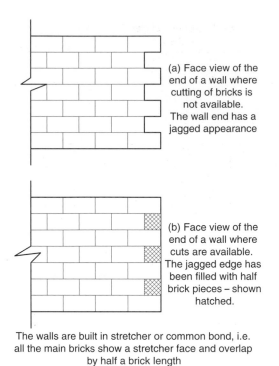

(a) Face view of the end of a wall where cutting of bricks is not available. The wall end has a jagged appearance

(b) Face view of the end of a wall where cuts are available. The jagged edge has been filled with half brick pieces – shown hatched.

The walls are built in stretcher or common bond, i.e. all the main bricks show a stretcher face and overlap by half a brick length

Fig. 1.6 Why bricks are cut.

is the half brick, which can be described as cutting a brick along a plane vertical to its bed, along the centre of its short axis. Figure 1.5 shows the dimensions of a whole brick and below it the dimensions of a brick cut in half – a **half batt**. Note that the half batt is not a true half of a brick length as there must always be an allowance made for the thickness of the mortar used in building, 10 mm.

This most simple of cuts makes the building of walls with straight, vertical ends possible. Figure 1.6 shows this quite clearly and more simply than words can describe. The wall is built in **stretcher bond**, i.e. the bricks are laid with their long axis along the length of the wall and the bricks in adjacent courses overlap each other by half a brick. The views shown are, (a) without using cut bricks and (b) using cut bricks. Not being able to use cut bricks

means that the end of the wall takes on a saw-toothed appearance, which is not the case if half brick cuts are available.

Another bond which is commonly used is **English bond**, more complex than stretcher bond and only used where the wall thickness is 215 mm thick – one brick or over. If we look at Figure 1.7, a drawing of two adjacent courses of this bond in a particular situation, we can see how another of the standard cuts can be put to good effect in maintaining the bonding of the bricks as well as allowing vertical ends to the walls. When we start to look at walls of one brick thick and upwards, a further development of bonding comes into play. Every bond in this category displays a grouping of bricks which repeats across a course. In some instances the pattern repeats across every course, in others adjacent courses display a mirror of that pattern. In one-brick walls in English bond the pattern is of two bricks, side by side and which are turned through 180°. This is called **sectional bond** and is shown hatched in all the figures which follow.

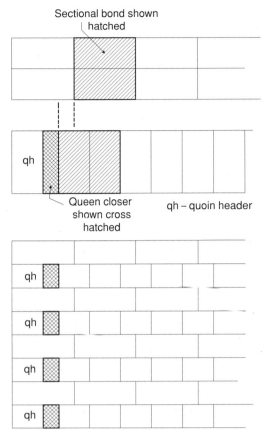

Fig. 1.7 The use of a closer in a one brick thick wall built in English bond.

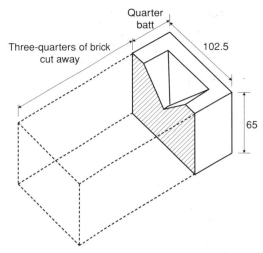

Fig. 1.8 Quarter batt.

pieces of brick cut into quarter, half or three-quarter lengths are referred to as **batts**, and other 'named' cuts as **closers**. They are all illustrated in Figures 1.8 to 1.13.

We will find out shortly in this chapter where and when to use these cuts when we look at how the various bonds are laid out course-by-course and situation-by-situation. What we must ask ourselves now is 'why do we bother to bond in any particular way at all?' Anyone who has played with old fashioned wooden blocks or their modern

The drawing shows the two adjacent courses at the end of a wall as well as the face view of the wall over a number of courses. The cut illustrated is called a **queen closer** and is formed by cutting a brick in a plane vertical to the bed and along the centre line of the long axis of the brick. Note that there is a whole brick in alternate courses at the end of the wall next to the queen closer. This brick is referred to as the **quoin header**. We will explain the terminology properly as we look at particular bonds and bonding in more detail.

Having established the need for cut bricks, let us look at the first of the standard cuts taken from a whole metric brick, beginning with that shown in Figure 1.8.

Having cut the brick in half, the bricklayer can also now cut it into quarters. Note that

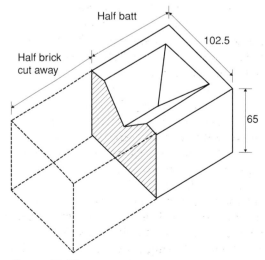

Fig. 1.9 Half batt.

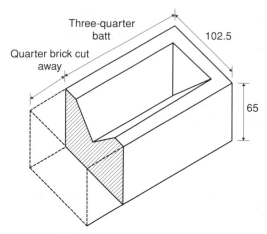

Fig. 1.10 Three-quarter batt.

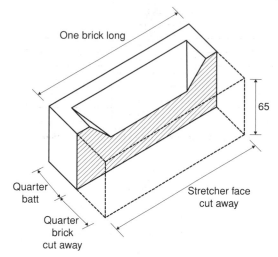

Fig. 1.12 Queen closer.

equivalent, Lego, will understand the need to connect vertical layers of these units together to increase the stability of an increasing height of built work. The broader the base and the more comprehensive the bonding of the layer, the higher the structure can be made. So it is with brick and blockwork walls. As we build we overlap the bricks in adjacent layers, thus ensuring that there is never a complete vertical layer joined only to the remainder of the structure with a layer of mortar.

The second criterion is to spread any vertical loading on a part of the wall to an ever-widening area of brickwork, thus dissipating the load. This is illustrated in Figure 1.14.

Convention on thicknesses of walls

Thicknesses of walls are not generally given in millimetres but in **multiples** of '**half a brick**' (which of course equals **102.5** mm) *plus* the intervening thickness of mortar (**10** mm) where appropriate. So a '**half brick**' thick wall would be **102.5** mm thick, a '**one brick**' thick wall

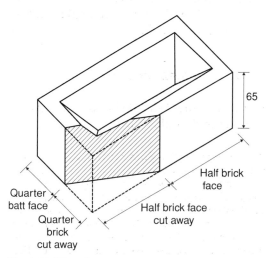

Fig. 1.11 King closer.

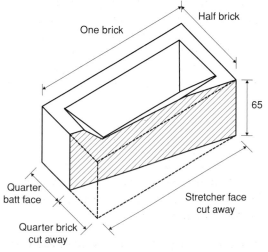

Fig. 1.13 Bevelled closer.

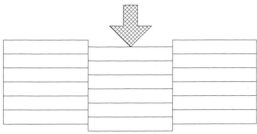

Column of bricks not bonded to wall either side

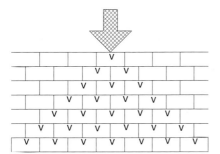

Transfer of load to more bricks in courses when
wall is fully bonded

Fig. 1.14 Loadings on walls which are not bonded and
those which are bonded.

215 mm thick, a '**one and a half brick**' thick
wall would be **327.5** mm thick, and so on[4].

This is a particularly helpful device when
drawing details, plans and sections, etc. as
well as in the preparation of contract documentation such as specifications of workmanship and material and bills of quantities. Many
of those documents can be pre-prepared in
standard form if this terminology is adopted
when describing brickwork, and then by the
simple addition of a statement regarding the
size of the bricks to be used the whole becomes
related to that simple, short statement of the
size.

Types of bond

We will illustrate bonds by showing what
takes place at three important points in any
wall construction:

[4] Note that a mortar joint of 10 mm is always added except with half brick thick walls.

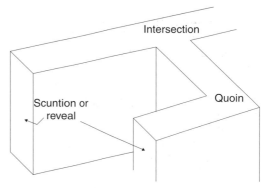

Fig. 1.15 Parts of a wall – scuntion, intersection and
quoin.

□ The end of the wall, called a **scuntion** or
reveal[5]. You may see an alternative spelling
of scuncheon.
□ The junction of two walls at right angles,
called an **intersection**.
□ The right-angled corner of a wall, called a
quoin.

Figure 1.15 illustrates this terminology.

For each bond we show two adjacent
courses. Where necessary explanatory notes
will follow each figure. Before beginning a
description of bonds it should be noted that
this is not a comprehensive list of bonds or
of situations. When looking at other texts the
reader will find numerous variations on the
bonds discussed here, additional bonds and
other situations which include intersections
and quoins at other than right angles and of
walls of different thicknesses.

Stretcher or common bond

The first and most simple bond, stretcher or
common bond, is illustrated in Figure 1.16.

[5] Mitchell's Construction series refers to this as a 'stopped
end' but this is an incorrect use of that term. A moulding, chamfer or other shape is frequently formed on the
edge(s) of timber, stone plaster etc. These shapes are
called 'labours' because their formation only involves
labour making the shape. Occasionally these labours
do not run the full length of the item in which they are
formed but are machined to form a definite 'stop'. This
is the proper use of the term 'stopped end'.

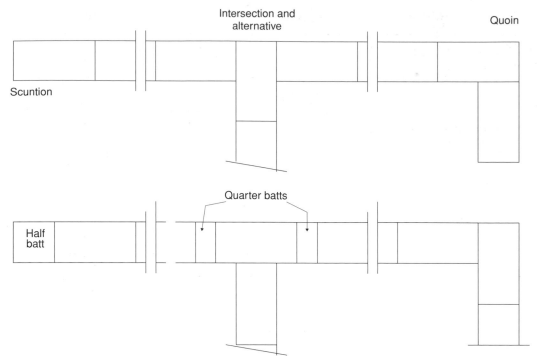

Fig. 1.16 Half brick thick wall in stretcher or common bond.

□ Only used for walls of **half brick** thickness, this is the only practical bond which can be used on a wall of this thickness, although we can build 'mock' bonds of other kinds. A little of that later.

□ Only shows stretchers on general face except for occasional closers and half batts used to maintain bond at quoins, scuntions and intersections.

Walls we will consider now will vary in thickness from 215 mm to 327.5 mm thick – 1 to $1^1/_2$ bricks thick.

We will begin with English bond and continue with Flemish bond, Scotch bond and garden wall bond. Finally we will show Quetta and Rattrap bonds, which are always 327.5 and 215 mm thick respectively. **Sectional bond** is shown hatched in all the figures which follow.

English bond (Figures 1.17 and 1.18)

□ The strongest bond

□ This bond maximises the strength of the wall

□ It is used on walls one brick thick and upwards

□ Note how the sectional bond changes as the wall increases in thickness.

Pattern on the face of the wall shows distinctive courses of headers and stretchers.

Flemish bond (Figures 1.19 and 1.20)

□ Not such a strong bond as English bond

□ It is used on walls one brick thick and upwards

□ Note how the sectional bond changes as the wall increases in thickness.

Decorative pattern on face of wall shows alternate headers and stretchers in each course with the headers centred under and over stretchers in adjacent courses.

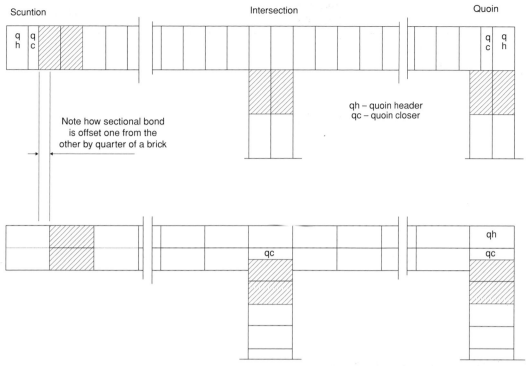

Fig. 1.17 One brick thick wall in English bond.

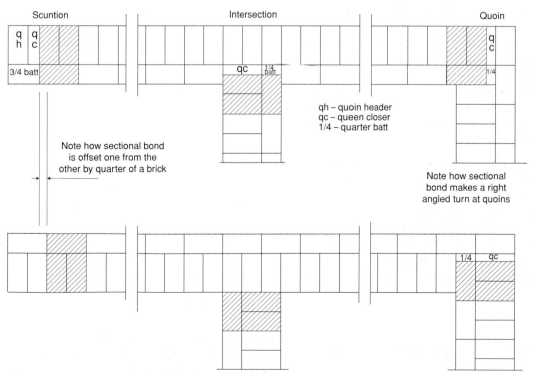

Fig. 1.18 One and one half brick thick wall in English bond.

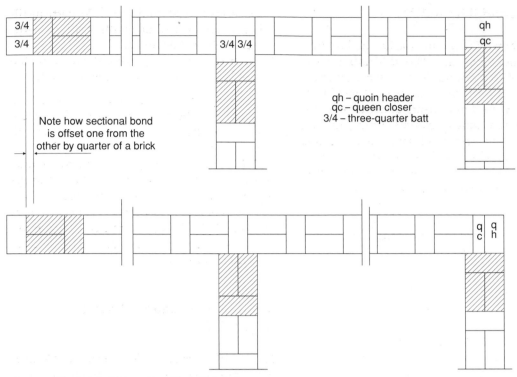

Fig. 1.19 One brick thick wall in Flemish bond.

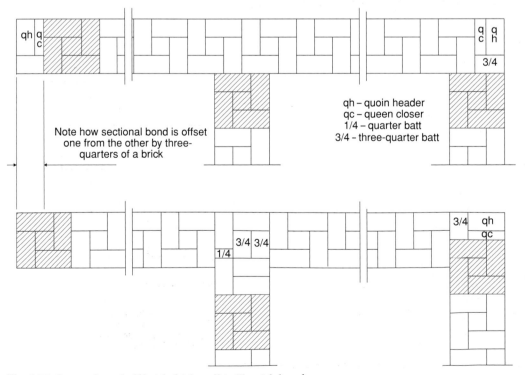

Fig. 1.20 One and one half brick thick wall in Flemish bond.

Quetta bond (Figure 1.21)

Note that on this drawing the hatched portion is a *void*, not the sectional bond. This bond was an attempt to produce a more weather resistant form of wall than the one brick solid wall commonly used in housing at the time, without using any more bricks or splitting the wall into two layers joined with wall ties. The idea never took off as the brick which protrudes from one vertical layer and touches the other merely serves to draw moisture from the outside to the inside face of the wall.

The bricks are all laid 'on edge', thus the wall ends up one and one third bricks thick. Building the bond with brick on edge poses a few problems:

- Bricks to be exposed on the face should have plain beds – no frogs, cavities or perforations
- If the wall is to be rendered or plastered then choose bricks with very shallow frogs or very small perforations
- If the wall is covered over in some other way, then the size of frog etc. doesn't matter

- Coverings used could include vertical tiling, slating or cladding with weather boarding or other sheet material
- All of these coverings would overcome the problem of moisture crossing the wall.

Walls with this bond have been used as the basis for part brick/part reinforced concrete walls with reinforcing rods set in the gaps and, when the brickwork is complete and the mortar hardened, concrete poured into the spaces. All-in-all, not an economic solution to keeping out the weather when the cost of over-cladding is added, but quite suitable for non-inhabitable buildings or garden or retaining walls, especially with the concrete fill and reinforcement for the latter.

Scotch and garden wall bonds (Figure 1.22)

Neither of these bonds is as strong as Flemish bond but they are used in situations where strength is not such an issue, e.g. garden walls.

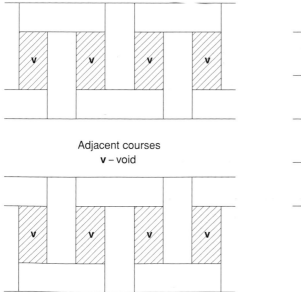

Adjacent courses
v – void

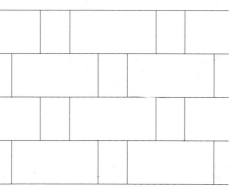

Pattern on face of wall

Note that the bricks are all
built 'on edge'. The bond
can also be built in blocks –
see later figures.

Fig. 1.21 Quetta bond – wall is always one and one half bricks thick.

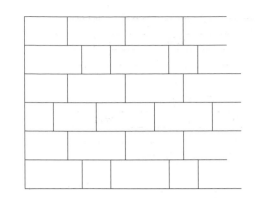

6
5
4
3
2
1

English garden wall bond   Flemish garden wall bond

Courses 1 and 5 show the typical header pattern of English bond and
the typical header/stretcher pattern of Flemish bond.
Courses 2, 3, 4 and 6 are all built entirely of stretchers on face.

Scotch bond only differs in the number of courses of 'all stretchers',
which is five.
Seven or more courses of stretchers are never used.

Fig. 1.22 Garden wall and Scotch bond.

Rattrap bond (Figures 1.23 and 1.24)

As in Quetta bond, the bricks are laid on edge, resulting in an interesting face pattern. The notes regarding Quetta bond apply equally to Rattrap bond, with an even lesser chance of keeping weather effects out of the building. A stronger bond than Quetta, it could be used for industrial or agricultural buildings and of course can be made more weatherproof by over-cladding. The bond is usually one stretcher to one header but it can also be built with three stretchers to one header, as shown in Figure 1.25. Note that it is always an odd number of stretchers but never more than three as this weakens the bond considerably.

Vertical alignment

Before leaving the patterns of bond, the vertical alignment of perpends must be studied. Two points in particular:

☐ Perpends are never built immediately above each other, i.e. in adjacent courses, on the face of a wall

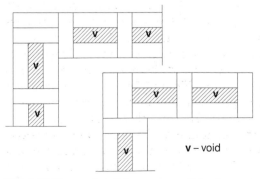

v – void

Fig. 1.23 Rattrap bond by one – wall is always one brick thick.

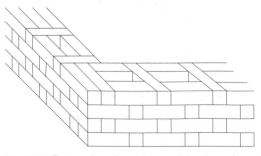

Fig. 1.24 Rattrap bond – orthographic view of quoin.

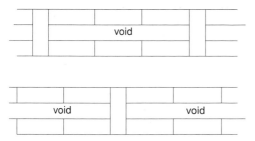

Fig. 1.25 Rattrap bond by three.

□ Perpends can be partly built above each other within the body or thickness of the wall
□ Perpends continuous through adjacent courses are known as **risbond** and the joints are known as **risbond joints**.

Figure 1.26 shows the face of an English bond wall and a Flemish bond wall. Note that

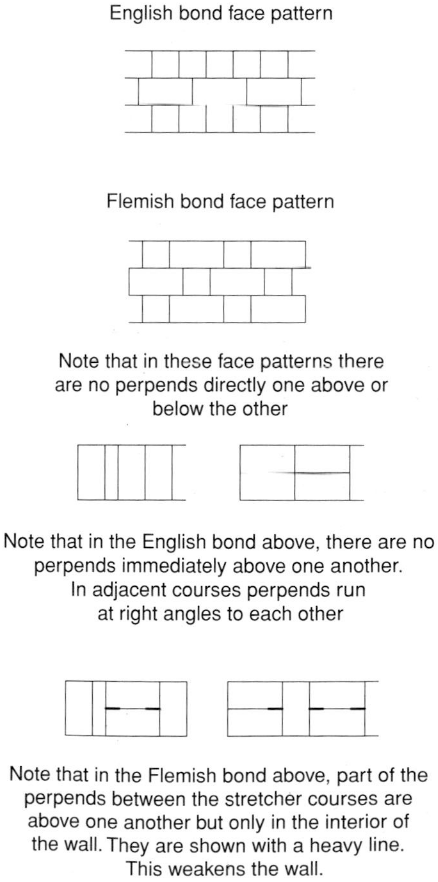

English bond face pattern

Flemish bond face pattern

Note that in these face patterns there are no perpends directly one above or below the other

Note that in the English bond above, there are no perpends immediately above one another. In adjacent courses perpends run at right angles to each other

Note that in the Flemish bond above, part of the perpends between the stretcher courses are above one another but only in the interior of the wall. They are shown with a heavy line. This weakens the wall.

Fig. 1.26 Risbond joint or *no* risbond joint.

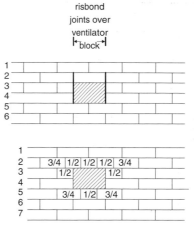

risbond joints over ventilator block

In this face drawing of the wall, the ventilator block has no risbond joints above or below it. Risbond has been avoided by using half and three-quarter batts. These can be coursed the other way round.

Fig. 1.27 Risbond joint at two course high wall ventilator.

neither shows any risbond, but that on the course plans risbond can be found within the Flemish bond wall.

A common source of risbond joint, fortunately of minor importance, is the building of ventilator blocks into half brick thick walls, as shown in Figure 1.27. As you now know, these walls are generally built in stretcher or common bond.

The reason risbond appears in the upper sketch of Figure 1.27 is the obsession the bricklayer has had with getting back to a bond which is simply laying each brick half over the one below, and he achieves this in five courses – but it is a visual disaster. A little thought would start this off in course 7 and get back to that in course 1 as shown in the lower sketch. And just to prove that bricklayers like this do exist and it's not all in the author's imagination, Figure 1.28 is a photograph of a facing brick wall with one course ventilators built into it. They mostly all looked like the one in the photograph. The photograph was originally taken in colour and the contrast across the different colours of bricks accentuated the risbond as well as showing up more clearly the poor general workmanship in this small area

Fig. 1.28 Risbond joint at one course high wall ventilator.

of brickwork. The ventilator is built into the bottom three or four courses of a two-storey gable wall of a house and the whole of the brickwork was to the same horrendous standard.

Honeycomb brickwork

This is a method of building a half brick thick wall similar to stretcher or common bond but leaving a space between every brick in each course, as shown in Figure 1.29. Therefore the bond can be by quarters or thirds, which describes the overlap of bricks in adjacent courses. With a quarter lap the void is half a brick wide, and with a lap of a third, the void is a third of a brick wide.

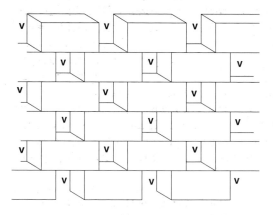

v – void

Fig. 1.29 Honeycomb brickwork.

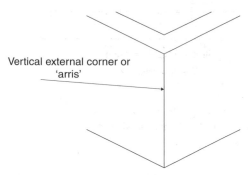

Fig. 1.30 Alternative use of term quoin.

This bond is used when building short support walls under ground floors which in turn are suspended over a void. The holes in the wall allow for a free flow of air in the underfloor void.

Quoins – an alternative definition

The term quoin has so far been used to mean the corner of a wall – the complete corner – from one side of the wall to the other. It can also be used to mean the vertical external corner or **arris** of the wall, as illustrated in Figure 1.30.

Another term associated with quoins is **quoin header**. Whenever a closer is used at a quoin or scuntion the whole brick nearest the corner is called a quoin header.

Half brick thick walls

It is generally believed that a wall showing a common or stretcher bond pattern is always half a brick thick and no other pattern can be used in a wall half a brick thick. This is *not* true. Half brick walls can show English, Flemish or garden wall bond on the face as a pattern – not as a true bond – and so deceive the unwary surveyor. How is it done? **Snap headers**.

Look back at Figure 1.27 which showed risbond joints around a ventilator and a possible solution. Such ventilators are frequently built into half brick thick walls and an appearance of a more decorative bond could be given by

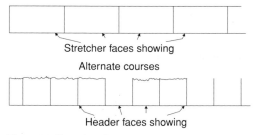

Stretcher faces showing

Alternate courses

Header faces showing

Fig. 1.31 Snap headers.

using snap headers in its construction. Indeed some of the cuts in the 'solution' would be termed snap headers. The term arises because any headers which show on the face of the wall are just 'headers' (whole bricks) snapped in two with the decorative header face showing and the rough cut end on the inside. This is illustrated in Figure 1.31.

Frog up or frog down

Cellular or single frog bricks can be laid with the depression on the bed facing up or down. The prime consideration in deciding which way to lay these bricks is the strength of wall required. Economics is the secondary consideration.

With the frog up, mortar for bedding the course above fills any frog or cells in the bricks below. Solid filling of the depressions gives a stronger wall – provided it is done properly and not skimped by bricklayers in a hurry.

With the frog down the mortar bed over the first course of bricks is nominally 10 mm thick. The next course above only beds into that mortar around the periphery of the frog, cell or perforation. Because there is a much smaller area of contact between adjacent courses, there is a much higher loading at these points. So for a given load, a wall built this way is stressed locally to a higher degree than a wall built frog up. The wall is weaker.

The economics

Laid frog up, bricks with large depressions will use much more mortar than when laid frog down. This costs money in two ways:

□ The cost of the extra mortar
□ The cost of labour spreading that extra mortar.

'Tipping'

No, not slipping the bricklayer a small appreciation of his services but a much more invidious practice by the bricklayer in a hurry.

To lay the bricks at all, a full bed of mortar is required. It is not possible to skimp on this as the bricks would not sit properly in the wet mortar if it were anything less than fully spread. However, it is possible to skimp on the mortar put into the perpends. This is done by the bricklayer who is hurrying to get enough bricks laid to earn a bonus, or is just lazy. As a brick is laid it is tamped or pressed down into the wet mortar which squeezes out each side of the wall. This surplus mortar is immediately cut away with the trowel. The next brick is picked up and that mortar on the trowel is 'wiped' off onto the header/stretcher arris of the brick. That end of the brick is then laid against the previous brick and from the outside of the wall, the perpend looks as if it was filled – but it is an illusion! This weakens the wall. The bricklayer should cover the header face of the brick and press this against the previous brick so that every perpend is filled solid with mortar.

Where very thick walls are built it is common to expect tipping at exposed perpends, and probably no perpends filled in the body of the wall at all. Many old **specifications**[6] called for each course to be grouted. **Grouting** is the filling of voids – in this case the perpends – with a mortar of a fairly fluid consistency. The lay reader will of course expect that the spreading of the next bed of mortar would fill all these perpends anyway, but this is not the case. Building mortar is of too stiff a consistency to run down into a 10 mm wide gap for a

[6] A specification is a document which was commonly prepared by the architect but is now done by the quantity surveyor. It lays down minimum standards for materials, components and workmanship to be used on a particular contract.

depth of about 65 mm. The inevitable drying action caused by contact with bricks stiffens the mortar even more and prevents it running anywhere. Even when a bricklayer pushes mortar into the perpends with the edge of his trowel, the filling is not complete.

Common and facing brickwork

So far we have only looked at the bonding of the bricks in a wall, without looking at why they are built into the wall. There are basically two ways in which bricks are built:

☐ As common bricks, *or*
☐ As facing bricks.

Common brickwork is used to describe brickwork which will generally have a 'finish' applied to it or whose external appearance is not of great importance. Examples of the latter might be the insides of garages, boiler sheds, etc., whereas the interior walls of a house would be plastered.

Despite the name and the fact that this work is rarely seen in the finished building, the layout and cutting of bricks to form bond must be done carefully and well. Just because it is 'common brickwork' is no excuse for shoddy workmanship.

If common brickwork is to have a finish applied, then the beds and perpends have the mortar raked back to provide a **key** for that finish. The finishes are either **plaster** or a mortar **render**, which are applied 'wet'. Raking back the joints allows the plasterer to press this wet material into the joints, thus giving a good grip or key.

Another way to finish common brickwork is to fill the beds and perpends full of mortar and strike this off flush with the face of the bricks. This is called **flush jointing**. Before striking the mortar off flush, the bricklayer will press the mortar down hard with a trowel. This gives a dense hard joint not easily broken down by weather etc. He or she may then rub the wall face down with a piece of sacking. This is called **bag rubbing**. Alternatively

he or she could brush it with a stiffish hand brush. Both techniques give a softer texture to the face of the wall. Flush jointing and bag rubbing or brushing are often used internally on walls as a precursor to some form of paint finish applied direct to the brickwork.

Facing brickwork

Facing brickwork is the term applied to brickwork where special attention has been paid to the external appearance, including the use of decorative or coloured bricks and decorative or coloured mortar beds and perpends.

Facing bricks can have one or both of the following features:

☐ A special colour either inherent in the base material or added to it or applied to the surface
☐ Texture deliberately impressed on the surfaces or as a result of the base material texture and/or the manufacturing process.

At least one stretcher face and one header face must have a 'finished' surface. Frequently one stretcher face and both header faces are 'finished'. This allows any wall of one brick thick to be built with two 'good' faces provided the lengths of the bricks are reasonably consistent and the bricklayers are skilful enough to stretch or compress the thickness of the mortar in the perpends to accommodate any inequality in dimensions of the bricks. Finishing only one header face allows any wall over one brick thick to be built with two 'good' faces.

Figures 1.32 and 1.33 illustrate how bricks require to be 'finished' on more than just the stretcher face to give a good appearance in facing brick work. Finishing half brick thick walls with a good face both sides is restricted this time by the accuracy of the width of the brick, and there are no internal perpends which can be adjusted in thickness to accommodate inadequacies in the dimensions of the bricks. Naturally any snap headers must be built with the cut face as a perpend so that both sides

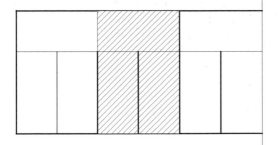

Plan of one course of English bond in the scuntion
of a one and one half brick thick wall.
The finished faces of the facing brick are shown in
heavy lines – remember only one stretcher face
and one header face are finished

Fig. 1.32 Scuntion of one brick thick wall built both sides fair face; bricks finished one stretcher and two header faces.

show finished faces. In this case the cutting is best done with a power saw. If you doubt that, just go back and look at Figure 1.28 and the perpends at the half bricks. The wavy edge of a brick cut with a chisel or a brick hammer is clearly seen.

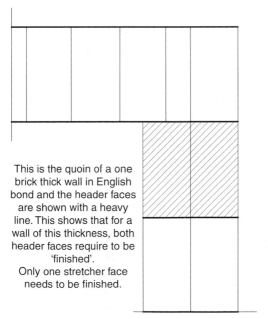

This is the quoin of a one
brick thick wall in English
bond and the header faces
are shown with a heavy
line. This shows that for a
wall of this thickness, both
header faces require to be
'finished'.
Only one stretcher face
needs to be finished.

Fig. 1.33 Quoin of one brick thick wall built both sides fair face; bricks finished one stretcher and two header faces.

Pointing and jointing

Naturally the finishing of the joints of facing brickwork is important, but any brickwork with joints exposed to the weather requires a good finish to the joints to prevent weather penetration. Pointing and jointing are done to enhance the weathering characteristics of the mortar in the beds and perpends since these are the most vulnerable part of the whole wall which might be exposed. Should the building mortar be unsuitable for exposure to the weather, then jointing is not an option; pointing must be carried out where the mortar can withstand any weathering effects. Apart from these practical considerations, aesthetics frequently play their part in choosing a pointing/jointing shape. Some of these shapes are shown in Figure 1.35.

On facing brickwork, flush jointing is occasionally done but bag rubbing or brushing are *never* done as this carries mortar over the face of the bricks, spoiling their appearance. Look at Figure 1.28 for mortar staining of the bricks.

It is more usual to strike the mortar off at an angle or some other shape to emphasise the joint and also to add to its weather shedding properties, as shown in Figure 1.34. The figure

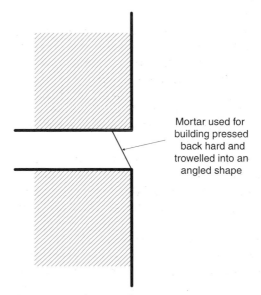

Mortar used for
building pressed
back hard and
trowelled into an
angled shape

Fig. 1.34 Weather struck joint in pointing or jointing.

shows a **weather struck** finish, the most common shape used to finish off a mortar bed or with which to finish **pointing**. The slope is designed to shed any water striking the surface. It is a shape preferred by bricklayers as it finishes off the upper edge of the course which is the bed set level with the builder's spirit level or the builder's line. Note that when the shape is worked on the mortar used for building it is termed **jointing**.

There are many reasons for, and occasions when, the mortar used for building facing brickwork is not suitable for the finished appearance of the wall: wrong resistance to weathering, wrong colour and wrong texture are the most frequent. It is generally considered too expensive to use coloured mortars for building. What is done then to keep the building mortar cost down and still have effective and expensive looking joints? Simply, build in the cheaper mortar, rake back the joints 10–15 mm and add coloured and/or textured mortar to finish the joints. This process is called **pointing**.

It is important to understand the difference between jointing and pointing:

□ **Jointing** means giving the mortar used for building a finished shape.
□ **Pointing** means raking back the building mortar at the joints and adding a separate mortar with a colour or texture and giving that a finished shape. The addition of the coloured or textured mortar is always done after the building mortar has hardened.

Different shapes can be worked on the mortar in both jointing and pointing operations, and a selection is shown in Figure 1.35.

Starting at the top left and working clockwise, **flush** or **flat** jointing or pointing has already been described. Despite its simple appearance it is not an easy option when building with facing brick as it is all too easy to smear mortar across the face of the brick and so spoil the appearance.

Recessed jointing or pointing used to be much more common on the Continent than in the UK. It is a good clean method of finishing

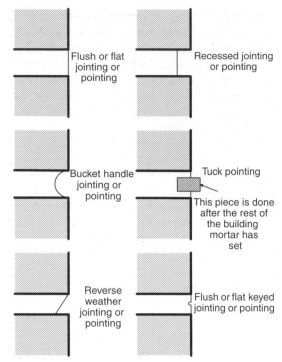

Fig. 1.35 A selection of joint shapes.

a joint although it does not shed water out of the joint, as weather struck jointing or pointing does.

Tuck pointing is very decorative and with the distinctive, exposed, regularly shaped, rectangle of (coloured) mortar draws the eye away from any irregularity in the size or building of the facing bricks. It is frequently used, therefore, for renovation work in old brickwork where the old mortar is raked out, the beds and perpends flush pointed and a groove struck, and after being allowed to set, the 'tuck' is formed, perhaps in a contrasting mortar. A tool for tuck pointing is illustrated in Figure 1.36.

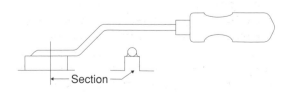

Fig. 1.36 Tuck pointing tool.

Keyed pointing or jointing is shown here worked on a flush bed or perpend. The key is simply a shallow line marked in the still plastic mortar and ruled straight down all the beds and across alternate courses in the perpends. The idea is to disguise any unevenness in the brickwork. Keying can be applied to recessed pointing or jointing.

Reverse weather jointing or pointing is a fairly obvious description but its effects are quite different to the regular weather struck joint. Being made from the lower edge of the course of brickwork, it tends to follow a more irregular line and so could, if below eye level, show up any such irregularities. However, in walls generally much of the work is above eye level so it has the effect of softening the line of the bed pointing.

Bucket handle or **grooved** pointing is easily achieved with a simple tool made by the bricklayers themselves from a bucket handle. Not just any old handle but the galvanised steel one from an old fashioned galvanised bucket. The handle is straightened out and cut into lengths (generally two or three) which are then shaped into a very open Z, as shown in Figure 1.37. Either end can then be used to press the mortar back into the joints, giving the required shape to the joint and density to the mortar.

It is important to realise that good pointing/jointing is all for nought if the bricks themselves are:

☐ Badly laid
☐ Of poor quality
☐ Of uneven sizes
☐ Chipped on edges or corners.

These faults are emphasised by attempts to point them up straight and true. The only pointing which might help would be tuck pointing where the base mortar is tinted to match the brick, the groove is struck straight and even and the tuck is placed in a contrasting coloured mortar. It is for this reason that tuck pointing is favoured by many professionals when carrying out re-pointing of old and worn brickwork.

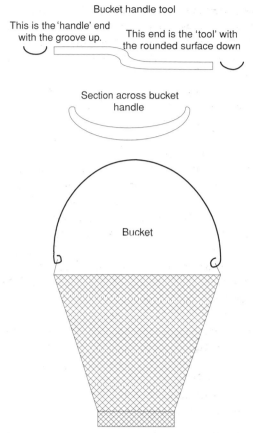

Fig. 1.37 Bucket handle pointing/jointing tool.

General principles of bonding

To recapitulate all we have learned about bonding brickwork we list here the seven principles which, if followed carefully, will ensure the proper layout of each course of brickwork in any thickness of wall. As you read each one you may wish to refer back to the illustrations of bonds, and note where each of them occurs. To start you off the figure numbers are given for the first principle.

Bonding 1

Correct lap set out and maintained by:

☐ closer next to quoin header (Figure 1.17, the upper course shows queen closers next to the quoin headers), *or*

☐ three-quarter batt starting the stretcher course (Figure 1.19, the upper course shows the scuntion beginning with a pair of three-quarter batts).

Bonding 2

☐ Perpends in alternate courses kept vertical (any figure showing face patterns of walls).

Bonding 3

☐ No risbond joints on faces of wall
☐ Risbond is allowed in the thickness but must be kept to the minimum 'allowed' by the particular bond.

Bonding 4

☐ Closers *only* used next to quoin headers.
☐ These can occur at quoins and scuntions.

Bonding 5

☐ Tie bricks at intersections and quoins must bond properly, i.e. by a quarter brick.

Bonding 6

☐ Bricks in the interior of a wall should be laid header wise across the wall as far as possible.

Bonding 7

☐ Sectional bond should be properly maintained.
☐ Front and back faces of walls should line through (Figure 1.38 illustrates this vital point).

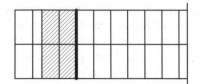

Sectional bond lines through back to front of wall as shown by heavy lines

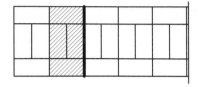

Bricks in thickness of the wall laid header wise, across the long axis of the wall

Fig. 1.38 Alternate courses in a two brick thick wall in English bond showing bricks in centre of wall and lining through of sectional bond.

BLOCKS

Block materials

Modern blocks are generally made from some form of 'concrete' and are the subject of BS 6073, Precast concrete masonry units. Blocks can also be made from fired clay, the subject of BS 3921.

Concrete blocks

There are three basic categories of concrete[7] used for block manufacturer

☐ **Dense concrete**
☐ **Lightweight concrete**
☐ **Autoclaved aerated concrete (aac).**

Dense concrete is usually described as concrete which has a dry bulk density[8] in excess of $750 \, \text{kg/m}^3$.

---

[7] Concrete is discussed more fully later in this chapter and in Appendix D.
[8] Bulk density is the mass of any material in unit volume, voids in the material being measured as part of the overall volume.

Lightweight concrete is therefore concrete which has a dry bulk density up to $750 \, kg/m^3$.

Aac is a special concrete made from a mixture of cement, lime and siliceous material which is heated under pressure. It is aerated before 'cooking' and has a dry bulk density from 350 to $750 \, kg/m^3$.

Dense and lightweight concretes

These concretes are made from three basic ingredients:

☐ Aggregates
☐ Binder
☐ Water

Additives such as colouring may be introduced.

☐ Aggregates are of two kinds, fine (sand) and coarse (normally gravel or crushed stone or other inert material)
☐ The binder is normally ordinary Portland cement (OPC)
☐ The water is necessary to trigger and complete the chemical reaction which makes the mixture 'set'.

Materials for aggregates in dense or lightweight blocks can be: natural gravel and sands, crushed stone, crushed stone dust, slags, furnace ash, exfoliated vermiculite, expanded perlite, crushed brick, crushed concrete, crushed clay blocks, wood fibre and sawdust or wood chippings. The wood material is usually treated with petrifying liquid. This involves impregnating the cells of the wood with minerals in solution – the petrifying fluid – so that they become almost a 'fossilised' replica of the original cells.

Autoclaved aerated concrete

☐ The aggregate is a mixture of sand and pulverised fuel ash (PFA). The proportions vary but grey coloured aac has a higher proportion of PFA than sand, and cream coloured aac has a higher proportion of sand.
☐ The binder is a mixture of OPC and building lime.
☐ Water is used to make the mixture set.
☐ Aluminium powder reacts with the alkali in solution to generate hydrogen, the bubbles 'aerating' the wet mixture before autoclaving.
☐ The mixture of alkaline and siliceous material combines under the heat and pressure to form calcium silicate, the same material as sand or flint lime bricks. The Victorians invented the name Tobermorite for calcium silicate.

All blocks of whichever concrete come in a variety of sizes. For a complete range of face sizes refer to BS 6073. Thicknesses generally available are: 50, 65, 75, 90, 100, 199, 125, 150, 200, 215, 250, 300 mm.

Dimensions of standard metric block

As in brickwork, blocks have coordinating dimensions within which the block and one layer of mortar are accommodated. The work size of the standard metric block is 10 mm less on face than the coordinating size, so 10 mm mortar beds and perpends are the norm.

Aac blocks are only ever manufactured as solid rectilinear shapes. Dense and lightweight concrete blocks are manufactured as solid, cellular and perforated shapes, as shown in Figure 1.39.

With a wide range of thicknesses available, walls built with blocks only have a single, vertical layer of blocks in them. In other words, the thickness of the wall is designed to suit one of the thicknesses of block available. It follows from this that blockwork is built in common bond. Blocks will overlap in adjacent courses. The overlap may be by thirds or halves. The choice of bonding by thirds or halves can depend on the setting out of the length of the wall to accommodate proper bonding at quoins

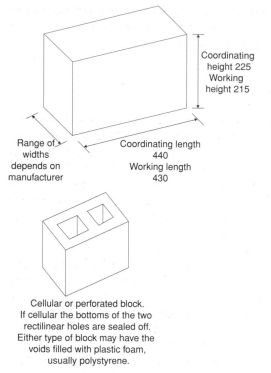

Cellular or perforated block.
If cellular the bottoms of the two
rectilinear holes are sealed off.
Either type of block may have the
voids filled with plastic foam,
usually polystyrene.

Fig. 1.39 Standard metric block dimensions and type by shape.

and intersections. The thickness of the block also affects this decision.

These last points are illustrated in Figure 1.40. Note the cuts at the quoin.

Unlike brickwork, block cutting is generally confined to cutting to a particular length to

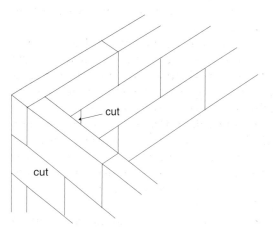

Fig. 1.40 100 thick block wall quoin.

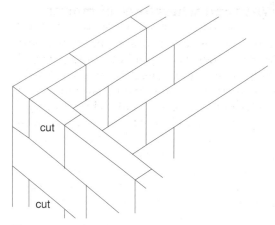

Fig. 1.41 150 thick block wall quoin.

accommodate the bond required. Dense concrete blocks can be cut with a bolster[9] and hammer or with a power saw. Lightweight aggregate blocks can be cut with a bolster and hammer, a power saw or chopped with a brick hammer. Aac blocks can be cut with a bolster and hammer (the bolster tends to sink deeply into the block without making much impression!), a power saw (this is a waste of power), a tungsten-tipped hand saw (good but labour intensive) or an axe or zaxe[10] (both a bit crude).

Figure 1.41 illustrates a quoin in a 150 mm thick wall. Note the cuts at the quoin. Although the 150 thickness is slightly more than one third of the coordinating dimension of a block length, there is no need to actually cut all the blocks at the quoin. The bricklayer simply makes the perpends in each course a little less than 10 mm between the quoins. He doesn't actually measure this as a fraction of a millimetre at each perpend. His skill allows him to make the adjustment a little at a time purely 'by eye' or 'by feel'. Only if there is insufficient length between the quoins to make sufficient adjustment, are blocks cut.

---

[9] A bolster is a cold chisel with a broad cutting edge, anything from 50 to 150 mm wide. 100 mm wide is a commonly used size in these circumstances.

[10] A zaxe is a tool used by slaters to cut roofing slates. It resembles a meat cleaver with a spike on the back edge.

Whys and wherefores of mortar

Mortar is what we use to build masonry. Mortar 'glues' bricks and blocks together, although the gluing effect is minimal. Mortar takes up irregularities in the shape and size of bricks and blocks, especially if built 'frog up'. By providing full contact between adjacent units, mortar spreads loads evenly throughout masonry. Mortar prevents weather penetration between bricks and blocks, prevents penetration of walls etc. by fire and sound, and can be a decorative feature of masonry.

Mortar is composed of:

☐ **Fine aggregate**
☐ **Binder**
☐ **Water**.

Additives such as colourings, **waterproofers**, **accelerators** and **air entrainers** are frequently introduced.

The most usual aggregate is washed building sand. The binder may be:

☐ ordinary Portland cement (OPC)
☐ building lime, *or*
☐ a mixture of both, *or*
☐ masonry cement.

Water is added to trigger and complete the 'setting' process of the binder(s). Water makes the mixture plastic, thus allowing the bricks and blocks to 'bed' into the layer of mortar.

The proportions of the principal ingredients are determined by the use to which the mortar will finally be put. The proportions are expressed as **parts** by **volume**. Where OPC or lime is used as the whole of the binder this is expressed as one volume. Where OPC and lime are combined, the OPC is expressed as one volume and the lime in its correct proportion to that one volume of OPC. Fine aggregate is the final volume.

For example, a mortar mix described as 1:2:9 is a mixture of one part OPC, two parts building lime and nine parts fine aggregate. This mix is termed a cement/lime mix. A cement mortar mix described as 1:3 is a mixture of one part OPC and three parts fine aggregate. The mix is described as a cement mortar mix as opposed to a lime mortar mix.

Cement

Cement is the best known ingredient of mortar and one most commonly mistaken for the mortar itself. Remember we build nothing 'in cement'. We put cement into certain mortars and concretes where its properties as part of the binder are used to advantage. There are occasions when the use of cement would not be appropriate.

The most common type of cement used is ordinary Portland cement (OPC) which is the subject of BS 12. The term 'Portland' came about because the colour of the cement – a greeny grey – resembled Portland stone.

Ordinary Portland cement was the invention of Joseph Aspden in 1824. To manufacture cement, clay and limestone are broken into small pieces, mixed and calcined in a long rotary kiln. The fuel can be oil, gas or powdered coal. A little gypsum powder is added to the clinker, which is then ground into a fine powder. The result is ordinary Portland cement. The cement can be filled into paper sacks of various weights, the most common being 50 kg, with 25 kg bags finding favour with jobbing builders and the DIY market. It can also be shipped out of the works in bulk, either in railway or road tankers. This material is loaded into silos at the plant or site where it is to be used.

The chemical composition of OPC is complex and varies according to the source of the minerals used in its manufacture. The hardening or 'setting' of mortar and concretes made using OPC is due to a chemical reaction between the OPC and the water used for mixing. Air is *not* required for setting to take place, nor does the mortar or concrete need to 'dry'. Because cements of this type will set under water and without air, they are often referred to as **hydraulic** cements.

White cement can be manufactured but in this instance the clay is white china clay and the limestone used is pure natural chalk. Coal is not suitable for firing the kiln. Powdered gypsum is added after the clinker has been finely ground.

Portland cement can be obtained in a **rapid hardening** variety. The setting process is speeded up by grinding the powder even more finely, thus increasing the exposed surface area of the cement particles. The surface area of the particles is given in square centimetres per gram of cement – the **specific surface**. The specific surface of rapid hardening cement ranges from 7000 to 9000 cm^2/g. The increased area in contact with water speeds up the setting process.

For **Portland blast furnace cement** (BS 146) 30%, by mass, of blast furnace slag is added to the clinker produced by calcining clay and limestone before grinding. Blast furnace slag is a material which demonstrates **pozzolanic** properties, i.e. a 'setting' action takes place in the presence of water.

OPC is a complex cocktail of chemical compounds, one of which – **tricalcium aluminate** – is attacked by sulphates in solution. By reducing the tricalcium aluminate content in the cement greater resistance to sulphate attack can be achieved – **sulphate resisting cement**.

Masonry cement is made by taking OPC and adding a filler and a plasticiser. The filler might be a material such as **pulverised fuel ash (PFA)**. PFA is the ash produced in vertical furnaces used in power stations which burn powdered coal. The ash is very fine, grey powder which is trapped in electrostatic precipitators and mixed with water before being dumped in lagoons. The PFA must be kept wet in order to stop it spreading out over the countryside in even the slightest breeze. The plasticiser is usually an air entraining agent complying with BS 4887.

Lime

Before the invention of OPC, lime was the most common binder used. Builder's lime is manufactured from carbonates of calcium and magnesium. The most common form is made from calcium carbonate – chalk or limestone – which is calcined in a kiln to give calcium oxide or 'quicklime'.

The equation for the process looks like this:

$$CaCO_3 \xrightarrow{heat} CaO + CO_2$$

The calcium oxide is **slaked** by immersing in water and forms the **hydroxide**, calcium hydroxide – $CaO_2(OH)$.

The burning of the lime used to be carried out next to sources of limestone and also where coal could be easily obtained to fire the kilns. The quicklime was extracted by hand which was dangerous and dirty work. Quicklime generates large quantities of heat when wetted, and on moist skin or in eyes will produce horrible burns.

Builders slaked the quicklime on site by covering it with water for at least three weeks. The excess water was drained off, leaving the wet hydroxide known as **lime putty**. The putty was run through a sieve to take out any unslaked lumps and then mixed with coarse sand to give a mortar.

Lime manufacture is now a factory process and the lime is supplied to the trade as a dry powder known as a **dry hydrate** – $CaO_2(OH)$. This can be mixed direct into mortars, proportioned by volume. The relevant standard is BS 890.

Limes based on calcium carbonates are the most commonly used for general building work. They are known as 'pure' or 'fat' limes. They are also termed **non-hydraulic limes**, as they require air in order to set. In particular they require the carbon dioxide in the air since setting is the action of the CO_2 converting the hydrate back into the carbonate, releasing water which simply dries out.

Another lime available is a magnesium-based lime which must have at least 5% of magnesium oxide before slaking. In practice the percentage is something in the order of 30+%.

Hydraulic limes are also made and these exhibit properties similar to OPC, as the limestone from which they are made is a grey limestone containing traces of clay.

BS 890 covers other limes – semi-hydraulic and eminently hydraulic. These are not of interest at the present time.

Sand

Modern mortars use washed building sand – not any old sand but an aggregate where there is a definite grading of the particle sizes such that the proportion of binder to sand at around 1:3 to 1:4 allows for complete filling of the voids between and coating of the sand particles by the binder. Sand is covered by BS 1200 – pit and river sands. Sea sand is *never* used.

Water

Water is the final ingredient in any mortar. It is no use specifying and procuring lime, cement and sand to the required standard if the water used for mixing is impure. For the majority of work it is sufficient to describe the water as 'potable' and to further reinforce that by specifying that the supply must be obtained from a main supplied with water by the local water authority. Natural sources of water such as rivers, streams ponds, etc. can be contaminated with excess sulphates, nitrates and other soluble chemicals which can weaken the mortar by destroying the cement or other binder, or can cause staining or even odours in the building when it is heated.

An example of the latter was found when a large office block in the centre of Edinburgh was having its heating system commissioned, although the fault in the end was not 'water'. A strong smell of old ammonia was detected on one floor and was finally tracked down to one or two rooms. When questioned, the subcontractor quizzed his employees and eventually found out that they were in the habit of taking plastic, soft drinks bottles of water into rooms with them. The water was occasionally drunk but more often was used to 'knock up' an already setting mortar. Occasionally, having drunk too much, the bottles were used for personal relief! One wonders if the workmen ever got the contents mixed up in other ways.

Which mortar mix?

Guidance on which mortar mix to use is given in the Building Regulations by reference to British Standard 5628. The editions quoted in the Regulations have all now been superseded and mortar is now included in Part 3, *Code of Practice for the use of masonry; materials and components, design and workmanship.*

Choice of mortar depends on the units being built, strength required and the season of the year with its attendant weather conditions. Further factors are:

☐ Location within the building
☐ Degree of exposure to weather, particularly driven rain.

We cannot, therefore, consider which mortar to use until we understand what we mean by exposure.

For external walls, location within the building is divided into three areas:

☐ To a height not less than 150 mm above finished ground level – from the damp proof course level down
☐ Between the level of the top of the damp proof course (DPC) and the junction of wall with roof
☐ Gable haffits – the triangular shaped piece of wall from eaves level to the ridge level.

Exposure is rated in accordance with the British Standard Code of Practice 8104, *Assessing exposure of walls to wind-driven rain.* It is based on a fairly complex series of data presented in the Code which can be selected and applied to any building according to its proposed geographical location. It is instructive to note that in this Standard the exposure of the wall of a house between DPC level and eaves can vary by as much as a factor of 2 to 3 and if there is a gabled wall, the top of the gable can vary by a factor of from 3 to 4.

There are two measures of exposure given in the Code, the unit of measurement being litres of rain driven onto the wall/square metre of vertical wall surface. Note 1 litre/m^2 is the equivalent of 1 mm of rainfall.

Exposure will be calculated as either annual or spell. **Annual exposure** is the total amount of water hitting the wall surface in one year and would be used to gauge the overall moisture content of the wall and the prospect of moss or lichen growth, etc.

Spell exposure is the total amount of water hitting the wall surface in relatively shorter periods of time with up to 4 days in the spell without any appreciable rain at all; it is used to determine peaks in rainfall and so enable the wall designers to take these into account. For example, if a wall were to have an annual exposure of 400 litres/m², that would indicate an annual rainfall against the wall of 400 mm – not a lot; but if a spell exposure rating of 400 litres/m² were indicated over a period of less than one day, the wall would have to be designed very differently. Such an exposure would indicate that a cavity wall would have solid water on the inner face of the outer leaf and that weeps at the bottom of the wall might be overwhelmed in their efforts to clear the water; more of this when we look at walls.

A number of factors are taken into account in the calculations:

- Airfield index – Imagine that on the site of the building there is a large open space without obstructions – an airfield for example – and that in the middle of this space the quantity of driven rain 10 m above the ground is measured at the orientation of the wall being examined.
- Geographical location – (see airfield index above). Maps have been prepared to a scale of 1:1 000 000 showing the increments as contours. Maps give the key to finding the **wind roses** from which the base for the exposure calculation is derived.
- Orientation – 12 directions are given on the **wind roses** (see airfield index above).
- Topography – ground slope and height.
- Terrain roughness – allows for the general terrain type up wind of the wall.
- Obstruction – nearby trees, hedges, fences and buildings, their distance and spacing.
- Overhangs and surface features – refers to eaves overhangs, canopies, string courses, etc. on the wall itself.
- Occasional features such as hills, cliffs and escarpments can also be factored into the calculation. The maps, and the factors of orientation, topography and terrain roughness which they introduce in the key to the wind roses, generally account for ranges of hills and mountains but cannot account for isolated features such as these. They must be dealt with separately.

Mortar used for building masonry is only one of many parts of British Standard 5628, *Structural work in masonry*. The Standard covers all types of masonry – brick, stone, concrete, etc. – and is a large and comprehensive document. As a result, the mortar mixes given in Tables 13 and 14 of the Standard cover all these masonry types. Our text here deals only with brick and concrete block and so the choice narrows slightly. The mixes and advice summarised in Table 1.2 should prove adequate with the limited masonry types under discussion.

'Fat' mixes

Mortar mixes suitable for building masonry units should be 'fat'. This is a difficult term to define, but imagine you are trying to spread out a mixture of plain sand and water into a thin layer. Such a mixture is gritty and stiff and if it is being applied to a porous surface the water is absorbed. To improve the spreadability of a mortar, to reduce the grittiness and loss of water on absorbent surfaces, one can do a number of things to the mixture:

- Include lime in the binder, *or*
- add an air entraining chemical, *or*
- use masonry cement as the binder.

Highly absorbent surfaces on bricks or blocks – especially in hot, dry weather – can take so much water out of the mortar that the chemical reaction between the water and the binder cannot fully take place and a weak bed and perpend are formed.

Table 1.2 Mortars for clay bricks and concrete blocks.

	Mix	Below DPC level	Above DPC level
A	1:1:5 or 1:1:6, or 1:5 using masonry cement	For all exposures and in/at any season	For all exposures and in all seasons
B	1:5 or 1:6 using OPC and a plasticiser	For all exposures and in/at any season	For all exposures and in all seasons
C	1:2:8 or 1:2:9, or 1:6 using masonry cement	—	For sheltered and moderate exposures and in spring and summer
D	1:8 using OPC with a plasticiser	—	For sheltered and moderate exposures and in spring and summer
E	1:3 using hydraulic lime	—	For sheltered and moderate exposures and in spring and summer
F	1:2 using hydraulic lime	—	For all exposures and in all seasons
G	1:3 using either OPC or masonry cement	For all exposures and in any season but to withstand a heavy loading	For severe exposures and at all seasons but to withstand a heavy loading

Adding lime to the mix

Using a cement/lime mortar has a number of advantages beyond providing the brick-layers with a 'fatty' mix. After it has set, the mortar:

- Is weaker than a pure cement mortar and will not damage the bricks or blocks
- Exhibits greater elasticity than a pure cement mortar, thus taking up thermal and moisture movement of the blocks or bricks.

If cement/lime mixes are not usually used, a masonry cement mortar or a cement mortar with an air entrainer are the next best, in that order.

Adding an air entrainer

Air entrainers are based on an industrial detergent and during mixing introduce lots of air bubbles into the water of the mortar. Because of the air bubbles the mix feels 'fatty', thus helping the laying and bedding of the bricks or blocks. Because the water surrounding the air bubble is held there by high surface tension, it is less likely to be taken up by an absorbent surface.

Using masonry cement

Using masonry cement as the binder gives a fatty mortar since the cement exhibits a combination of characteristics akin to an OPC and lime mix. The set mortar is also softer and more elastic than a pure cement mortar.

General rules for selection of mortar

- Adequate strength to resist crushing loads in the beds of the masonry
- Weaker than the masonry units so that shrinkage of the mortar does not crack the units
- Sufficient elasticity to absorb movement of the units
- Good 'spreadability' with proper water retention on absorbent surfaces
- Appropriate for the season of the year
- Capable of withstanding the exposure to be experienced.

Mortar additives

Additives frequently introduced are:

☐ **colourings**
☐ **waterproofers**
☐ **accelerators**
☐ **air entrainers**
☐ **frostproofers**.

Colourings

OPC, which is grey/green in colour, can be tinted dark red, green, buff, brown, dark grey, etc. by the addition of powdered tinting agents. To obtain pale colours silver sand and white cement must be used in the mortar.

Waterproofers

In some instances the term 'waterproofers' is a misnomer. Some additives are only water repellants.

Water repellants prevent water penetrating the pore structure provided there is no pressure behind the water. Under hydrostatic or hydraulic pressure, mortar, even when treated with a repellant, will allow water to pass through its pore structure. An old form of water repellant was the inclusion of linseed oil. This was dispersed as an emulsion in hot water and then added to the mixing water. Modern water repellants are based on synthetic resins which coat the pore sides, preventing water from entering unless under pressure.

To make mortar completely waterproof, one has to fill every pore or seal off every pore into a closed cell structure. Again, modern chemistry has given us synthetic resins which are capable of blocking off the pore structure, even under considerable pressure, but be aware that they are not all 'equal'. For example, there are waterproofers based on EVA (ethyl vinyl acetate), available in a liquid and a powder form. Both are simply added to the mortar mix in the recommended proportions, but the liquid form is quickly washed out of

the pore structure whereas the powder form is more permanent in its effect.

Accelerators

Accelerators are chemicals added to a mortar to speed up the reaction between the binder and the water, known as 'setting'. Calcium chloride was widely used for this purpose but too high a proportion 'killed' the set completely and if in contact with ferrous metals caused corrosion. If a very fast set is required it is better to use quick-setting Portland cement.

Air entrainers

Air entrainers are based on industrial detergent and are 'wetting' agents. Numerous small air bubbles are formed during mixing. While this can allow a reduction in the amount of water used when making concrete, with a mortar the effect is to reduce the loss of water to dry or porous building units. This has a number of effects:

☐ The mix appears 'fatty' for laying the units
☐ Water is retained in the mortar beds and not absorbed so readily by dry, porous units
☐ Water is not lost so readily in dry hot weather
☐ There may be a lowering of strength.

Frostproofers

Frostproofers are nothing more than accelerators. Acceleration of the 'set' generates more heat, thus 'proofing' the mix against immediate frost attack. An accelerated setting time also means the mortar has a shorter time span during which it can be frozen.

Mixing in additives

Additives in liquid form are best put into the mixing water. Mixing water is generally kept in a steel or plastic drum next to the mixer.

So to a known quantity of water the correct proportion of additive may be mixed in. No matter how much mortar is mixed, it will then contain the correct amount of additive. Most dry additives, i.e. in powder form, are added in proportion to the mass of OPC or other binder. A proper container should be manufactured and calibrated to ensure the correct proportion of additive is placed in the mixer with the dry ingredients.

Mixing mortar

Mixing of the ingredients is generally done in a mechanical **concrete mixer**. The term 'cement mixer' is frequently used but is quite incorrect.

The dry ingredients are placed in a drum and mixed to a uniform colour. Water is finally added and mixing continues to give a uniform consistency. *Never* mix beyond that point. Mixing for longer than is necessary to give a uniform consistency will entrain an excessive amount of air into the mix, which will weaken the mortar.

Mortar should never be remixed, nor should old mortar be introduced into a fresh batch. Old mortar should not be 'freshened up' or 'knocked up' by adding more water.

Once the water is added to the batch of ingredients, the chemical reaction of 'setting' commences. It takes up to about 3 hours to reach an 'initial' set. This is the point beyond which the mortar should no longer be used for building. In practice mortar should be used within two hours of mixing.

Schematics of drum-type concrete mixers and pan mill mixers are shown in Figure 1.42, and a photograph of a drum mixer is shown in Figure 1.43.

Black ash or old soft bricks were often used as aggregates for mortar, instead of washed building sand. These were placed in a pan mill which had heavy rollers rather than paddles. These rollers first ground the brick or lumps of ash to a powder, after which the binder and water were added and the rollers squeezed the mixture out to the sides and so mixed it to an even consistency.

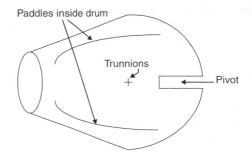

Rotating drum-type concrete mixer. The drum rotates on the pivot and tilts up on the trunnions for mixing or down for emptying

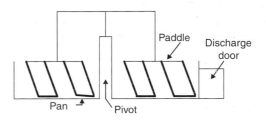

Pan-type mortar mixer. Either the pan rotates and the paddles are fixed or vice versa

Fig. 1.42 Schematic of drum and pan mortar/concrete mixing machines.

Mortar during this period was frequently a lime mortar or a cement/lime mortar, the lime being bought in as 'slaked lime' and in lumps. Pieces of slaked lime could be ground down in the pan mill with the ash or brick and water, giving a 'fatty' mortar. Black ash is no

Fig. 1.43 Drum concrete mixer.

longer used as an aggregate. Corrosion of ferrous and galvanised fittings has proved to be a problem because of aggressive minerals from the black ash, notably sulphates which also cause deterioration of the OPC binder. Soft, broken bricks are no longer used as an aggregate, the supply of washed building sand being cheap and readily available. Pan mill mixers with rollers are no longer used but horizontal pan mixers and their modern equivalent, the turbo mixer, are used. Both types use paddles to 'stir' the ingredients. The paddles may rotate with the pan fixed or the paddles may be fixed and the pan rotates.

To ensure correct and consistent proportions in the mix a gauging box should be used. This is a four-sided box which is placed on a clean board and filled once with binder. The box is lifted up, replaced on the board and filled, say, five times with sand. The six volumes are shovelled into the mixer and the result is a 1:5 mortar. Figure 1.44 illustrates the technique.

Mortar can be mixed by hand but only for small jobbing work. Consistency between batches is difficult to achieve with hand mixing. Gauging boxes should be used and the ingredients mixed dry on the board with a shovel, turning the pile over at least three times. Water is added and the whole mixed to an even consistency. Loss of binder frequently occurs when hand mixing and, in consequence, many operatives put more OPC into the mix just in case it is lost. The loss is carelessness and better mixing technique should be employed, with *no* additional binder added as the mix might well end up being too strong.

Good or bad weather

Conditions affecting the building of masonry are hot dry weather, frost and snow, and rain.

Hot, dry conditions

□ Hot, dry conditions dry out bricks and blocks and built masonry.
□ Absorption of water from mortar increases.
□ Dampen down bricks and blocks and cover stacks. *Don't* soak the bricks and blocks.
□ Use an air entraining agent or plasticiser in the mortar.
□ Cover built masonry with wet hessian to prevent premature drying out of the mortar.

Frost and snow

□ Bricks, blocks and building sands should be covered with waterproof covers *before* frost and snow occur. A low heat can be applied under the waterproof cover.
□ Materials which are frozen *must not* be used for building.
□ Even if materials are not frozen, building must *stop* when the air temperature reaches 4°C on a falling thermometer.
□ Building can recommence when the temperature reaches 4°C on a rising thermometer.
□ Mortar which freezes has no free water to react with the binder to achieve a 'set'.
□ When mild conditions return the mortar will be found to have dried out rather than 'set' and will therefore be weak and crumbly.

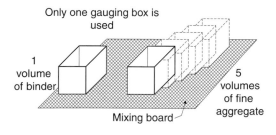

This system of mixing by volumes can be used when hand mixing where mixing actually takes place on the mixing board
OR
the board is placed in front of the concrete mixer and the measured materials are shovelled into the drum or pan for mechanical mixing, and either discharged onto the board or into a wheelbarrow

Fig. 1.44 Use of gauging box for proportioning ingredients in a mortar.

☐ Masonry affected has to be taken down; the bricks and blocks can be cleaned and rebuilt.

☐ It is important that the site has an accurate thermometer displayed outside the site office during the winter.

☐ If a clerk of works[11] has been appointed to the site he will keep a written record of weather conditions including twice daily temperature readings plus max./min. readings every 24 hours.

---

[11] A clerk of works is the architect's representative on site. He is employed, usually on large or high value sites, as a full time, on site, representative. He usually has wide authority to approve or condemn workmanship etc. and can be of inestimable assistance to all the professionals who only visit the site at intervals.

Rain

☐ Rain makes bricks and blocks unsuitable for building, the excessive moisture content leaving mortar in beds and joints so sloppy that it runs out – particularly the binder, leaving a weakened mortar behind and trails of binder down the face of the wall.

☐ If frost follows rain, then the water in the mortar and the units may freeze.

☐ Masonry just built should be protected from penetration by rain. Waterproof covers should be placed on top of the wall and weighted down. A good overhang beyond the face of the wall may be necessary. A full covering on the windward face may be necessary.

2

Substructures

Excavation generally	34
Topsoil	35
Subsoils	36
General categorisation of subsoils and	
their loadbearing capacities	37
Foundations	37
The principal considerations	38
Simple foundation calculations	39
The mass of buildings	39
Mass, load and bearing capacity	40
Foundation width and thickness	41
Reinforced concrete foundations	44
Failure of wide, thin, strip foundations	44
Trench fill foundations	45
Critical levels and depths	46
Level	46
Finished ground level	47
Bearing strata	48
Depths and levels	48
Step in foundation	49
Setting out	49
The site plan	49
Where do we put the building?	49
Equipment required for basic	
setting out	49
Setting out procedure	50
Excavation	53
Marking out the excavation	53
Excavation for and placing concrete	
foundations – and not wasting	
money doing it	53
Building masonry walls from foundation	
up to DPC level	57
Ground floor construction	59
Detail drawings	59
Wall–floor interfaces generally	62
Precautions	62
Solid concrete floors	62
Single and double layer concrete floors	
with hollow masonry wall	62
Hung floors	64
Hung timber floors	64
Hung timber floor alternatives	66
Hung concrete floors	67
Blockwork substructure	71

Buildings can be divided into two parts: the part generally below the ground floor and which extends down into the ground, and the part above the ground floor, which is termed the superstructure. This chapter is, therefore, quite a long one as it deals with ground conditions into which a building is set, right up to and including ground floor construction.

The sequence of events which will be discussed in this chapter begins with an examination of the ground conditions on a site, the excavations which must be done, the concrete foundations which must be placed, the walls built off these foundations, a variety of ground floor constructions, and treatment of the ground inside the building. Besides this, the chapter describes how the builder will place the building on the site with a good degree of accuracy, and details of techniques in general for that work are described in Appendix B.

Excavation generally

Generally speaking, ground conditions are divided between two types of soil: vegetable soil or **topsoil** and **subsoil.**

Over much of the world's land surface vegetation grows, and the layer of soil in which it grows is what we refer to as topsoil. Although the basic 'ingredients' of topsoil are mineral in origin, it contains a very high proportion of organic matter as well as bacteria, insects and other creatures such as worms.

Below the layer of topsoil there is a thin layer of material which is neither topsoil nor subsoil but is considered to be topsoil for our purposes in building construction. It is a transitional layer where the two forms of soil meet. The subsoil layer lies below and has no organic constituents in its make-up, although it might be home to living creatures that burrow below the topsoil.

Topsoil

Topsoil must be removed before a building is built. This is a requirement of all Building Regulations. The reasons are threefold:

(1) Topsoil has no loadbearing capacity and therefore cannot hold a building up, nor has it sufficient cohesion, depth or mass to allow the anchoring of a building in it.

(2) Topsoil contains organic matter which if left under the building would rot and cause a health hazard and/or give off methane gas which could cause an explosion, or hydrogen sulphide which has an unpleasant odour as well as being a health hazard. Rotting vegetation also attracts vermin, particularly insects, and this could be a source of disease etc.

(3) Vegetable roots, bulbs, corms, seeds or tubers left behind could sprout under the building and cause damage to any treatment applied over the area of the building, particularly damp proof layers. Roots can penetrate drainage and ducting systems, blocking or disrupting supplies or discharges.

The author remembers his father laying a concrete floor for a garage. The concrete was laid

Fig. 2.1 Site with topsoil removed – concrete foundation in foreground.

over a deep layer of compacted ashes obtained from a neighbouring farm. Some time after the floor had set hard it showed evidence of localised cracking and eventually sprouts appeared through the cracks. The spot was dug out and underneath were found a few potatoes buried in the ashes. It seems incredible that a potato could force its way through a concrete slab but this was during World War II and cement was in short supply, so perhaps it was not a particularly strong concrete.

Figure 2.1 shows a building site with the topsoil removed and concrete foundations for a house in the foreground.

The thickness of the topsoil layer will obviously affect the cost of the building. The more there is to remove, the costlier the exercise becomes. So how thick is that layer? Generally the layer of topsoil is fairly uniform across much of the UK, at around 200 mm where the ground has been cultivated, down to around 100–150 mm for naturally occurring land with good vegetable growth. Of course there are large areas where there is only heather growing, large areas of heathland, and large areas with peat and little growing on it. Also there are sites where there has been demolition and filling in, commercial dumping or filling in of depressions, holes, quarries, etc. There could be little or no topsoil in these areas, although there may well be vegetation growing. We will not go into foundations appropriate to these extreme conditions but will deal only with a simple topsoil layer over a stable subsoil.

One of the problems with presenting drawn information is that many of the drawings are prepared with no particular site in mind, and the designer therefore has to assume some 'standard' which can be followed by anyone trying to interpret the drawings. This idea of having a standard is something which will crop up on other occasions as we progress through the text. The 'standard' thickness adopted for topsoil is usually 150 mm, so that is what we will adopt in any drawings in this text.

When only one building is being erected it is usual to remove the topsoil from the immediate area of the building, but with mechanical excavators so readily and cheaply available it is much more usual to follow the 'big site' practice and remove the topsoil from the area of the site. The topsoil is most often required for the garden ground around the building and so must be preserved in some way. Any excavated material is called **spoil** and can be disposed of in a number of ways:

□ It can be removed from the site to a place where it is either disposed of, or stored, or used.
□ It can be immediately spread around the site being worked on. This is accompanied by the process of bringing the surface to a particular level and so is usually termed **spreading and levelling**.
□ It can be put into heaps called **spoil heaps** and re-excavated at a later date for use on the site being worked on. Spoil heaps of topsoil should not be higher than approx 2.50 m as storage in higher heaps can be detrimental to the soil in the longer term.

Subsoils

Subsoils can be divided into three groups:

□ Those capable of carrying the load of simple **low-rise buildings**[1] without the need for special techniques or precautions

---

[1] A low rise building is one generally not exceeding two storeys in height plus a pitched roof.

□ Those capable of carrying the loads imposed by much larger buildings with appropriate foundation techniques
□ Those requiring special techniques or precautions for even the most simple of structures.

We will concentrate on the first category.

But first the subsoil. Subsoil does not start at a line dividing it from the topsoil. There is a gradual transition from one to the other and generally occupies a layer 50–75 mm thick. This transitional layer is not generally suitable for building on and so the rule is that foundations are placed *in* the subsoil, not on top of it.

So, is there a standard thickness of subsoil? No there is not, simply because in theory there is no real end to the subsoil, but for practical construction purposes it can hardly extend into the hotter depths of the earth's crust. Is there a standard kind of subsoil? Again no; subsoil varies across the world, different kinds of soil being encountered at different depths and for different thicknesses.

No matter what the thickness or kind of subsoil, the most important consideration for anyone placing foundations in it is the **loadbearing capacity**, which can be defined as the force acting on a unit area which will cause the foundation to just fail. So we need to have a margin of safety and arrive at the **safe loadbearing capacity**. Where we have only our eyes and experience to guide us, a good rule of thumb is to halve the loadbearing capacity we might assess from visual examination and use this as the safe loadbearing capacity.

Loadbearing capacity depends on a number of factors, not all of them applicable to every kind of subsoil:

□ Kind of subsoil
□ Thickness of the layer of that subsoil
□ Kind and thickness of layers underlying the subsoil
□ Degree of compaction of the subsoil layers
□ Moisture content and general water level
□ Degree of containment of the layers
□ The presence or not of underground flowing water.

Ever since textbooks on construction have been written, attempts have been made by various authors to define what they conceive as standard types of subsoil. They have succeeded by and large and modern books generally categorise subsoils in the manner given below, together with an approximation of the loadbearing capacity given in newtons or kilonewtons (kN)[2]. For ordinary work these loadbearing capacities are quite adequate provided they are converted into safe loadbearing capacities as mentioned above. Only when very thin layers of subsoil are encountered or odd placement of low capacity layers is met, does the ordinary builder require expert guidance through sampling and analysis of the subsoil. The techniques used are well beyond the scope of this text but are generally known collectively as 'soil mechanics'. The specialists in this field can sample subsoil on a building site and test it to provide a safe loadbearing capacity, and can even go on to design the foundations for a particular building. They can also predict the probable cause and type of any failure which might occur due to overloading, variable weather conditions moisture content, etc.

General categorisation of subsoils and their loadbearing capacities

Naturally Building Regulations have a lot to say about loadbearing capacities. This usually takes the form of a reference, including tables which give widths of strip foundations in a variety of soil types for a range of loads. In the tables soils are generally classed as gravel, sand, silty sand, clayey sand, silt and rock. All of these must then be assessed according to their condition – compact, stiff, firm, loose, soft and very soft. Only rock allows a foundation the same width as the wall. The

others require a minimum width – given in the table – depending on the subsoil and its condition. A range of wall loadings is given, and the various widths allowable are noted underneath.

These tables are not too different from the tables which have been promoted in a variety of textbooks over the past 40–50 years, all based on Building Research Station Digests 64 and 67. As a simple example, one table quoted in a set of Regulations gives a width of 400 with a wall load of 40 kN per metre in a subsoil of compact sand or gravel. As will be seen when we consider simple width calculations a little later, this means that the safe bearing capacity of the subsoil is considered to be 100 kN/m². And this is despite the fact that maximum loadbearing capacity for these subsoils is quoted in many texts as >300 and >600 kN/m² respectively! By the rule of thumb quoted above, the safe bearing capacities of the soils would be >150 and >300 kN/m² respectively, which would indicate that the figures quoted in the Regulations for foundation widths and safe bearing capacities have a much higher margin of safety.

Foundations

Now that we have had a brief look at soils we really must consider what we mean by foundations. There is clearly a wide variation in the loadbearing characteristics of the soil categories in the tables mentioned above, and we need techniques to allow us to build in the majority of situations these will present.

First, it has to be said that foundations are generally made by pouring wet concrete into holes in the ground. The shape involved can be simple or complicated and everything in between. We are really only concerned with simple shapes – long rectilinear pieces of concrete cast in the ground and presenting a horizontal surface on which walls can be built.

Concrete is discussed in more detail in Appendix D and this should be read if the discussion in this chapter gets away from you.

---

[2] A newton is the force exerted by a mass of one kilogram acting with the force of gravity, g. So if we know that a mass of 1 kg is laid on the ground, the force exerted by that mass will be 1×9.81 m/sec² = 9.81 newtons. The kilonewton is derived from this and is equivalent to 1000 newtons.

The better the loadbearing capacity, the simpler the foundations may be. If we were dealing with a stable rock formation, then there might be no formal foundation constructed. Instead the rock would be levelled off and the walls or frame built straight off the rock surface. The majority of foundations are of two main types:

- **Strip** foundations[3]
- **Deep strip** or **trench fill** foundations.

In strip foundations a relatively wide trench (500–750 wide – wider than the walls) is dug and a layer of concrete poured in, to a thickness of at least 150. Walls are built off this layer of concrete. In deep strip foundations a narrow trench is dug (only slightly wider than the walls) and filled with concrete almost to ground level. The walls are then built starting almost at ground level. We will study these two types of foundation in greater detail later in this chapter. The other extreme would be soft ground to some depth overlying a firm strata with good loadbearing characteristics. In a situation like this it might be best to drive **piles** into the soft ground until they sit in the **bearing strata**. Piles are long columns of material which can be driven into the ground, or they can be cast in concrete and combinations of concrete and steel tubes sunk into holes 'drilled' into the ground. Once in, the tops of these piles are linked together with **beams** of concrete and the walls then built on the beams.

This is a very simple explanation of a complex process which has more than one way of being achieved. The piles themselves can be made of a variety of materials – concrete both **precast** and **cast** in situ, steel of a variety of sections, and even timber. On the other hand, the poor layer of soil might not be too bad and provided we can build sufficient area of foundation, we can spread the building load thinly over this poor soil. In this case we are looking at a **raft foundation**. This involves placing a layer of concrete over the area required to support the building. Note that this area could be much larger than the building but would be at least equal to the building's footprint. Then we could get into the more esoteric types of foundation: for example we could make a large box of concrete, divided internally into closed cells, and 'float' this on the site of the building. The individual cells would have valves which allow air to be bled off and water or silt to enter. Using these valves the box is sunk into the ground to the required level and the building built on top[4].

Obviously some of these solutions are very expensive and it would depend greatly on the total value of the building being built, as well as how urgent the need for that building, before making a final decision on whether or not to build at all.

The principal considerations

Foundations are placed in the subsoil to:

- Provide a base on which the building can be built so that it will not sink into the ground
- Prevent a building being lifted out of or off the ground
- A combination of both of the above.

The regulations further stipulate that the foundations must not be at risk from movement in the subsoil. This movement can be caused by two mechanisms:

(1) The subsoil takes up or loses moisture and as a result it swells or shrinks, causing the building to be displaced. Plants take up a great deal of moisture from the ground and in transpiration this passes into the atmosphere. Trees are major users of water and many species such as poplars and willows take up more than most. Thus they can take moisture from the subsoil during drought conditions, which can cause the subsoil to shrink. Trees are best kept well away from buildings – or vice versa – where shrinkable clay is encountered.
(2) Moisture in the subsoil freezes and causes the building to be displaced.

---

[3] Figure 2.1 showing a building site has a house foundation in the foreground which is a strip foundation.

[4] The BP refinery at Grangemouth is one of the largest of its type in Europe. Many of the buildings and petrochemical plant are built this way.

The first mechanism is a feature of **shrinkable clays**, which exhibit extreme swelling and shrinkage as their moisture content varies. These clays are distributed mainly in a wide area across south-east England[5]. Until recent years it was considered adequate to place the bottom of any foundation 1000 mm below the finished ground level to get below the point at which the moisture content will vary either by drying out or by absorption of additional water. The sides of the foundation and any walls built below ground could also be affected and there had to be a crushable layer of material at the sides of the foundation and the wall below ground level. However, ever drier and hotter summers have placed some doubt on the efficacy of the 1000 mm depth and many professionals now argue for a greater depth.

The second mechanism is a feature of all subsoils and the only way to avoid **frost heave**[6] is to keep the bottom of the foundations below a depth to which frost can penetrate. This is generally held to be 450 mm. Some parts of the UK regularly have frosts penetrating to this depth, while it happens very occasionally across wide areas. Local knowledge is not the only guide in this matter; the Meteorological Office has information on weather patterns across the UK and has a website at www.met-office.gov.uk/construction/pastdata5.html or www.met-office.gov.uk/construction/pastdata3.html

Simple foundation calculations

In the construction forms – brick and timber frame – which we will study, there is sufficient mass inherent in the form to obviate the need for the foundation to keep the building in the ground. So, we will be principally concerned with providing adequate support.

This does not mean that within the construction form we can ignore the need to tie portions of the construction together or to

[5] For further information on the distribution of soil types across the UK, contact the British Geological Survey by visiting their website at www.bgs.ac.uk
[6] So-called because water when it freezes expands and so causes the ground to expand and lift upwards.

the foundation. To determine the amount and type of support we need to know:

- The mass we intend to support, for which we need to know the construction form to be adopted
- The bearing capacity of the subsoil on the construction site.

The mass of buildings

Mass of buildings depends on the construction form adopted, the building type and the end use of the building. Calculation of mass is not difficult and a few steps will now be outlined.

To give the reader some idea of the range of masses, consider some contrasting buildings:

- Two-storey house with brick cavity walls, timber floors, timber roof with concrete tile covering. This is the medium to heavy weight type of low-rise building. Wall masses where upper floors and roof have to be taken into account range from 2500–3000 kg/m of wall.
- Timber frame house with brick or rendered blockwork cladding, timber floors, timber roof with concrete tile covering. This is a light to medium weight type of low rise building. Wall masses where upper floors and roof have to be taken into account range from 1500–2500 kg/m of wall.
- Timber frame single or two-storey 'club house' with timber or 'plastics' weather board cladding, pitched roof with bituminous shingles. This is a light weight construction. Wall masses where upper floors and roof have to be taken into account range from 800–2000 kg/m of wall.
- Only an unlined timber garage or garden shed would be lighter.

The mass of a building is transferred to the foundations through the walls. The external walls carry the bulk of this load and we normally only calculate foundation size for this amount. The building load on the foundations comes about because of the following:

□ Walls which in turn support floors and roofs.
□ Roofs bear **snow loads** (rain also adds to the load but not as much as snow and wind so it is ignored).
□ **Wind loads** affect the walls and roofs of buildings but mainly the roofs where the loads can be positive as well as negative – positive loads pushing against the walls or down on the roof, and negative loads trying to suck the walls out or the roof. When the wind blows, both types of load are experienced by the walls and roof.
□ Floors bear dead and live loads – dead loads are those imposed by the masses of other parts of the structure such as partitions built off the floor and things like furniture and fittings; live loads are those imposed by masses of the occupants, the dog and the cat, and these occupants moving about or jumping up and down.

Those walls which support both floor(s) and roof will have the heaviest loadings. Normally we calculate only for these walls and make all other foundations the same. Where there might be a considerable saving in excavation and concrete, more individual calculations might be done.

Mass, load and bearing capacity

The mass in the walls is normally calculated on a 'per metre' length of wall basis, measured in **kilograms** (kg). The load for a given mass is measured in **newtons**, i.e. 1 newton $= 1\,\text{kg} \times g$ (the acceleration due to gravity – 9.81 m/sec^2). It is more usual to employ the derived SI unit of **kilonewtons** (kN), i.e. 1000 newtons.

Bearing capacities are also given in kilonewtons per square metre. *Safe* bearing capacities should be taken as *half* of that figure if only a visual examination of the subsoil has taken place. This allows not only a factor of safety but also for variations in strength due to disturbance and varying moisture content.

A simple calculation will give the foundation area required to support a building's walls or, for a length of 1 metre of wall, will give us the width of the foundation. Consider the following.

A building has a mass at the foot of the walls of 10 200 kg/metre length of wall. The walls will be 265 thick. The subsoil identified visually has a bearing capacity of 300 kN/m^2, so a safe loadbearing capacity (SLC) would be 150 kN/m^2. To begin the calculation of the width of the foundation required we take the mass of wall – 10 200 kg/m – and convert that to the imposed load per metre:

$$10\,200 \times 9.81\ \text{N/m} = 100\ \text{kN/m}$$

The width of foundation required is:

$$\frac{\text{Load per metre } 100\ \text{kN/m}}{\text{Safe bearing capacity/m}^2\ 150\ \text{kN/m}^2}$$

$$= 0.66\ \text{m}$$

$$= 660\ \text{mm}$$

In the short discussion earlier regarding subsoil types and loadbearing capacities, a foundation width of 400 was given for a wall with a load of 40 kN. Applying the above formula we can calculate the SLC:

$$\frac{\text{Load per metre } 40\ \text{kN/m}}{\text{SLC}} = 400\ \text{mm or } 0.40\ \text{m}$$

$$\text{So: SLC} = \frac{40}{0.40} = 100\ \text{kN/m}^2$$

This seems somewhat at odds with the loadbearing capacities quoted for the soils in excess of 300 and 600 kN/m^2 and our own contention that we should halve these for visual assessment of safe loadbearing capacity.

The thickness of the walls was given above and yet played no part in the calculation. It was included deliberately to show that foundation width is not necessarily related to wall thickness. In this case the foundation will project beyond the faces of the wall by approximately 200 and this does have a bearing on the thickness of the foundation – as the next section will illustrate.

Foundation width and thickness

First, we make no apology for repeating some of the important points already mentioned in the text. The apocryphal tale comes to mind of the sergeant major who was congratulated on how well his recruits were trained and had this to say, 'First Ah tells 'em wot it is Ah'm gonna tell 'em, then Ah tells 'em, then Ah tells 'em wot Ah told 'em'.

So for at least the second and not the last time in this chapter:

- Foundations must be placed *in* a bearing stratum so that the full bearing potential of the stratum is exploited.
- Note the zone between two strata where *mixing* can occur and which is not suitable for foundations.

Both these points are included in Figure 2.2.

- The bottom of the foundation must be **deep enough** to avoid **frost heave**. This depth can vary with local weather conditions but **generally 450 mm** is sufficient in the UK and is illustrated in Figure 2.3.
- In shrinkable clays, the bottom must be deep enough to avoid heave due to changes in moisture content of the clay, as shown in

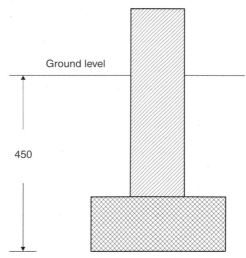

Fig. 2.3 Minimum depth to avoid frost heave.

Figure 2.4. In all but the most severe conditions 1000 mm is sufficient.

- Very narrow trenches are difficult to work in while building walls and once the depth exceeds 700–800 it becomes almost impossible. So the cladding of foundation sides and backfilling with crushable material shown in Figure 2.5 is only possible when the

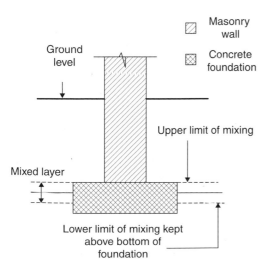

Fig. 2.2 Foundation penetrating mixed layer into subsoil.

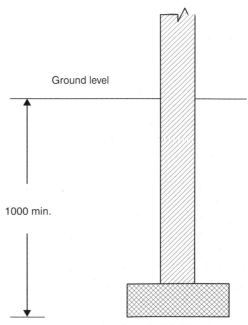

Fig. 2.4 Minimum depth in shrinkable clay.

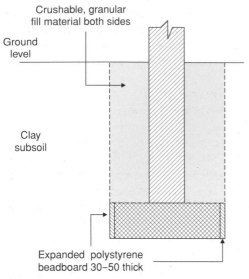

Fig. 2.5 Protection from heave in shrinkable clay.

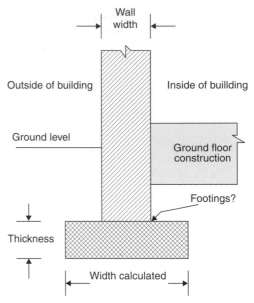

Fig. 2.6 Foundation with width calculated – thickness?

foundation is of a reasonable width. (See deep strip foundations later in this chapter.)

☐ In shrinkable clay, the sides of foundations and the walls on them must be protected from subsoil movement. Expanded polystyrene[7] is frequently used, and filling round the walls is done with granular, crushable material which will in all probability have to be imported onto the site. The use of this material is illustrated in Figure 2.5.

☐ When using bearing capacities quoted in tables and textbooks combined with a visual inspection only, the capacities should be halved as discussed earlier.

We will study two types of foundation: **strip foundations** and **deep strip** or **trench fill** foundations. The method of calculating width is the same for both types. Thickness depends primarily on type but for main wall, strip foundations is *never less* than 150 mm. Figure 2.6 shows a typical section through a strip foundation.

---

[7] The product used is a 'beadboard' made by compressing beads of expanded polystyrene so that they adhere into a board or sheet material. It is the same material which is frequently moulded into packaging for fragile merchandise.

Width is calculated as before and **thickness** depends on:

☐ wall width
☐ whether the concrete is to be **plain** or **reinforced**
☐ whether or not there will be **footings**.

Assuming that the concrete will be plain and there will be no footings, thickness must be equal to the projection of the concrete from the face of the wall, as shown in Figure 2.7.

Note that walls are always centred on the foundation and that any additional width of concrete in a plain foundation which extends beyond the point where the dotted line reaches the bottom, would, in theory, tend to break off. To be effective the extra width would have to involve further techniques to strengthen the concrete.

For a given wall thickness the width required for a foundation increases as:

☐ wall load increases, *and/or*
☐ bearing capacity of soil decreases.

As the width increases so does the thickness, as illustrated in Figure 2.8. There comes a time

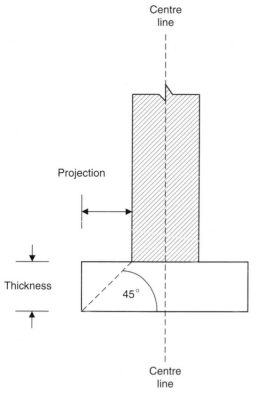

Fig. 2.7 Calculation of thickness of plain foundation.

when it is more economic to consider a course of action other than further increasing thickness. Dictated mainly by economics, two possibilities are open to the designer:

☐ Use footings
☐ Reinforce the concrete.

Footings are formed by building courses of brickwork, each course being half a brick wider than the one above, as shown in Figure 2.9. The additional width is *always* evenly distributed on each side. The footings courses are usually laid brick on edge. The net result is the spreading of the load from the wall over a widening area of brickwork and then onto a relatively thin, wide layer of concrete. The thickness of the concrete equals the projection as before.

There is an economic limit to the number of courses of footings one would place. That limit varies as the relative costs of brickwork and the alternatives change with time. Quantity surveyors should be capable of calculating guide figures appropriate at the time a building is being designed.

The second alternative is reinforcement of the foundation.

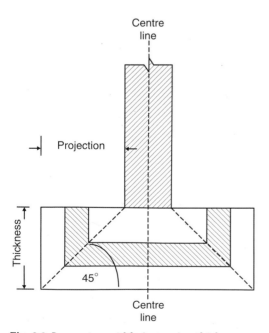

Fig. 2.8 Increasing width, increasing thickness.

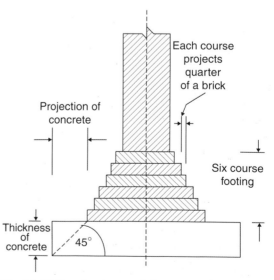

Fig. 2.9 Wall on footing on plain concrete foundation.

Reinforced concrete foundations

The use of reinforcement bars is generally confined to the foundations for larger structures. Fabric is generally adequate for reinforcing strip foundations to low rise buildings.

Fabric is supplied in sheets, available in several standard sizes which may be cut to suit the width and length of the foundation being laid. Individual pieces must overlap with adjacent pieces by at least one 'cell'. Corners and intersections should fully overlap. Laps should be tied together with 1.6 mm soft iron wire ties. If a large number or quantity of foundations is being laid, it can be more economic to have the fabric supplied in roll form, the width being set to suit the foundation width.

Failure of wide, thin, strip foundations

There are two modes of failure:

□ **Bending** as shown in Figure 2.10
□ **Punching shear** as shown in Figure 2.11.

To avoid failure by bending, fabric must be placed near the bottom of the concrete, as shown in Figure 2.12. Concrete is not strong in tension but the fabric is and so the elongation of the bottom layer is prevented or controlled. Concrete is strong in compression and generally does not require reinforcement in the top layer to prevent failure by crushing. However, the narrow wall concentrates its load over a small area of concrete and thus the wall could punch its way through the strip foundation. Fabric in the bottom alone cannot resist this and so another layer is placed in the top, as shown in Figure 2.13.

In theory, the bottom layer has to be the full width of the strip foundation and the top layer can be narrower. In practice, unless there is an excessively wide foundation, both layers are the same width. The financial effect is minimal. The same type of fabric should be used for both layers, thus preventing errors

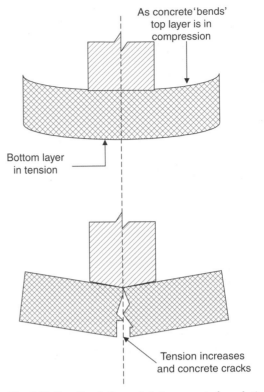

Fig. 2.10 Bending failure of plain concrete foundation.

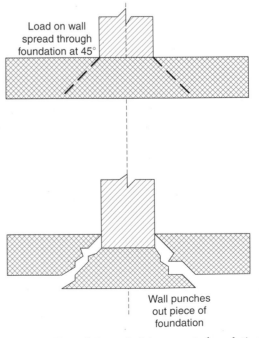

Fig. 2.11 Shear failure of plain concrete foundation.

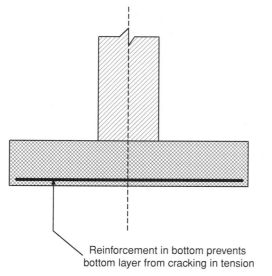

Reinforcement in bottom prevents
bottom layer from cracking in tension

Fig. 2.12 Reinforcement of bottom layer of concrete foundation.

in placing; don't think things like that never happen – they do with monotonous regularity.

Trench fill foundations

Trench fill is sometimes referred to as deep strip but to avoid confusion this text will refer only to trench fill.

In cohesive soils (which generally have an adequate loadbearing capacity) it is possible to require a foundation to be little wider than the wall to be built. Money can be saved

then by digging a very narrow trench with a mechanical excavator equipped with a narrow 'bucket'. Trench depth must still be maintained so as to reach into the bearing strata and to avoid frost and/or clay heave.

Because of the cohesive nature of the soil, the sides will remain stable until concrete is poured in, provided there is the minimum of delay between digging and filling and no intervening adverse weather conditions. Concrete is filled in to a point just below the finished ground level.

The operatives do not need to work *in* the trench as is the case with strip foundations. Support for the foundation is not just from the bottom of the trench but can also be from the friction of the sides of the foundation in the ground. This is illustrated in Figure 2.14.

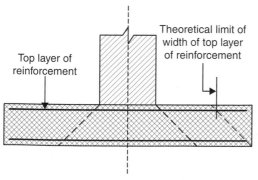

Top layer of
reinforcement

Theoretical limit of
width of top layer
of reinforcement

Fig. 2.13 Top layer reinforcement of concrete foundation.

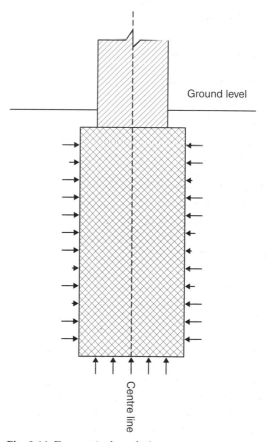

Ground level

Centre line

Fig. 2.14 Deep strip foundation.

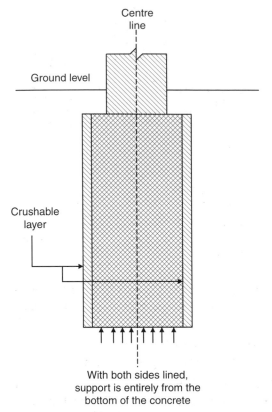

Fig. 2.15 Deep strip foundation in shrinkable clay.

Trench fill foundations do not fail by bending or punching shear. Shrinkable clay is often suitable for trench fill foundations. In this case the sides of the foundations should be lined with a 'crushable' material such as expanded polystyrene sheet. The additional work is illustrated in Figure 2.15.

Critical levels and depths

From our study so far it should be obvious that there are a number of critical levels and heights which depend on:

☐ The lie of the ground, i.e. how it slopes etc.
☐ How it will be finished around the building
☐ The nature of the subsoil
☐ The need to protect the foundation from **heave**.

Level

The word level can be applied in two ways:

☐ We can refer to a 'level' when we refer to an absolute height above a **datum** or to a height relative to some other object.
☐ We can describe a surface as 'level' when we describe a horizontal surface.

The most common form of datum is an Ordnance Survey (OS) benchmark. (See also Appendices A and B, which deal with Maps and Plans and Levelling.) These marks carved into corners of buildings etc. are related to sea level at Newlyn in Cornwall. This is the datum for all OS levelling in the UK. The mark comprises two parts: the deeply incised horizontal line, the centre of which is the actual level referred to on OS maps, and the broad arrow symbol underneath the line. The arrow is a British Government symbol used on everything from prison clothing to army weapons.

Figure 2.16 is a sketch of a typical benchmark. Photographs of various benchmarks are shown in Appendix B.

Temporary benchmarks are frequently used in building work, and two examples are shown in Figure 2.17. They may relate to OS levels or to an arbitrary value. Levels are given in metres to two places of decimals, e.g. 84.56.

As shown in Figure 2.17, a wooden peg driven into the ground and set round with

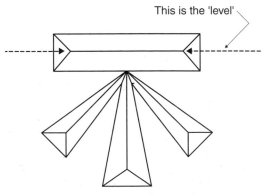

Fig. 2.16 Sketch of benchmark symbol.

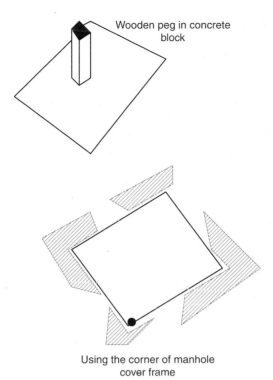

Wooden peg in concrete block

Using the corner of manhole cover frame

Fig. 2.17 Temporary benchmarks.

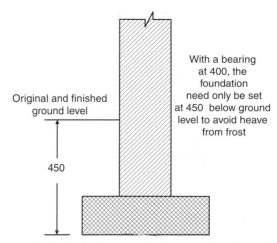

Original and finished ground level

450

With a bearing at 400, the foundation need only be set at 450 below ground level to avoid heave from frost

Fig. 2.18 Finished ground level at same level as original ground level.

concrete can be used as a temporary benchmark. Even a length of steel angle iron can be treated this way. The pegs are usually painted in contrasting stripes and the top is usually painted red. Where there is a danger that the benchmark might be disturbed, a low fence or rail can be built around it. The corner of the manhole (usually a cast iron cover) is a favourite with many surveyors but the corner of the frame should be marked, with paint, not the actual lid.

Finished ground level

By 'finished ground level' we mean the height of the ground above a datum after all excavation and/or filling work and the spreading of topsoil has been carried out around the building. The ground around a building does not always need to present a horizontal surface, and the finished surface can be higher or lower than the original ground level.

When original and finished ground level are the same, as shown in Figure 2.18, the excavation depth is set by:

☐ Need to reach a bearing
☐ Need for cover to prevent heave.

Frost cover dictates that the bottom of the foundation is 450 below **finished** ground level, as shown in Figure 2.19. There should be no difficulty reaching a bearing with that minimum depth. The foundation bottom must

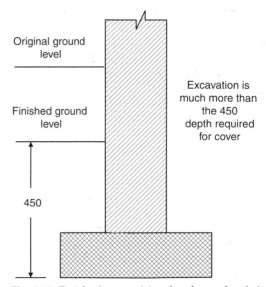

Original ground level

Finished ground level

450

Excavation is much more than the 450 depth required for cover

Fig. 2.19 Finished ground level at lower level than original ground level.

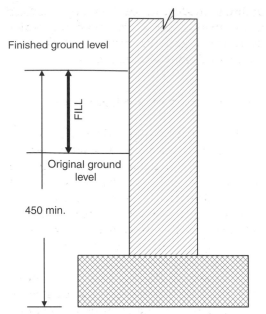

Fig. 2.20 Finished ground level at higher level than original ground level.

reach down to bearing strata, as shown in Figure 2.20. Cover is generally no problem. Note the large amount of filling.

Bearing strata

A bearing may not be found at a convenient depth below finished ground level. There will be occasions when the foundation must be deeper than the minima previously illustrated, and also site surfaces are seldom truly horizontal, nor do bearing strata follow the same angle of slope.

Assuming a level site, strata may be horizontal, as illustrated earlier, or strata may slope slightly, moderately or steeply, as illustrated in Figure 2.21.

If the site surface slopes, strata may be level or strata may slope with or against the site surface, as illustrated in Figure 2.22.

Depths and levels

Depth of foundation below finished ground level is controlled by:

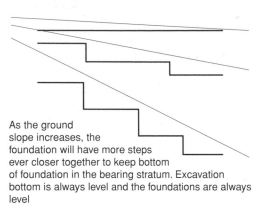

Fig. 2.21 Bearing strata sloping in relation to ground level (1).

- □ The need for frost cover or to avoid clay heave
- □ The need for the foundation to be *in* the bearing stratum
- □ The slope of the finished ground

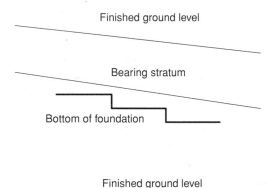

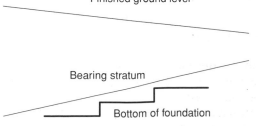

Fig. 2.22 Bearing strata sloping in relation to ground level (2).

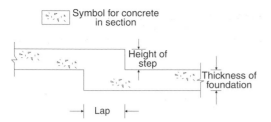

Fig. 2.23 Step in foundation – the rules.

- The slope of the bearing stratum
- The need to be economic – no excessive excavation; without **stepped foundations** there would be too much excavation.

Step in foundation

In very steeply sloping strata, the steps may not just be frequent but the height of the individual steps may be great. For plain concrete, the step height should be restricted to the equivalent of the foundation thickness, otherwise the laps, and perhaps the whole foundation, should be reinforced with steel fabric. The step is illustrated in Figure 2.23.

The lap must equal the thickness, twice the height or 300, whichever is greatest. The height of the step should be set to a multiple of the masonry unit height – for standard metric bricks this means multiples of 75.

The vertical faces of the step in the foundation need support while the concrete sets. This can be done simply by placing bricks across the foundation after allowing the lower layer to set a little, or for large steps timber formwork has to be erected and fixed to pins set into the bottom of the trench.

Setting out

The site plan

If the reader has not already done so, now is the time to read Appendices A and B, on Maps and plans, and Levelling.

A plan of the site usually to a scale of 1:500 is drawn showing the existing features of the site. To this plan are added the proposed buildings, roads and services, any ancillary works such as car parks, retaining walls, screen walls and landscaping work; new ground levels may also be shown.

Where do we put the building?

The site plan so carefully and laboriously prepared is just so much waste paper if the men on site cannot place the building where the plan shows it to be. Foundations must be placed at the correct level and ground floors at the correct height above finished ground level. All depend on the correct transfer of the proposals from plan to the physical site in a proper, well ordered manner.

When planning the site layout, the architect or engineer is frequently constrained by 'building lines', i.e, the front of the building has to be sited on an imaginary line drawn on the site surface. This line should be shown on the site plan, with its position relative to known fixed points clearly marked.

Other examples of lines from which reference can be made when siting a building are:

- A road or pavement kerb line
- Extension of the frontage line of existing buildings.

A typical site plan is shown in Figure 2.24.

There may be no such building line and the corners of the plan of the building may be marked by dimensions taken from at least two known fixed points.

Equipment required for basic setting out

- Dumpy level, tripod and staff
- 50 × 50 wooden pegs 600–750 mm long
- 2 lb hammer and a claw hammer
- 65 mm plain wire nails
- Two steel measuring tapes 30 or 100 m long
- Builder's line and builder's level
- Measuring rods
- Crosscut hand saw
- 94 × 19 wrot boarding.

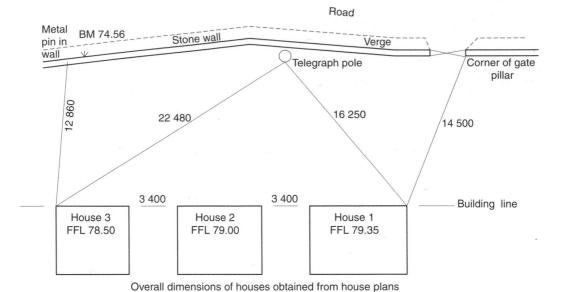

Road

Metal pin in wall   BM 74.56   Stone wall   Verge   Corner of gate pillar

Telegraph pole

12 860

22 480   16 250   14 500

3 400   3 400   Building line

| House 3 FFL 78.50 | House 2 FFL 79.00 | House 1 FFL 79.35 |

Overall dimensions of houses obtained from house plans

Fig. 2.24 1:500 site plan.

Setting out procedure

Walk round the site and identify the fixed reference points used for the original survey. Look for nails, pins, hooks or loops, marks, etc. to or from which measurements were taken. Confirm these points by re-measuring.

Initially the ends of building lines and corners of walls of new building works are marked by placing a wooden peg in the ground and placing a nail in the top of the peg, as shown in Figure 2.25. This may appear a very primitive way to mark out a spot in the middle of a building site, but it has been used for many decades and no better way has been found.

If there is no building line, mark the corners of the building(s) by placing pegs in the ground. Figure 2.26 illustrates how it is done. Using a steel tape from each of the known fixed points, measure the distance given on the site plan to the end(s) of the base line or building corner. Where readings coincide, drive in a peg. Measure again across the top of the peg and put in a nail.

In the case of a building line, with the position of the ends fixed, stretch a builder's line from nail to nail, and using a steel tape

measure along the line to establish building corners. Mark each corner with a peg and nail, as shown in Figure 2.27.

Having found two corners of the proposed building, set out and mark the others. Use

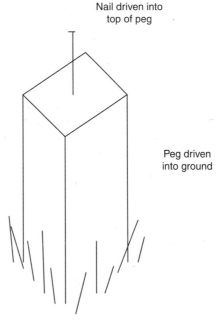

Nail driven into top of peg

Peg driven into ground

Fig. 2.25 Peg with nail, used to mark corner of a building on site.

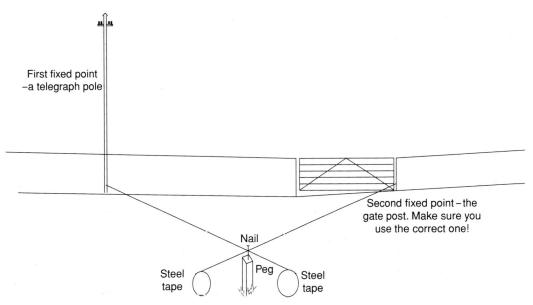

Fig. 2.26 Finding position of peg from two points.

the Pythagoras theorem in any combination of right angled triangles, as shown in Figure 2.28.

Having placed pegs at all four corners, check their positions by measuring the diagonals as shown in Figure 2.29. The lengths of the diagonals should be within 6 mm of each other on a medium size house.

We now know where the corners of the building's walls will be. When we place foundations we excavate and place concrete in the bottom. Both excavation and concrete are wider than the walls. The pegs marking the corners will therefore be displaced and we lose our markers. This is shown in Figure 2.30.

We need to improve on these first basic markers so that we have reference points which:

□ Are not disturbed by subsequent work
□ Indicate vertical alignment as well as horizontal
□ Allow repeatability of corner marks, measurements and levels.

We need **profile boards**, as illustrated in Figure 2.31. Profile boards are set up at each corner of the building. Each set 'looks' two ways along adjacent walls, as shown in Figure 2.32. To 'see' the outline of the building's concrete foundation and walls, builder's lines may be stretched from profile to profile. With all that string around, there is no room to get in to excavate for the foundation, but see how excavations are marked out below.

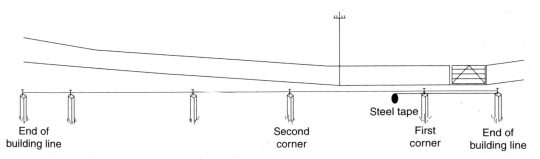

Fig. 2.27 Pegs on the building line marking building corners.

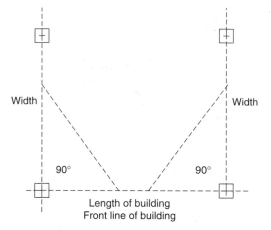

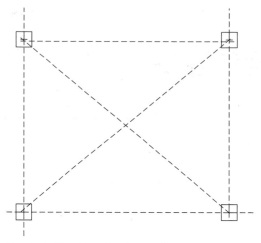

Fig. 2.28 Using the Pythagoras theorem to find other corners of building.

Fig. 2.29 Checking the diagonals.

Of course you may end up working for a large company which can afford electronic position and distance measurement (EPDM) equipment, and then there is no need for all these pegs and profiles. One or two strategically placed pegs with a nail in the top, well protected, are left from the first survey and subsequently used to set out every bit of work on the site from roads and sewers to house drains and the buildings themselves. (See Appendices A and B.)

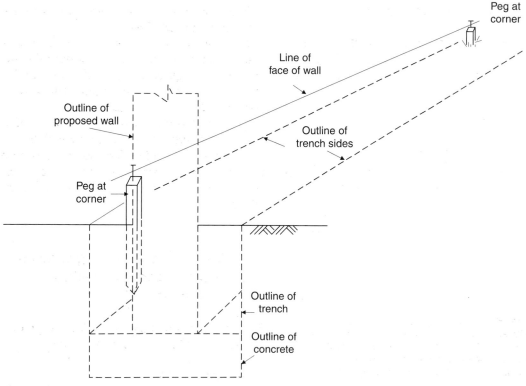

Fig. 2.30 Losing markers for building corners when excavating.

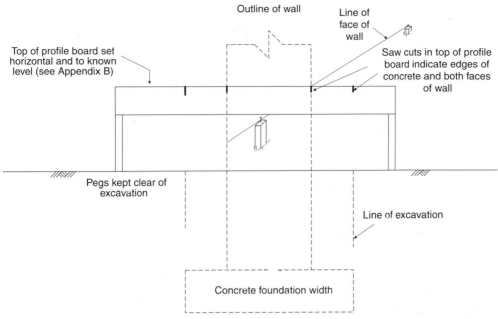

Fig. 2.31 The profile board.

Excavation

Marking out the excavation

Only put up builder's line for the outline of excavation/concrete; people with picks and shovels or an excavator could break the

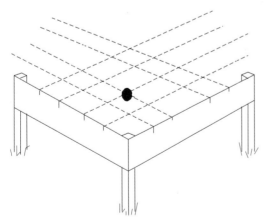

With builder's line stretching from notch to notch in opposite profile boards, the intersections of the strings mark the corners of various parts of the substructure. The black dot marks the corner of the wall to be built and should be exactly over the original peg and nail.

Fig. 2.32 The corner of the building wall.

strings. To avoid this, sprinkle old cement, lime or sawdust on the ground, following the builder's line as closely as possible. Now remove the builder's line.

If during work the mark on the ground is erased, string up the line and mark again. After a first rough excavation, the line can be restrung to check for alignment. This can be done as often as is necessary.

By using the other saw cuts in the profile board, wall outlines can be established. By using the known level of the top of the profile boards, the floor levels, depth of excavation and concrete thickness can be set and maintained.

Excavation for and placing concrete foundations – and not wasting money doing it

Excavation is now largely carried out by machine. Generally it is done in two stages:

☐ Bulk excavation
☐ Trimming to size (may be done by hand).

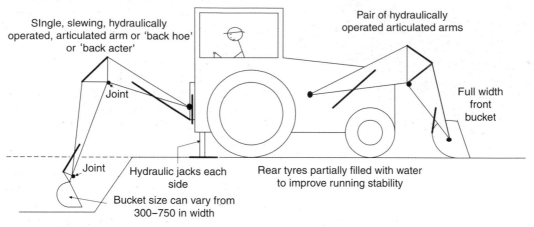

Heavy lines indicate hydraulic rams to operate arms etc.

SIngle, slewing, hydraulically operated, articulated arm or 'back hoe' or 'back acter'

Pair of hydraulically operated articulated arms

Joint

Full width front bucket

Joint

Hydraulic jacks each side

Rear tyres partially filled with water to improve running stability

Bucket size can vary from 300–750 in width

Fig. 2.33 JCB-type excavator.

Accuracy is important so as not to waste money:

☐ Too much excavation is a waste of effort and of concrete
☐ Too little may mean too little concrete or not enough cover against frost or clay heave.

With modern machinery it is common practice to remove vegetable soil from the entire site and store in spoil heaps for later use. Work therefore commences from subsoil level. If major adjustment to ground levels was required this would be done after removing the vegetable soil and before setting up profile boards for foundation excavation.

The most common excavator used is of the JCB type – a tractor-based machine having front and rear buckets on hydraulically operated arms (see Figure 2.33).

From the preliminary site investigation, the level of the bearing strata will have been established, vegetable soil has been removed, ground levels have been adjusted, and profile boards have been erected giving a datum for the three dimensional work about to take place. With the outline of the concrete marked on the ground an excavator commences work, digging out between the lines to form a trench. The digger driver must dig down *into* the bearing strata – *not too little, not too much*. He

requires some assistance and some means of gauging how deep he has dug, so has an assistant called a **banksman** who uses a device called a **traveller** to guide the digger driver to the correct depth of excavation, as illustrated in Figure 2.34.

The photograph in Figure 2.35 shows two men using a traveller between two profiles set up to gauge the excavation depth for a road.

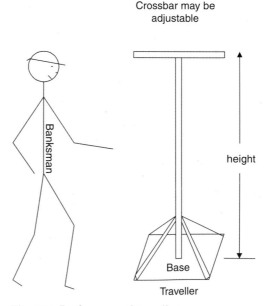

Crossbar may be adjustable

Banksman

height

Base

Traveller

Fig. 2.34 Banksman and traveller.

Fig. 2.35 Banksmen and traveller between profile boards.

The man on the left is holding the traveller, the man on the right is viewing the traveller cross bar in relation to the profiles in front of and beyond the traveller. They are checking the subsoil level to see that enough has been excavated before the first layer of hardcore goes down as a road base. They will alter the traveller (shorten it) to check that the hardcore bed is the correct thickness.

Travellers are made of timber, and the critical dimension is the height. The depth at which the concrete must be laid has already been determined. We know the level of the top of the profile boards. The height of the traveller is made *slightly less than the difference* between these two levels. The traveller is used by placing it at the bottom of the excavation and having the banksman sight across the profile boards past the cross bar on the traveller,

like the matchstick banksman in Figure 2.36. If the traveller protrudes above the line of sight more digging is required. If the traveller is below the line of sight too much digging has been done.

The bucket on the excavator will be one of a standard range of sizes, narrower than the trench being dug. On some subsoils, the buckets used will have a toothed cutting edge. These teeth disturb the bottom of the trench excavation.

Setting the traveller to less than the full height for the first pass means that the disturbance does not penetrate below the required level. Setting the traveller now for the full height allows the **disturbed layer** to be removed either by hand or with a plain edged bucket. This is called **trimming** and is illustrated in Figure 2.37.

Excavating into the ground for any purpose can be dangerous due to the possible collapse of the sides. The deeper one goes, and the more friable the subsoil, the more dangerous it becomes. In certain circumstances it is impossible to excavate at all unless one uses **timbering**. This is the short-hand term for **shoring, strutting and waling** – a range of systems to support the sides of excavations to prevent collapse and protect workmen down in the excavation. Appendix F describes the techniques involved. Shoring, strutting and waling is **temporary work** and is, by convention, never shown on drawings. It will not be

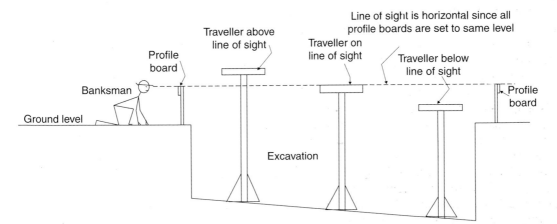

Fig. 2.36 Traveller in use in an excavation.

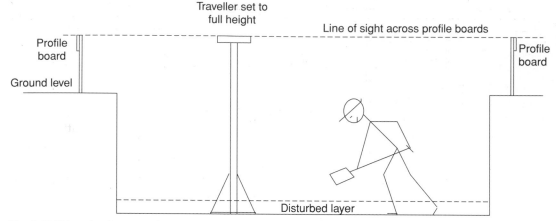

Fig. 2.37 Trimming bottom of excavation.

shown in the drawings included in this text, except for Appendix F.

After excavation the trench should:

☐ Be straight
☐ Have a horizontal bottom
☐ Have the bottom at the correct depth in the bearing stratum
☐ Have an undisturbed bottom.

Straightness is tested by restringing the builder's line on the profiles and checking the sides of the excavation with a builder's level, as shown in Figure 2.38.

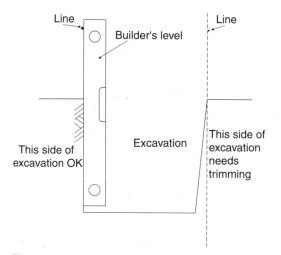

Fig. 2.38 Trimming sides of excavation.

Laying the concrete

Once the excavation sides and bottom have been trimmed, the concrete should be laid, before the bottom of the trench can be disturbed again or become sodden with rain.

Special precautions are taken to deal with ground water. Either the water is drained off in specially dug trenches from the low points in the excavation, or **sumps**[8] are dug at the low points of the excavation and mechanical pumps take out the water and dispose of it outside the excavations.

The concrete must be laid to the correct thickness:

☐ Too thin means a weak foundation, failing in bending and shear.
☐ Too thick means a waste of money on too much concrete.

The upper surface must be horizontal and allow full courses of masonry up to DPC level. Once you start to pour concrete into the trench, you lose sight of the bottom of the

[8] A sump is a hole dug below the lowest point of an excavation. The hole need not be excessively large – a 600 sided cube might be adequate for house foundations. The end of an extract hose fitted with a strum box (a coarse perforated metal filter) is fitted on the end to prevent small stones being sucked over and damaging the pump.

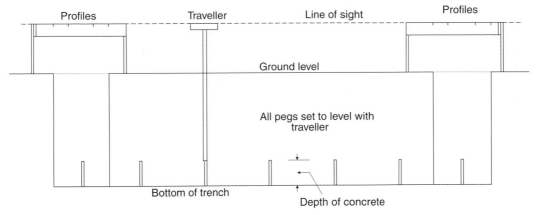

Fig. 2.39 Using the traveller to set concrete foundation thickness.

trench and all measure of concrete thickness so provision must be made to allow this measurement to be made or maintained. The profile boards provide us with a unique reference line to control both thickness and the level of the upper surface. Reduce the height of the traveller so that it extends from top of profile to top of concrete. Drive pegs into the bottom of the trench such that the traveller intercepts the line of sight across the profiles, as shown in Figure 2.39. Pegs are spaced out about 1 metre apart down the centre of the trench.

Building masonry walls from foundation up to DPC level

With the foundation concrete laid, we are ready to start building the walls. Start by finding the position of each corner using a builder's line on the profile boards, and a builder's level. The builder's level is applied to the adjacent strings and this will fix the wall's corner, as shown in Figure 2.40.

Corners

At each corner, bed a brick in mortar as shown in Figure 2.41. Adjust its position in both directions until it falls directly below the crossover of the builder's lines. This is the corner of the wall.

Building the corner out

As the position of each corner is established, additional bricks are added in that first course for a distance of five or six brick lengths, as shown in Figure 2.42.

Keeping the course properly aligned

Alignment of this first course of bricks is done using a builder's level, with reference to the

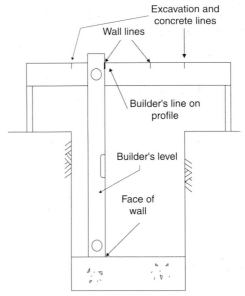

Fig. 2.40 Finding corner of wall on concrete foundation.

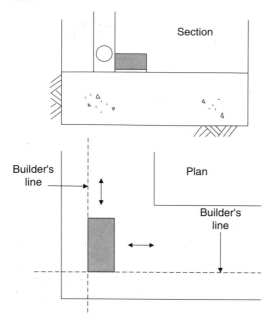

Fig. 2.41 Setting the first brick of the corner of wall on the foundation.

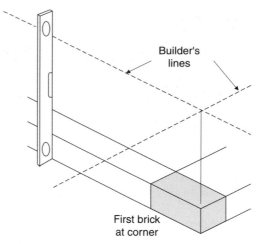

Fig. 2.43 Keeping the first course of bricks aligned on the foundation.

builder's line strung on the profile, as shown in Figure 2.43.

Building up the corners

With the first course started at the corners and properly aligned, the rest of the corner is built up keeping it 'plumb' (vertical) by using the builder's level. The brickwork is 'racked back' on each wing, as shown in Figure 2.44.

Filling in

With corners built at each end of a length of wall, a builder's line is strung from the

brickwork about four or five courses up, as shown in Figure 2.45. Bricks are laid between the corners, filling up the four or five courses.

Going up

With the first few courses infilled, corners are built up again and infilling continued, as shown in Figure 2.46. Brickwork is built to a maximum of 1.50 m (20 courses) high in any one working day, irrespective of wall width. This height is referred to as a **lift**. Building any higher on unset mortar brings the risk

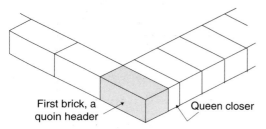

Fig. 2.42 Setting up the first five to six bricks at corners on the foundation.

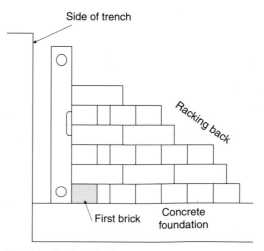

Fig. 2.44 Racking back.

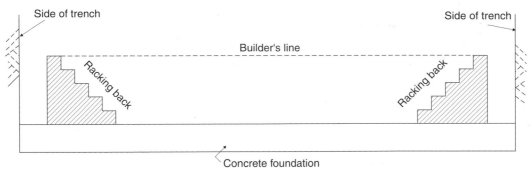

Fig. 2.45 Two corners built and racked back ready to fill in.

of the wall settling down unevenly and even collapsing. Left overnight, another lift can be built next day.

Ground floor construction

We are now approaching the point where drawings in the text will in the main be considered to be 'construction details'. As such, the details will show the construction of more than one piece of the building. For example, we cannot show how a ground floor is constructed without relating that to the construction of the foundation and the substructure walls; we cannot show how a roof is tiled without showing the carpentry structure involved

at key parts of the roof, and so on. So as we look at a variety of ground floor constructions we must also examine how or if they affect the way in which the rest of the substructure is constructed. So we really look at the interface between a floor and a wall.

Detail drawings

The most efficient way to 'describe' the interface between two or more pieces of construction is to draw an appropriate detail. When the detail is complete it must contain:

☐ **annotation**
☐ **illustration**
☐ **dimension**.

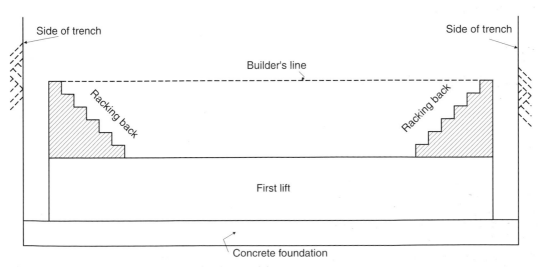

Fig. 2.46 Corners extended and ready for the next lift.

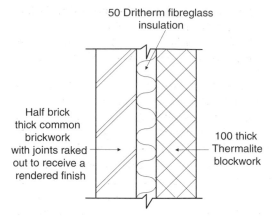

Fig. 2.47 Wall detail – symbols for materials in section.

Annotation

An annotation is a note describing what has been illustrated, e.g. Figure 2.47, and expanding on what is seen, e.g. a symbol for insulation cannot distinguish between, say, polystyrene and glass fibre; it requires annotation to make the distinction.

Illustration

An illustration is a drawing to scale or a sketch to an approximate scale or in proportion. Different line widths or strengths are used to emphasise important features. The wrong line thickness or strength sends confusing messages, as can be seen in Figure 2.48. Correct drafting symbols and conventions must be used.

Dimension

All key dimensions must be given so that no-one need scale off any dimension. Scaling dimensions from a drawing is not done if at all possible. If one would have to scale off, the drawing is missing key dimensions and should be returned to whoever produced it for them to put in the key dimensions. Look at Figure 2.49 for an example of key dimensions.

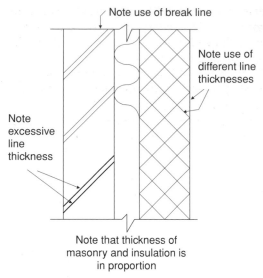

Fig. 2.48 Use of correct line thickness.

There is no need to give every single dimension. Some can be left to be calculated. That is quite legitimate and indeed expected.

An example of a detail

The detail for a wall–ground floor interface must include information, where appropriate, on the foundation, the wall, the floor, DPC, DPM (damp proof membrane), joining DPC and DPM, wall insulation, floor insulation,

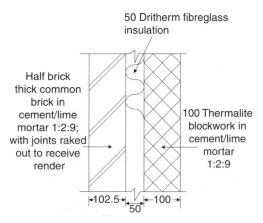

Fig. 2.49 Wall section with dimensions and annotation.

finishes to wall and floor, ground levels and ventilation. This is all part of the annotation and includes information on material, workmanship and dimensions.

DPCs and DPMs are discussed in Appendix I, together with weeps and ventilation of underfloor voids.

The following is a simple example relating to a non-domestic situation.

A one brick thick wall is supported on a concrete foundation in 'normal' soil. The floor will be a solid concrete floor with a power floated finish. No insulation is required. Because bearing strata and finished ground level are not mentioned here, we can assume that what is required is a 'standard detail', i.e. what things would look like with the minimum amount of work required, a 'level' site and existing and finished ground levels the same.

The detail is shown in Figure 2.50:

- All key dimensions are given
- Dimensions are *always* in millimetres
- The illustration includes proper symbols for materials:
  - Earth
  - Concrete
  - Common brick – pairs of diagonal lines
  - Facing brick – pairs of diagonal lines with an opposite single line
- Hardcore and backfill – 'stones' and diagonal pattern respectively
- DPC and DPM – both shown as a pair of fine lines linked with blocks of black.

The illustration is very 'out of scale', e.g. the Visqueen sheet is only 250 microns thick.

Annotation is added, e.g. concrete mixes and finish, fill material, facing brick type, etc. On this detail the annotation is a bit over the top. It would be only one detail on a large sheet showing the ground floor plan of foundations, substructure walls and partitions, plus several other details. So it is going to be tedious to repeat a lot of the fine detail every time we produce a detail. Much of the common information would be consigned to a column on the right of the whole drawing under notes, and might include one statement on each of the following:

- Facing brick type
- Building mortar
- Pointing mortar
- Pointing or jointing style
- Hardcore specification
- Backfill specification
- DPC specification
- DPM specification
- Concrete finishes.

The concrete mixes referenced as A, B, C, etc. already refer to a printed document, the preambles, issued to contractors when they tender for the works. This would cut down on the information surrounding the detail and would be quite acceptable.

For this text we will still give as full a specification for each figure as we feel is necessary to avoid ambiguity or to make a specific point regarding the construction used. Where it might confuse or clutter up the drawing we will omit it, but please remember when drawings are done professionally for a contract, the information has to be on the drawing either as annotation or in notes on the drawing.

The detail in Figure 2.50 is of course only appropriate for non-domestic building with a very light floor load. There is no insulation of walls or floor. The walls being one brick thick

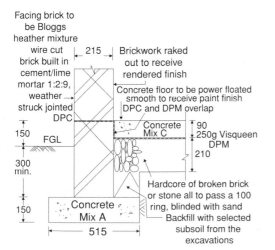

Fig. 2.50 Simple non-domestic substructure/ground floor interface.

are not of a sufficient weather resisting standard for domestic construction. The floor is a 'single layer' concrete floor and the thickness of concrete shown is the minimum allowed under the Building Regulations.

Wall–floor interfaces generally

When studying the variety of ways in which ground floors are constructed, it is usual to consider how they interface with:

☐ Outer walls
☐ Internal loadbearing walls
☐ Internal non-loadbearing walls
☐ Intermediate support.

Ground floor construction divides into:

☐ Solid concrete floors – laid direct on an inert fill and not supported by any type of wall
☐ Hung floors – constructed over a void and supported by walls and/or beams and columns. Hung floors can be constructed from timber, concrete or steel or a mixture of these.

Precautions

Being next to the ground, these floors and the walls require special precautions to prevent moisture from the ground penetrating up into the structure:

☐ A DPC – damp proof course – is placed in the walls
☐ A DPM – damp proof membrane – is placed under or within the floor structure.

DPCs and DPMs must be joined where walls and solid floors meet, otherwise moisture will pass between them. The joint is a particularly vulnerable part of that form of construction.

Voids in many forms of construction require to be ventilated. This is to allow any build up in moisture levels to dry out quickly before damage can occur. Voids under hung floors must be ventilated for this reason. We will look at how this requirement is met later in this chapter.

DPCs and DPMs are discussed in Appendix I, together with weeps and ventilation of underfloor voids.

Solid concrete floors

Single and double layer concrete floors with hollow masonry wall

Domestic standards require floors and walls to have maximum values for the passage of heat, known as **U-values**. We will study that concept in Appendix K. To meet these standards it is normal to utilise materials with a high insulation value and/or incorporate non-structural insulation into the walls and floor.

To meet the weather resistance standards, masonry walls for domestic construction must generally be of **cavity** construction. (Walls have a chapter to themselves, Chapter 3, where the arguments in favour of cavity walls for domestic construction are put forward.) This means building two vertical layers of, say, masonry with a gap between. The vertical layers are usually half a brick thick or 100 mm of blockwork. Asymmetrical layers can be used but should total at least 200 mm. The minimum thickness for any one layer is 75.

Two layers of masonry – termed **leaves** or **skins** – must be made to act together and provide mutual support and stability. This is achieved by using **wall ties** spaced at 3 or $4/m^2$.

Figure 2.51 shows a domestic wall/floor interface. The same wall and foundation construction is shown in Figure 2.52. The solid concrete floor in this instance is of **sandwich** construction, the layers of concrete being the minimum thickness allowed under the Building Regulations.

A few points must be noted:

☐ The construction shown is not the only way to provide insulated solid or sandwich floors or insulated walls.

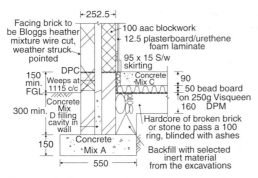

Fig. 2.51 Hollow wall/single layer concrete floor suitable for domestic construction.

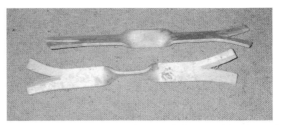

Fig. 2.53(a) Galvanised steel twisted wall ties.

Fig. 2.53(b) Galvanised steel wire butterfly wall ties.

☐ Materials specified are only one or two of a wide range available.

☐ Figures 2.53(a), (b), (c) and (d) show a selection of the many kinds of wall ties available to tie two masonry leaves together. Figure 2.53(a) shows a couple of galvanised steel twisted ties which are rigid. The twist is positioned in the cavity between the masonry leaves, and any moisture from the outer leaf trying to cross over the tie drops off the edge of the twist. Figures 2.53(b), (c) and (d) show pairs of butterfly twisted wall ties, stainless steel strip chevron ties and stainless steel wire wall ties, respectively. Note that each has a drip designed into it and an arrangement at either end to grip into the mortar beds.

Fig. 2.53(c) Stainless steel chevron wall ties.

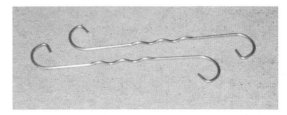

Fig. 2.53(d) Stainless steel wire wall ties.

☐ DPM and DPC still overlap between wall and edge of concrete behind the foam plastic upstand. Different materials for insulation, DPCs and DPMs will be studied in their own right.

Note that the annotation has been minimised on the assumption that there will be adequate notes appended to the whole drawing.

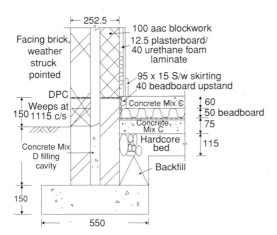

Fig. 2.52 Hollow wall/double layer concrete floor suitable for domestic construction.

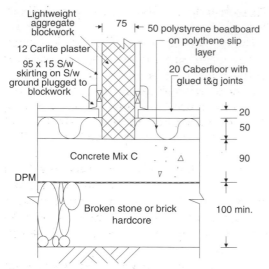

Fig. 2.54 Non-loadbearing block partition built off concrete subfloor.

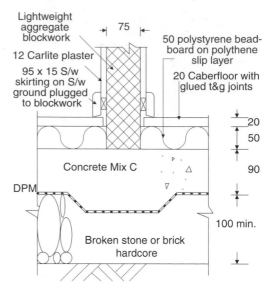

Fig. 2.55 Loadbearing block partition built off concrete subfloor.

Partitions

While single and double layer concrete floors do not have intermediate support walls, they can interface with partitions in different ways. The interface depends on the type of partition and the strength of the floor.

The simplest interface is a **non-loadbearing partition**, shown in Figure 2.54. This is simply built off the upper surface of the concrete slab. The floor can be single layer or sandwich construction. Note different insulation and floor surface options.

With a **loadbearing partition** one must either provide a foundation or, as shown in Figure 2.55, thicken and possibly reinforce the floor slab locally. A foundation might be required for a very heavily loaded wall. Using the floor slab can be economical if the upfill is deep, even if the whole floor slab has to be reinforced.

Hung floors

Hung timber floors

The timber hung floor is popular with occupiers as it has a natural resilience which

contributes to the comfort of the occupier and which is absent in concrete floors. If there is a greater than average depth of filling to be done under the floor, or if the floor level is well above finished ground level, then it will be cheaper to construct than a solid concrete floor. Otherwise, costs depend on the current relative material and labour values. Quantity surveyors should be able to advise on the current cost conditions for any configuration.

The principle on which the floor is constructed is quite simple:

☐ The inside of the building is filled with inert material up to finished ground level and a DPM laid
☐ The floor is built over the filling, leaving a minimum gap of 150 mm between the DPM and the nearest piece of timber
☐ The floor is constructed of timber planks or boarding fixed to **joists** supported by the outside walls.

The joists are timbers laid *on edge* and these are fixed **at centres**, i.e. they are evenly spaced out across the area of the floor at precise intervals. The timber or man-made board is nailed down to the joists to provide a flat, usable surface. Figure 2.56 illustrates the principle.

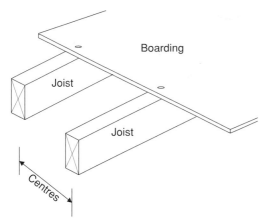

Fig. 2.56 Joists and boarding schematic.

The size of the joists used depends on:

☐ The live and dead load on the floor
☐ The span of the joists
☐ The spacing or **centres** of the joists
☐ The strength of the timber used.

The type and thickness of floor boarding depends on:

☐ The live and dead load on the floor
☐ The spacing or centres of the joists.

These criteria are discussed in Appendix C, Timber.

The void between the DPM and floor must be ventilated. Support by the outside walls for the floor must be above the level of the DPC. The means of support varies but generally the most economical option is the provision of a **three course corbel** and **wall plate**. Ground floor joists are *never* supported by building them into the wall.

Where the floor spans longer distances than the joists will bear, intermediate support must be provided. This is most commonly done with **sleeper walls** built in **honeycomb brickwork**. Other options are considered later in this chapter.

Unlike the solid concrete floor, there is no need to join the DPC and DPM. Any moisture rising in the wall up to DPC level is free to evaporate from both faces of the wall. Evaporation from the inside face is carried away by the through ventilation of the void.

Details usually show both the outer wall and intermediate support wall interfaces. Figure 2.57 shows a typical set of details. They show 'standard' assumptions on the relationship of finished ground level, original ground level and bearing stratum level.

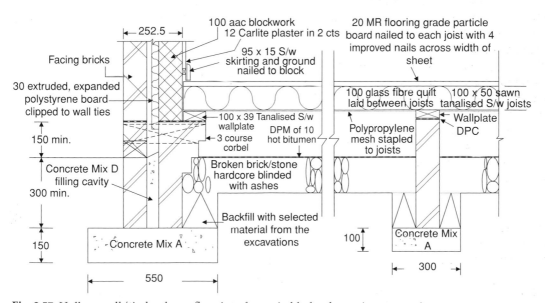

Fig. 2.57 Hollow wall/timber hung floor interface suitable for domestic construction.

Weeps are gaps deliberately left between the ends of bricks to allow moisture to drain through the wall.

Wall plates are continuous lengths of timber, generally treated with preservative, which are supplied by the carpenter and bedded horizontally in mortar by the bricklayer. Joints in running length and at corners should be **half lapped**[9]. Wall plates fulfil a number of functions:

- [] They provide a level surface on which to fix other timbers – in this instance floor joists.
- [] They provide a base on to which the joists are nailed, thus fixing the joists in position at the required centres.
- [] They spread the load of the joists over the supporting masonry.

The following materials and techniques already mentioned are described in more detail later in this chapter or elsewhere in this text as indicated:

- [] DPCs and DPMs (Appendix I)
- [] Weeps (Appendix I)
- [] Air bricks and ventilation of ground floor voids (Appendix I)
- [] Sizing of joists
- [] Timber types, **stress grading**[10] (Appendix C) and sizing of joists
- [] **Floor boarding**, types and fixing[11] (Appendix C)
- [] Alternative methods of joist support
- [] Alternative **solum** finishes.

The solum is the area of ground and the finishes applied to it, inside the external walls and any partitions built off foundations. It is not any particular finish – only descriptive of a finished state, the ground prepared to receive a finish or any stage between. One can speak of the **solum** ready to receive hardcore, the **solum** ready to receive concrete, the **solum** ready to receive hot bitumen, and so on.

---

[9] Joints in timber are described in Appendix C
[10] Stress Grading is described in Appendix C
[11] Floor boarding is discussed in Appendix C

Hung timber floor alternatives

Before looking at the alternatives to the timber hung floor shown in Figure 2.57, it is important to note that it represents the most economical solution to the provision of a ground floor for domestic construction as well as fully meeting the requirements of the Scottish and English Building Regulations.

The next few figures illustrate alternative forms of:

- [] Joist support at main wall
- [] Joist support by sleeper wall
- [] Solum finish:
  - ▪ which *can* support a sleeper wall
  - ▪ which *cannot* support a sleeper wall.

It is quite feasible to 'mix and match' the alternatives.

Figure 2.58 shows alternative means of support for joists at the outer wall/floor interface.

The inner leaf of the wall below DPC level is built one brick thick (A). Note that the 252.5 cavity wall is still centred on the foundation.

Intermediate support for the sleeper joists is by a honeycomb brick wall on its own foundation (C).

The solum finish is 65 mm thick concrete with a DPM under (B). The DPM is usually polyethylene at least 250 microns thick. The concrete is there primarily to hold the DPM in place. Note how the DPM turned up at the edges of the concrete. At 65 mm thick the concrete cannot be loadbearing but it would give:

- [] A useful surface for a **crawl space**[12]
- [] Protection from penetration by vermin, i.e. rats and mice.

A 10 pitch DPM would do neither of these.

Any loadbearing partition wall would have to be supported by its own foundation.

Figure 2.59 shows a commonly accepted **solum** in many textbooks – the 100 thick layer

---

[12] A **crawl space** is any area under a floor or in a roof which allows a workman sufficient access – even on hands and knees – to enter and work, maintain or repair components in the space.

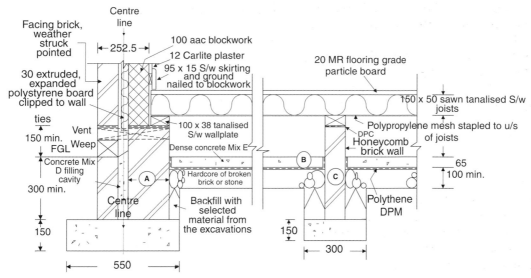

Fig. 2.58 Alternatives A, B and C of joist support.

of dense concrete. A DPM *must* be provided (D). The only advantage therefore is to provide a greater degree of support than 65 of concrete can provide. A lightly loaded sleeper wall could therefore be built directly off this layer of concrete.

The support for the floor joists is entirely by half brick thick honeycomb walls supported by the 100 thick concrete layer (E). As the regulations require a DPM under the concrete, no DPC is required in the sleeper walls.

Hung concrete floors

Hung concrete floors suitable for domestic construction are generally formed in one of two ways:

☐ Using precast concrete beams and filler blocks
☐ Using permanent steel formwork or shuttering spanning over beams or joists with a thin in-situ concrete slab poured over. The

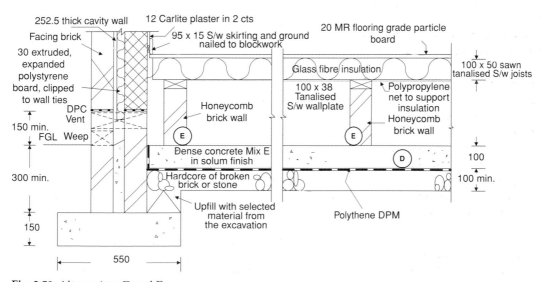

Fig. 2.59 Alternatives D and E.

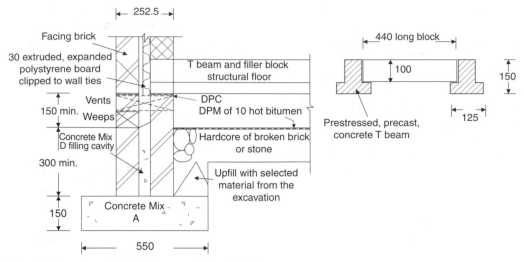

Fig. 2.60 Precast concrete T-beam and filler block floor.

joists may be of concrete or wood and very occasionally of galvanised cold rolled steel.

The first of these floors uses inverted T beams of prestressed, precast concrete, placed at centres across the required floor area. The space between the beams is filled with blocks. It is illustrated in Figure 2.60.

Generally the blocks are ordinary building blocks and particularly lightweight or aac concrete blocks. Some proprietary systems use specially cast blocks which may be shouldered to fit into the 'T' beam. They may be especially strong; they may be hollow or solid. Alternatives for the filler blocks are shown in Figure 2.61. Blocks can also be made of polystyrene; still quite strong but giving improved levels of insulation to the floors.

Because the beams are of prestressed[13], precast concrete:

☐ They can be built into outer walls without the worry of ends rotting away, and so save

[13] Prestressing is a means of increasing the efficacy of the reinforcement in a concrete member. Basically, when concrete is poured into a mould the reinforcement has been placed in position. In prestressed concrete that reinforcement is tensioned before the concrete is poured in. The rebar is of high tensile steel wire. The setting concrete grips the wire and when the mould is struck the wire remains in tension, literally pulling the concrete together and better able to carry a high load.

on the need for **corbelling** or formation of **scarcements** for joist support
☐ They can be cast with a modest depth but in long lengths and so can span across most houses without the need for sleeper walls.

Spans available generally do not exceed 8 m. If intermediate support should be required, a dwarf support wall of half brick thick honeycomb brickwork should be provided. The wall must have its own foundation and should *not* rest on any oversite concrete layers as the self weight of the floor is quite considerable.

A DPC is still required in the walls, and the beams should be bedded in mortar immediately above that DPC. A solum finish is required, including a DPM. Choose any from those described for hung timber floors. The underfloor void must be ventilated in the same way as the void under hung timber floors. Weeps must be provided in the outer leaf of cavity walls.

The cavity must be filled up to ground level with weak concrete. Once the beams and concrete blocks are in place, this rough floor provides a useful working platform for any work which is to follow. This was particularly the case in Scotland where a trestles and battens scaffold was generally built inside the building rather than having an external scaffold, a situation that has changed considerably since the 1980s.

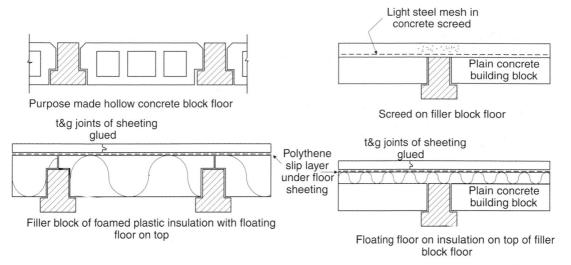

Fig. 2.61 Alternative filler block floor construction.

Once the building is weathertight, a floor finish can be applied over the beams and blocks. A number of options are available, from simple screeds to complex built layers of insulation and board or sheet flooring.

So the beam and filler block floors provide the structural base upon which a 'finish' is placed. The finish can be:

- ☐ Timber based sheet
- ☐ A fine concrete screed.

Insulation may be necessary in the floor and can be placed below either timber or screed. Timber sheet is laid 'floating' with the sheet clear of wall edges to allow for moisture movement. The sheets are wedged together until the glue in the joints has set.

While it is not impossible, it is inadvisable to attempt to fasten battens (and nail floor boarding or sheet to them) on this type of floor. Generally the beams are of very dense concrete, making nailing or plugging difficult, while power nailing is apt to shatter the concrete. Nor should one fasten to the blocks as movement of the timber will displace the blocks and this will result in a noisy floor when the occupants put their weight on a batten fixed to a loose block. Timber floors therefore must 'float' and be allowed to move separately from the concrete base.

The thickness of any screed placed over the sub-floor depends on the purpose to which the floor will be put and the type of finish placed on the screed. Insulation may be placed over the floor under the screed, in which case a minimum thickness of 65 screed would have to be laid. In addition, some thin screeds would benefit from the inclusion of a light steel reinforcement – galvanised chicken wire would be adequate to prevent serious cracking due to temperature and/or moisture movement[14]. A floor incorporating heating cables or pipes must have a minimum screed thickness of 65.

Formwork

Two types of formwork can be used for the in-situ concrete slab hung floor and these are illustrated in Figure 2.62:

- ☐ A solid, profiled sheet of hot zinc coated steel

[14] It is not generally appreciated that concrete products actually expand and contract. They do so under the influence of temperature changes in the same way as all other materials. The movement caused by varying moisture content is even greater, the concrete expanding as water is taken up and shrinking on drying out.

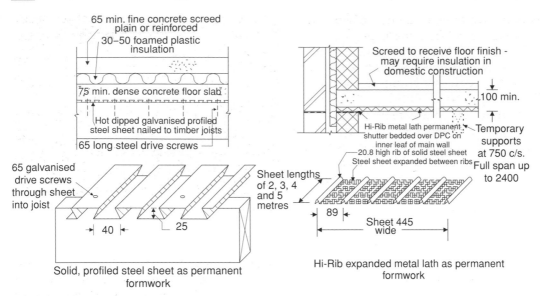

Fig. 2.62 Permanent formwork for hung concrete, cast in-situ floors.

□ A sheet of hot zinc coated steel which has ribs of solid sheet adjacent to areas of expanded sheet.

Either way, the formwork has two functions:

□ As formwork, and for that purpose it might appear that the solid sheet would better retain the concrete; however, the mesh will retain a dry mix quite well and also 'keys' to the concrete much better.
□ To act as reinforcement for the floor slab.

On no account should any separating layer be placed over the formwork before the concrete is poured, as this would destroy the essential 'key' between concrete and metal.

Even if the floor is a lightly loaded domestic one, the span of the steel sheet or lath is limited by the wet weight of the concrete. Wet concrete has a mass of approximately 2.5 tonnes per cubic metre, so a square metre of formwork must support 250 kg of wet concrete at 100 mm thick. Permanent supports can be placed at intervals dictated by the span, to support the wet concrete, or they can be placed as dictated by the finished dry slab and temporary intermediate supports used to hold up the formwork until a full set takes place. Temporary supports are generally as simple as scaffold planks on short timber or dry masonry props. In any event it is unlikely that spans from outer wall to outer wall can be achieved on anything but the narrowest of buildings.

The profiled sheet shape shown in Figure 2.62 is a continental profile and until recently had no equivalent in the UK. The expanded metal lath material with the deep rib is manufactured in the UK under the name Hi-Rib. The strands of metal in the expanded metal lath portion of the sheet are angled to give a better key in the concrete.

The thickness of concrete poured varies according to load and span, but would normally be a minimum of 75 mm. Additional reinforcement meshes and thickness of concrete can be introduced to cope with almost any domestic or light commercial load. These add to what is an already expensive and heavy structure.

Consolidation of the concrete slab is generally carried out either by tamping with a screeding board, or by using a poker vibrator. Tamping is the best method to use with Hi-Rib as the slabs are too thin to allow use of a poker vibrator. Using the vibrator so near to the lath would literally push the cement and water through the lath and leave the larger particles of aggregate behind. The vibrator could be used on the solid profiled steel sheet but care would have to be taken not to separate the cement and sand from the aggregate.

The use of such floors is unusual in the UK as they are generally much more expensive than the alternative forms of construction. They are promoted quite heavily in continental Europe where they are used for floors which are subject to continuous damp, such as in bathrooms – particularly wet rooms, shower rooms, kitchens and laundry rooms. In all these instances they are frequently confined to small areas with short spans to load-bearing partitions. Remaining areas of floor are constructed using cheaper alternatives.

Blockwork substructure

Finally, before leaving this chapter, we should show a form of substructure which is very popular now with many house builders. The technique uses blocks which have extra height, laid on their faces to give a solid wall whose thickness is the height of the blocks.

A detail is illustrated in Figure 2.63, which shows the substructure prepared to support a timber frame building which has a facing

Fig. 2.64 Photograph of substructure as detailed in Figure 2.63.

brickwork cladding. The solid wall is built off strip foundations to just below or at finished ground level. The advantages of this solid wall are:

□ Blocks, even laid on their face, are very much quicker to lay than bricks.
□ There is no need for cavity fill, thus saving on an otherwise additional operation which always risks knocking over recently laid brickwork.
□ Costs of blocks can be considerably less than bricks when bricks have to be transported some distance.
□ Using special aac or lightweight aggregate blocks, the work is not so strenuous.

We will return to this detail when we look at timber frame wall construction in Chapter 3.

Figure 2.64 is a photograph of a fairly similar construction, the main differences being

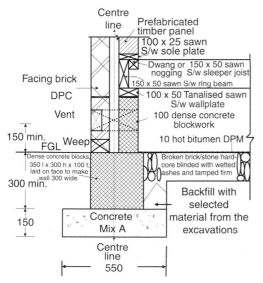

Fig. 2.63 Solid blockwork wall and brick/block cavity wall as a substructure for a timber frame panel construction.

Fig. 2.65 Sleeper walls in the above substructure.

the solum finish (polythene DPM with 65 concrete over) and the final wall cladding, which was rendered blockwork rather than facing brickwork. The photograph clearly shows the block on flat solid wall up to what will be finished ground level, and the start of the cavity wall with one course of blockwork laid up to what will be DPC level. Above that DPC will be the ground floor, ring beam and timber frame construction, while on the line of the outer face will be the blockwork cladding.

The photograph in Figure 2.65 shows the sleeper walls built inside the building perimeter. Note that they are of blockwork and so cannot be conveniently built 'honeycomb'.

To provide ventilation across the complete solum, large gaps are left in the sleeper walls, and these are bridged over with concrete beams, in this case prestressed concrete beams, 75 thick. The sleeper walls have their own strip foundations – one can be seen in the opening of the nearest wall.

The building was being built in an area that has an abundance of concrete blocks which can be produced and delivered to site much more cheaply than common bricks. The only place bricks were being used on the whole site was in some work in the main sewer manholes under the cast iron covers, and even there only around 200 Class 2 engineering bricks could have been used on the whole site.

3

Walls and Partitions

General	73
Requirements	74
Walls – environmental control	75
Heat loss and thermal capacity	75
Resistance to weather – precipitation	75
Air infiltration	77
Noise control	79
Fire	79
Dimensional stability	79

Walls of brick and blockwork	81
Insulation of external walls	84
Timber frame construction	88
Traditional timber frame	88
Modern timber frame construction	91
Loadbearing and non-loadbearing internal partitions	96
Expansion joints	99

General

Why do we build walls? What do walls do after they are built? How do we build walls to fulfil these functions?

The layman thinks of building walls mainly to provide shelter, but they fulfil a number of other functions as well:

- To support other parts of the structure – upper floors and roof
- To provide mutual stability with other building parts
- To give privacy to the occupants from outside the house and between compartments in the house
- To modify the micro-climate inside the building:
  - Keeping it hotter or cooler than the outside ambient temperature
  - Raising or lowering the relative humidity
  - Keeping the effects of wind and/or precipitation from entering the house
  - Modifying the sound entering or leaving the building.

This chapter will concern itself principally with the structural aspects of domestic walls, but cannot completely ignore the other facets mentioned above. The chapter will also concern itself with two basic types of construction:

(1) Masonry walls of brick and blockwork
(2) Timber frame walls.

And in two situations:

(1) Main supporting and enclosing walls for a house
(2) Internal supporting and/or dividing walls within a house – the partitions.

Walls can be categorised in different ways:

- By **load**:
  - Structural – designed to carry vertical loads plus their own weight
  - Stabilising – designed to resist horizontal or oblique forces and carry vertical loads plus their own weight
  - Non-loadbearing – designed to carry only their own weight.

□ By **construction type**:
  ▪ Unitised construction – built with units such as bricks or blocks bonded to avoid vertical planar continuity of joints which might otherwise become the locus of cracking and structural weakness
  ▪ Homogeneous – built using plastic materials which subsequently **set** or dry out, such as clay, earth (adobe) or in-situ concrete; clay and adobe can be reinforced with straw or reed or small branches (**cattied**); concrete can be reinforced with steel rods, bars, mesh and fibres or plastic or vegetable fibres such as bamboo cane.
  ▪ Framed – posts and beams of timber, steel or concrete are erected on site and covered with boarding, a masonry skin, metal sheet, plastic sheet etc. The skin plays no role in supporting the load on the frame. The frame alone carries imposed loads caused by other parts of the structure, occupancy, wind and snow, etc. The skin may provide a measure of **racking resistance**.

Racking resistance

The simple frame, A, shown in Figure 3.1, will tend to pivot at all 12 joints when loads are

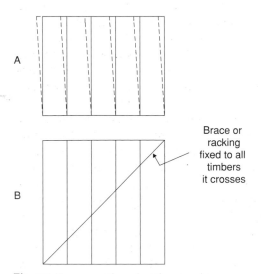

A

Brace or
racking
fixed to all
timbers
it crosses

B

Fig. 3.1 Frames with and without racking.

applied – wind, workmen and occupants leaning on the walls. This movement is called *racking*. Frame B has a brace and will not move.

In domestic construction, timber frames are clad with a skin of plywood or structural particle board – stressed skin panels. The panels are made in a factory and glues are extensively used. Panels may be for a part or whole wall length but will be for complete storey height walls. Door, window and other openings are preformed in the panels. In calculating the loadbearing capacity of the panels the **skin** or **sheathing** is taken into account. The skin provides all the racking resistance.

Requirements

Requirements for walls vary according to use and situation. Some or all of the following might be required. All but one or two subdivisions are covered by the statutory requirements of the Building Regulations:

□ Structural stability
□ Environmental control
□ Dimensional stability
□ Fire resistance.

Each of the above can be further broken down. Structural stability has three distinct facets:

□ Resistance to overturning
□ Resistance to crushing
□ Resistance to buckling.

Environmental control includes:

□ Resistance to weather
□ Resistance to air movement
□ Thermal resistance
□ Thermal capacity
□ Noise transmission
□ Fire resistance.

Dimensional stability:

□ Thermally induced movement
□ Movement due to changes in moisture content
□ Structural movement – bending, buckling, compression, deflection
□ Chemical reaction.

Fire resistance

□ Combustibility – how easily a material will ignite and sustain ignition
□ Surface spread of flame – the rate at which flame will spread across one material to ignite another surface
□ Stability; integrity; insulation.

Walls – environmental control

Heat loss and thermal capacity

Thermal transmittance is measured in watts per square metre per degree Kelvin and is a measure of the rate of heat transfer through a piece of the structure – wall, floor or roof – known as a U-value. The Building Regulations lay down maximum values for this rate of heat transfer. It is calculated from known values for thermal resistance of individual materials and their thickness in the construction plus the inside and outside surface resistances. We will look at a simple method of calculating a U-value for a wall in Appendix K.

Thermal capacity is the quantity of heat required to raise unit volume of a piece of construction by unit temperature. Dense materials have a higher thermal capacity which means they take longer to absorb heat, but having absorbed heat will give out that heat for a longer period. So heavy construction once heated up can act as a store of heat over a period when heat input is not available. A lightweight construction does not have this store effect but of course the occupied area heats up much more quickly and the occupants feel comfortable more quickly.

Resistance to weather – precipitation

Principally, resistance to weather means the ability to resist the effects of precipitation. There are basically two ways to stop water penetrating a wall: provide an impervious surface, which can be very thin, or provide a relatively thick and finely porous cladding or complete wall.

Note that water penetrating a wall does so as **solid** water which dries out by evaporation, i.e. as **water vapour**. Drying out therefore can take a lot longer than the initial wetting.

Thin impervious claddings can be in large or relatively small pieces. The use of large pieces means a shorter total length of joints. The use of small units such as tile or faience gives a large total length of joints. Jointing material is usually a mortar and so is porous. It is at the joints that water penetration will take place. Water penetrating these joints will spread rapidly in the porous masonry backing and will only be able to evaporate from the joint line – the more joints the quicker the wall will dry out. If there is a frost before the wall dries, there is a risk of the water freezing, and the thin facing material will burst off the wall face. This can happen not just with tile etc. but also with some facing bricks which have a dense impervious face. With the introduction of ever stricter U-values, the risk of the wall freezing becomes greater.

Porous wall finishes are not necessarily the answer unless the materials used give a very fine and evenly distributed pore structure. The ideal situation is to have a pore size such that take up of solid water is inhibited, is evenly distributed over the face and thickness of the wall and the water is as free as possible to evaporate from the surface. Aac blocks exhibit these qualities but are rarely acceptable as a finish on the outside of buildings.

Water is not retained in the wall so the risk of bursting on freezing is reduced. The fine porous structure can take up some of the expansion of the ice formed on freezing without being damaged.

Although beyond the scope of our study of domestic structures, it is as well to point out that if one wishes to use thin impervious materials, the joints should be arranged to **prevent** water penetration, e.g. use interlocking metal panels or timber or plastic 'weather boarding'.

Within the scope of our study, the use of finely porous materials would include the use of suitable facing bricks with a compatible mortar mix, *or* the use of common brick or concrete block which is covered with a 20 mm layer of rendering in the form of either a wet or dry dash finish.

Wet and dry dash finishes comprise a minimum of two layers of a suitable mortar spread evenly over the wall surface. The actual mortar mix depends on the type of surface to which it is applied, the suction of the surface and the degree of exposure to be experienced. All renders have a fine porous structure.

A dry dash finish comprises two coats of mortar, the second applied after the first has set. Before the second coat has set, 6 to 10 mm natural or crushed stone gravel is thrown or 'dashed' against the soft mortar and pressed gently in; it adheres to the surface, giving a decorative effect.

Figure 3.2 is a photograph of four renders. From the top:

- A new wet dash render
- An old (weathered) wet dash render
- Two new dry dash renders.

A wet dash render or 'roughcast' comprises a coat of mortar applied to the wall and allowed to set. A second coat of mortar to which 6 mm gravel has been added is applied on top of the first coat. The second coats of these wall finishes frequently contain 'tints' or colouring agents, and to achieve a 'white' finish, a white OPC is used together with a pale coloured sand and white chippings of crushed quartz or marble. The first coat is always scored to provide a mechanical key for the second coat.

Two-coat work of this type is expected to finish 20 mm thick overall. This can be achieved quite easily on good flat, straight brick or blockwork. If the wall is badly off the straight with lots of bumps and hollows, it is normal to apply three coats, the first coat more or less filling up the hollows and the two successive coats being the dry or wet dash as already described. This first coat is termed a **straightening coat**. The next coat is a **rendering coat** and the final coat is a **dashing coat**.

The efficacy of a wet or dry dash finish should never be underestimated. Under the harsher climatic regimes experienced in western, northern and in particular the coastal

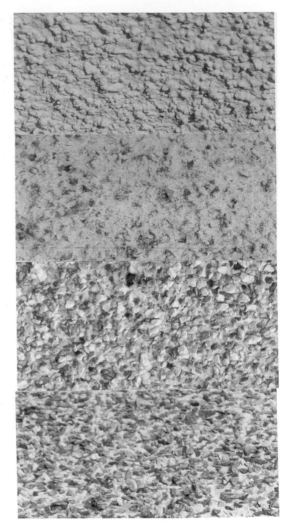

Fig. 3.2 Dry and wet dash renderings.

areas of the British Isles, equivalent wall constructions at a height of about three or four storeys for the outer leaf of a cavity wall would be:

- one brick thick facing brick, *or*
- half brick thick common brick with a 20 mm two coat wet or dry dash finish (*no* additives necessary).

There are a few points to note:

- The rendered wall is cheaper to construct
- The rendered wall is considerably lighter in weight and so a lighter foundation might be used

Table 3.1 Mortar mixes suitable for external rendering of masonry walls.

Background and type of finish	Undercoats		Final coat	
	Mix for severe exposures	Mix for moderate or sheltered exposures	Mix for severe exposures	Mix for moderate or sheltered exposures
Dense, strong and smooth/moderately strong, porous backgrounds				
Wet dash	H or A	H or A	H or A	H or A
Dry dash	H	H	H	H
Moderately weak, porous backgrounds				
Wet dash	A	A	A	A
Dry dash	H or A	H or A	H or A	H or A

☐ The wall as a whole will be half a brick less in thickness so a narrower foundation can be used

☐ In harsh climates with higher levels of insulation, the chances of facing brick spalling off are greater than the chance of damage to a rendered wall.

So don't despise regional preferences for wet and dry dash finishes. They have been used all these centuries for very good reasons.

Waterproofers are sometimes added to the outer coat to improve the resistance of the render to weather penetration, but where the masonry behind is of adequate thickness or is the outer leaf of a properly constructed cavity wall, there should be no need for such an additive.

Waterproofers

In the 1950s the author was shown an 'old roughcaster's' method of waterproofing. About $\frac{3}{4}$ litre of boiled linseed oil was added to a bucket of hot water and whisked to emulsify it. This was added to the mixing water for the mortar used for the dashing coat of a wet dash finish. The quantity of OPC in the batch being mixed was a full 112 lb bag – approximately 50 kg.

More modern materials include PVA (polyvinyl acetate) and EVA (ethylene vinyl acetate). PVA will wash out of the render in

a year or two while EVA is fully waterproof and should last the life of the render.

EVA can be purchased as a powder or a liquid. Recent research has shown that liquid EVA blocks most of the pores, thus helping to retain water within the wall rather than letting the wall 'breathe'. Powdered EVA is more successful. It does not block the pores but lines them, thus repelling solid water trying to enter while retaining a fully open surface for evaporation.

Silicone-based additives, like PVA, are not long lasting, their efficacy wearing off in a matter of a few years. Silicone-based coatings for application after the render has set are equally short lived but of course reapplication is always an option.

Old editions of Building Regulations gave lists of mortar mixes suitable for all coats of such wall finishes. An up-to-date version of one list is given in Table 3.1, with the recommendations for use as a dry or wet dash wall finish. Mortar mixes suitable for rendering were given as:

☐ A 1:1:5 or 6 or 1:5 using masonry cement
☐ H 1:$\frac{1}{2}$:4 or 4$\frac{1}{2}$ or 1:4 using masonry cement.

Air infiltration

There has been a long period when any air infiltration through the fabric of a building was seen as a bad thing, and great effort was put into reducing it to zero – or as near as one could get. In the few instances where this was achieved, the occupants discovered that they could experience one or two of a series of phenomena:

- A build-up of lived-in type smells – mostly unpleasant
- A lack of fresh air, a stuffiness
- An increase in the relative humidity in the house
- Dampness and mould growth
- Timber rot
- Increased insect activity
- Greater susceptibility to colds and respiratory infections.

By keeping air from penetrating the building, moisture laden vitiated air was allowed to remain inside to the detriment of the building, its contents and the occupants. The draughty conditions of the old housing stock, where every joint allowed a gale of wind to come through, were not desirable, especially when the cost of heating the air lost was considered. But the old buildings had a plentiful supply of fresh air coming in. What is required is a balance.

Any resistance to the passage of air will also exclude damp air, thus helping to keep the interior of the building dry during wet weather. Masonry structures of brick or block are difficult to 'seal' against infiltration, failure usually taking place at their interfaces with other components such as hung floors, roofs, doors and windows. The structure itself if rendered and/or plastered will be quite windtight. Timber frame construction being made with panels covered with a ply or particle board makes very tight structures, only failing at interfaces with floors, roof, windows and doors. In any construction form sealing these interfaces can be done, but it requires careful supervision of the workmen on site who rarely see the need for all the small detailed work and will bodge or omit it if you are not looking.

In masonry walls, openings can be made 12–15 mm larger all round, the frames suspended on metal brackets in the opening and then sealed all round with one of the expanding foams now on the market. In timber frame construction, the openings are made 6–8 mm larger all round and then sealed round the frames with silicone-based mastic.

Improving the seal between ground floor and walls is largely a matter of choosing a solid concrete ground floor rather than a timber hung floor. There are then no underfloor voids to ventilate and no chance of draughts at the skirting boards. If a timber floor finish is required, lay the timber floor on the insulation (plastic foam sheet) on the concrete floor slab, or nail it to battens laid over the slab and place insulation (glass fibre or mineral wool quilts) in between the battens. Trying to fix these battens to the slab results only in a creaky floor. Float the floor on insulation and fix it only to the battens.

The joint of roof to walls is difficult, but again choice of structure, design and insulation can be combined. There are two good choices:

- Do not use the roof space for rooms or water storage tanks with a cold roof[1] as such coomb ceilings are impossible to seal. If coomb ceilings must be built use a warm roof[2] construction and seal the insulation down to the wall head.
- A cold roof can be used over a horizontal upper floor ceiling but the ceiling board must seal to the walls and the insulation must seal to the wall head.

Much else can be done in detail to reduce air infiltration but attention to these major areas can bring it down from a UK average of 7–9 air changes per hour (**ach**) to 2 or 3 ach. Even lower values can be obtained but mechanical ventilation must be introduced to get over the problems of retaining foul air within the house. Heat recovery in conjunction with mechanical ventilation then becomes a serious consideration, for if mechanical ventilation is installed for the whole house – not just an extract fan in a bathroom and a kitchen – the

---

[1] A cold roof has insulation inserted immediately behind the ceiling finish, the space above being left 'cold'.

[2] A warm roof has insulation placed immediately under the roof finish: thus the whole of the roof void is kept warm by leakage through the ceiling finish.

loss of heat with all the air can be an expensive exercise. Discussion of these options is well beyond the scope of this module but trials on some housing association houses in Orkney about 10 years ago confirmed that it was possible to build and achieve low air change rates without re-educating the work force, and that heat recovery could work really well in a normal domestic situation but did require a better understanding of how the system could be used to advantage by the occupants. The workmen of course were all regular employees of the contractor(s) and were used to building tight houses in what is a very windy environment. Some occupants took the advantage in monetary terms by saving energy – others took the advantage by enjoying much higher ambient temperatures over the whole house than they could otherwise have afforded. One cannot say that one group was right and the other wrong.

Noise control

Noise is a complicated topic and far beyond this basic text. Suffice to say that the details shown will provide satisfactory sound levels in adjacent compartments of a building. These details rely mostly on a high mass to achieve a suitable reduction on sound levels. The walls we will include in our discussion will generally perform within any requirements of the Building Regulations when built in a single dwelling.

Fire

Structures are usually designed to have **stability**, **integrity** and **insulation**. In the event of a fire:

☐ Stability postpones collapse to allow occupants to escape and fire fighters to tackle the blaze.
☐ Integrity is designed to postpone the passage of flame, hot gases or fumes into 'safe' areas where they might harm occupants escaping and fire fighters working, and also prevents ignition of neighbouring areas.

☐ Insulation prevents conduction of heat which would otherwise ignite neighbouring areas and might even scorch escaping occupants or fire fighters.

The walls we will discuss will generally meet the requirements of the Building Regulations when used in the context of single or terraced housing.

Dimensional stability

Thermally induced movement is a phenomenon which is fairly well understood by most lay people but in terms of building structures is far from obvious as structural features generally either have low or similar coefficients of thermal expansion. Where thermal movement causes problems it is generally in conjunction with some other parts of the building, such as the services where metal and plastic pipes etc. expand and contract at higher rates than the structure generally, causing the pipes etc. to fail, but structural damage occasionally takes place.

Moisture movement is quite critical in terms of building structure. Timber and masonry have quite high expansion and contraction rates as they take up moisture and subsequently dry out. Of course, if parts of a structure are kept in stable conditions, no movement will take place and no damage can occur. As details are studied, methods of avoiding the effects of this movement will be highlighted.

A common example of damage caused by movement due to moisture content is shown in Figure 3.3. It is the cracking of masonry down the side of an opening spanned by a concrete lintel. The lintel on shrinking pulls the masonry away under the wall rest, causing a crack. Two pieces of DPC under the lintel rests would allow it to move independently of the masonry.

Structural movement, i.e. associated with deflection, buckling and compression, are either static or progressive. If progressive the building collapses, but if static, the building comes to rest and may be perfectly safe but

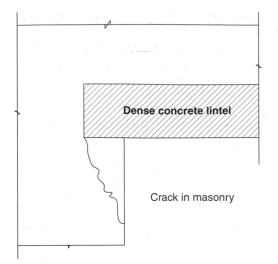

Fig. 3.3 Cracking in lightweight concrete wall with dense concrete lintel.

will show unsightly cracking or gaps between deflecting members and other surfaces. These gaps and cracks, if on the external face of the building, can allow weather to penetrate and damage the building.

Chemical reactions are sometimes difficult to predict, especially with new materials or new techniques, but an example might be the effects of sulphate attack on OPC-based products. Mortars and concretes subjected to prolonged attack by strong solutions of sulphates swell up. This is a case of the irresistible force and the (almost) immovable object. For example, walls resting on foundations being attacked will start to lift out of the ground and the effects in terms of cracking of walls will be seen wherever there is a weakness in the wall above, e.g. at quoins, scuntions of openings, where there is a lack of wall ties in cavities, lack of proper bond, etc.

A good example comes from an at first baffling consultancy where the author was called to look at a house which was displaying cracking in unusual places as well as at a corner of the building close to some windows. In fact, what was unusual about all the cracking was that it was taking place on the outer leaf of the cavity wall but not on the inner leaf. There was also a hung timber ground floor which was rising and falling, especially under a door

into an integral garage. As it rose it jammed the door into the garage.

The householder was unwilling to let the team go down into the void under the floor, which at that point was quite deep. Using a bore scope, the cavity in the walls was examined. A builder was hired to dig out the foundations on the outside and trial bores were taken of the subsoil under the concrete. Nothing unusual was observed during this process except a certain scarceness of wall ties round the corner area of the wall up to first floor level. In one of the excavations the builder actually found a bundle of wall ties ready bound up as supplied by a merchant. A bore scope examination of the underfloor void showed nothing obviously wrong. Wall ties were inserted to bring the numbers up to the required standard and tell tales were fitted across the cracks so that the movement could be monitored.

Six months passed and still the building cracked. Winter was coming so the cracks were pointed up to exclude the worst of the weather. Another six months and still the building cracked. By now the possibility of the foundations overturning or sinking had been ruled out, but there was nothing seen which could be the cause.

Then we were allowed to go into the underfloor void and we found that the floor going up and down was due to two things. The floor joists were built into the walls – but not the one with the big crack – very tightly and it was very humid under the floor. The joists were moving and since they were restricted longitudinally, they hogged – they bowed upwards, lifting the floor. What was making it humid under the floor was the fact that although there were vents built into the outer leaf of brickwork, there was no hole through to the inside of the wall – no ventilation to allow the space to dry out. What was surprising was that there was no sign of rot in any of the timbers.

A builder came in and cut the vents through, and with great difficulty eased off the grip of the walls on the timbers. The floor problem was immediately sorted out, but we were still left with the wall cracks. They continued to extend. After the builder had gone, the under-floor void was monitored and on the second or third visit we saw that the hot pitch DPM looked as though it was bubbling in one or two places. We didn't think too much about it but on the next visit these bubbles and more had burst and a white crystalline effusion was growing from the gaps. It was pure sulphate of something! Digging out some of the DPM revealed that the upfill below the DPM was spent coal waste, very rich in sulphate, and here it was piled into a deep substructure corner against a brick wall and over the inner edge of a concrete foundation.

Now we understood what was happening. The OPC-based mortar and concrete in the inner leaf and the foundation was expanding. This in turn lifted the inner leaf. There was a dearth of wall ties in the ground floor wall, so while the inner leaf kept rising out of the ground, the outer leaf wanted to stay where it was except that above the first floor level it was tied to the inner leaf. So, upper and lower parts parted company. The new ties inserted were flexible so had had no effect on this movement in the short time since their installation. The only cure was to take out the coal waste, remove the damaged brickwork and concrete and replace with sulphate resisting cement based products. No movement has been observed since this was done but it was an expensive underpinning exercise while it lasted.

Walls of brick and blockwork

Cavity walls have already been introduced in our study of substructures but not in any great detail, and certainly without some background as to their current popularity.

The Industrial Revolution brought with it the need to build cheap housing quickly to provide for a burgeoning workforce moving from rural to urban areas. In many instances this meant building in brickwork, and the favoured method was to build one brick thick walls to support the roof and occasionally an upper floor. From a structural point of view this was quite adequate but the methodology was inadequate on practically every other front.

One of the first of these inadequacies to be tackled was the problem of water penetration. Rain blown against a one brick thick wall will penetrate the full depth of the wall if it is of sufficient volume and for a sufficiently long period of time. Even if penetration does not take place during the rainfall period, moisture in the outer layers of the wall will percolate to the inner face, making the inside of the house damp. Together with a lack of ventilation and substructures without DPC or DPM, this dampness caused many of the health problems associated with that type of housing.

To overcome these problems, DPCs and DPMs were introduced into the substructure and the **cavity wall** was introduced. Instead of a single layer of brickwork one brick thick, two layers were built each a half brick thick and the layers were joined together with wrought iron wall ties placed at 3 per square yard. The layers are called **leaves** or **skins**. The prime purpose of the cavity is to prevent moisture reaching the inner leaf of masonry.

Wrought iron was a good choice of material for wall ties, for although it rusted, it was not as liable to rust to destruction in a short time as the later invention – mild steel. With the introduction of mild steel came the need for better protection from corrosion and this was applied by hot dip galvanising. Alas, the standard of galvanising applied was found in the 1960s to be insufficient and one or two building failures led to an assessment of the amount of hot zinc applied to metalwork in general. An updated British Standard was promulgated and no further failures have occurred although monitoring of walls with the old ties has to continue.

The first cavity was set at 2 inches, a little over 50 millimetres. Subsequent work on cavity walls has revealed that the cavity can be up to 75 wide, with ordinary wall ties at $4/m^2$. The wall ties are there to stabilise the separate layers of brickwork laterally, one with the other, and to spread some of the vertical loadings from one layer to the other. Because they bridge across the cavity there is a danger that they could provide a route by which water could be carried across the cavity. For this reason their design incorporates a **drip** or series of **drips.** Water will always form droplets on the lower edge of anything, and so it is with the drips formed on the wall ties. The ties are enclosed in the cavity and so in a space which has no disturbance from wind or weather. The water trying to cross runs down the drip and when sufficient has gathered will form a droplet which falls to the bottom of the cavity. The inner leaf remains dry – but only if the wall tie remains clean and the drips are built in the correct way up. The latter possibility has been obviated by designing the ties in such a way that a proper drip is always presented no matter which way the tie is built into the wall.

As cavity walls form the major part of all domestic construction work in some way, it is appropriate to widen our knowledge of them, but first what are the objections to solid walls as a whole?

Solid walls of masonry can be used for domestic construction. However, they must be a minimum of 250 mm thick:

☐ In blockwork this means using a 250 block
☐ In brickwork this means building a wall one and one half bricks thick – 327.5 mm using a metric brick.

Such walls would be very 'cold' and would require insulation to be added. Insulating walls in general will be covered as the chapter evolves.

The reason for that minimum thickness requirement is solely the exclusion of weather, particularly precipitation. Walls 200 thick are adequate to bear all normal domestic structural loadings, so to keep out the rain we are building very heavy blockwork and a massive amount of brickwork.

Early mass industrial housing was frequently built as cheaply as possible, which meant the provision of one brick thick walls. In very exposed areas a one brick thick wall will allow rain to penetrate, and in other less exposed areas dampness will be present on the inner face of the wall. Rendering the outer face and applying strapping, lath and plaster on the inner face did nothing to prevent these conditions arising, although these methods were frequently employed in remedial work.

Only the adoption of a radically new form of construction would be of any help, and it came in the form of the cavity wall. First built as two 'skins' or 'leaves' of half brick thick brickwork with a 2 in cavity between, the leaves were tied together by the insertion of wrought iron straps of metal at a rate of about 3 per square yard.

The idea of the cavity was a simple one – like all good ideas. Water penetrating the outer layer of masonry would simply run down the inner face until it reached the foundation. The inner leaf never got wet. The wall ties ensured structural integrity and mutual support for the leaves. The wall ties were shaped to prevent water running across the tie and wetting the inner leaf in. Figure 3.4 illustrates the principle.

Cavities have been set at a standard width of 50 (or 2 in) since their introduction. Wider cavities can have advantages in terms of being able to hold more insulation, but going beyond 75 makes the use of ordinary wall ties a problem structurally. A wall with a wider cavity has to be specially designed, perhaps having special wall ties and so on. In addition to the structural problems, a wider cavity allows the air in the cavity to start circulating.

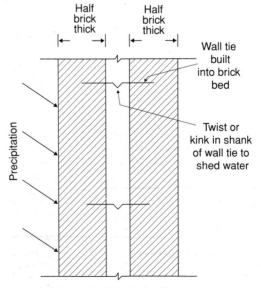

Fig. 3.4 Schematic of cavity wall.

Warmed air rises, cold air descends and this circulation adds to the heat loss through the wall. Of course, reducing the effective width by putting in insulation will reduce this effect.

Cavity walls of masonry construction are now built in any combination of brick or concrete block, reconstructed stone and natural stone, and any type of brick or block can be employed in either leaf. Where the brick or block in the outer leaf cannot withstand the effects of weather it would be coated with a dry or wet dash. The leaves don't have to be of equal thickness as long as the sum of the thicknesses equals at least 200 mm, although a leaf of less than 65 mm thick will not be very satisfactory from a structural point of view. Beyond that thickness, ordinary wall ties are not really effective and a better arrangement has to be made to connect the two masonry leaves. Figure 3.5 shows the open end of a cavity wall built with concrete block leaves, each 100 thick. A wire butterfly wall tie can just be seen.

Wall ties are no longer made of wrought iron (which resisted corrosion well) but of galvanised mild steel, stainless steel or plastic.

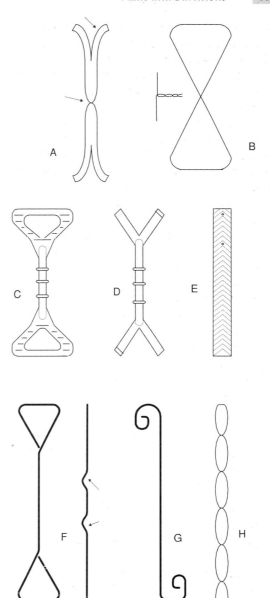

Fig. 3.6 Selection of wall tie types.

Fig. 3.5 Block cavity wall.

Some metal ties are made of wire, some of sheet or plate. Plastic ties are injection moulded from polypropylene. All have split or shaped ends to build into the mortar bed, and shanks with various forms of drip. Figure 3.6 shows sketches of several different types of tie:

A – A tie of galvanised mild steel, the ends **fishtailed** and shank twisted to provide a drip. Also available in stainless steel

B – A tie of light gauge wire formed into a double loop and the ends twisted together to form a drip. Very flexible tie, used in party wall construction where solid ties would transmit too much noise

C – A 'plastic' tie. The plastic used is poly-propylene, which is very tough and strong. The ties are very flexible and have no corrosion problems

D – Another shape of plastic tie. The fish tail ends each have a small local thickness to improve grip in the mortar bed

E – A 'chevron' tie. Available in galvanised steel or stainless steel. Both made from thin sheet stamped with very shallow grooves in the chevron pattern. Galvanised version prone to corrosion failure. Flexible tie with a poor grip in the mortar bed

F – A wire tie, a heavier gauge of wire, available in galvanised mild steel or stainless steel

G – Another shape of wire tie in stainless steel

H – An unusual tie of stainless steel strip, twisted into a continuous helix.

One of the difficulties of bridging the cavity with wall ties is that the ties can gather mortar droppings. The droppings can allow water to run or be drawn across the tie onto the inner leaf. If the quantity of water drawn across is excessive, a damp patch appears on the inside face of the wall.

To prevent mortar gathering on the ties it usual to specify that a batten will be used to prevent droppings resting on the ties, and it will be drawn up by cords through the cavity just before the next course of ties is built in, as shown in Figure 3.7. Alternatively, a straw rope can be laid over the ties and drawn up through the cavity before the next course of ties is laid, although that material is ever more difficult to source.

Insulation of external walls

While the adoption of the cavity wall and the ever-growing variety of wall ties have solved the problems of water penetration of walls,

Batten laid on wall ties, cords to one side

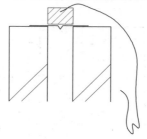

One leaf built to next course for ties, mortar droppings gather on batten, cords put over leaf just built

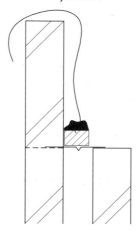

Other leaf built to next course for ties, batten pulled out of cavity by cords, bringing up mortar droppings with it

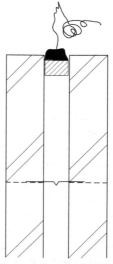

Fig. 3.7 Keeping the cavity clear of mortar droppings.

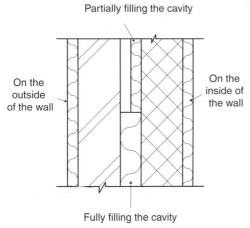

Partially filling the cavity

On the outside of the wall

On the inside of the wall

Fully filling the cavity

Fig. 3.8 Possible placement of wall insulation.

they did nothing to improve the thermal performance of the cavity wall – which is much better than a one brick solid wall.

Initial attempts centred on the use of lightweight concretes or aac to make blocks used only for the inner leaf. These are not sufficient in meeting today's much higher standards but can be used in conjunction with layers of insulation material. We will see how they can contribute when looking at U-value calculations.

Insulation material can be placed in one or more of three positions in a masonry cavity wall, as shown in Figure 3.8:

- ☐ On the outer face
- ☐ On the inner face
- ☐ In the cavity
  - ▪ filling the cavity
  - ▪ partly filling the cavity (always on the inner leaf).

Insulation can be placed in the cavity as the wall is built, or it may be injected or blown into the cavity.

Injection of insulation into the cavity will be examined a little later, as it is generally an exercise carried out on an existing wall.

Insulation applied to the outside of a wall is generally done as part of a refurbishment of an existing structure. It involves specialist fixings and a new weather resistant finish. These systems are not examined in detail in this text.

One, for example, uses a composite layer of insulation: perforated building paper, a stainless steel mesh and a two-coat render. The insulation is polystyrene bead board to which is stuck a layer of building papers[3] with vertical slots, and through the slots is woven a stainless steel wire mesh. This first layer is fixed to the masonry wall with a fastener not unlike a framing anchor (see Appendix G) with a large plastic washer. Once fixed, a two-coat cement/lime render is applied which effectively seals off the insulating layer from the weather.

This leaves the study of systems built or injected into the cavity, or built onto the inner face of a wall.

Materials available are:

- ☐ Glass fibre
- ☐ Mineral wool
- ☐ Foamed plastics:
  - ▪ polystyrene beadboard or as loose beads
  - ▪ extruded polystyrene
  - ▪ urea formaldehyde foams
  - ▪ phenol formaldehyde foams
  - ▪ polyurethane foams.

Glass and mineral wool materials comprise glass or mineral wool fibres laid down into a quilt or mat, a slab or batt, each of differing grades or densities. The fibres are coated with a resin-based adhesive; the lighter the product the less adhesive is used. Quilts are available in rolls, generally 1200 wide, and these may be further 'split' inside the pack into 2 × 600 mm or 3 × 400 mm, which suit standard **stud** centres. Quilts can be supplied 'plain' or 'paper covered' one or both sides. The 'papers' can be plain, bitumen bonded, foil faced, etc.

Batts are supplied as flat sheets in a variety of sizes and to a variety of densities. Higher density batts have better loadbearing characteristics but poorer thermal performance. Those to be built into a cavity are of low

---

[3] Building paper is specially formulated to resist the passage of 'solid' water but allow the passage of water vapour. Resistance to the passage of solid water is only effective in the long term if the 'wet' face of the material is left uncovered, i.e. it cannot be pressed up against the wet inner leaf.

density, high thermal performance and have a width of 440 mm – six courses, which is generally the vertical spacing between wall ties. Those batts designed to fill a cavity have the fibres aligned across the width to assist with the drain down of any 'solid' water which penetrates the outer leaf.

The foamed polystyrene and polyurethane are supplied as sheet material, which can be applied to inner wall surfaces or as a partial or full cavity insulation. A variety of densities is available. The extruded polystyrene is very tough and for the same thickness has a better thermal performance than **beadboard** of the same density. The extruded board is commonly referred to as **blueboard** and it has a good resistance to water vapour and is waterproof. The beadboard is quite tough but is more compressible than the extruded board of the same density. It does have some load-bearing ability and special grades are manufactured and commonly used under floor screeds and floating timber floors. It is not waterproof and is very permeable to water vapour. It should be familiar to anyone who has unpacked a computer, TV or white goods, its main use being in moulded packaging, although the grade used is much more friable than the one used as insulation.

Polyurethane board is very friable and is usually supplied faced on one or both sides with bituminous felts, foils and papers. It is slightly more expensive than polystyrene but when first made has a higher thermal resistance due to the foaming process leaving the pores filled with carbon dioxide rather than air. However, after a period of time the carbon dioxide leaches out and is replaced with air, and there is no thermal advantage. The other plastics are more expensive and so have largely been overtaken by polystyrene.

Filling the cavity with sheet or batt insulation presents one problem – how to stop water crossing the insulation to the inner leaf. With the batts, the fibres are aligned vertically in the cavity; provided adjacent sheets of the material are pressed firmly together, water will not cross at the joints even at the wall ties. The joints in the batts must be kept clear of mortar droppings.

Extruded polystyrene and urethane foam sheets are generally waterproof and water can only penetrate at the joints. To stop this a rebated or tongued and grooved joint is formed on the sheets. Polystyrene beadboard is very porous but it is thought that it is not so loose as to allow water to drain downwards, the water tending to cross the insulation to the inner leaf. For this reason, some professionals are of the opinion that it should only be used for partial filling of a cavity.

Partial filling of the cavity is usually done with foamed plastic sheet. The problem here is to ensure that the sheet remains firmly against the inner leaf. As the masonry skins are built, the sheet will be placed against the inner leaf between the courses with wall ties, the ties passing through the horizontal joints of the sheets. Most tie manufacturers now make patent plastic 'clips' which fasten to the shank of the tie and can be pressed firmly against the sheet, holding it in place. Figure 3.9 illustrates such a device.

The urea formaldehyde and phenol formaldehyde plastics are chiefly used as 'foam in place' insulation or sealers. They form a waterproof layer between the two leaves of the wall, therefore water penetrating the outer leaf will concentrate on the face of the insulation. As insulation in building they are used to insulate existing cavities by injecting through holes drilled through the mortar joints of the outer leaf.

Problems of dampness crossing the cavity have been experienced and have been largely due to mortar droppings left on wall ties. In the 'clean' cavity, the foam seals round the shank of the wall tie. If there are mortar

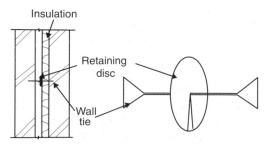

Fig. 3.9 Retaining clip for partial fill insulation in cavity.

droppings the foam will not form a seal, and water is concentrated at these weak points and penetrates the foam and the inner leaf. This can occur even if the unfilled cavity had not shown signs of being bridged prior to injecting the foam.

Using loose foamed polystyrene beads is an effective way of filling a cavity with insulation, generally as a retrofit or upgrading exercise. Holes are drilled through the masonry joints and beads are blown into the cavity through a fairly large nozzle. At the nozzle another smaller pipe injects an adhesive into the stream of beads. Once the adhesive has set, any cutting of holes through the wall in the future does not result in all the beads running out and having to be replaced. Because the insulation formed has a very open texture, water penetrating the outer leaf drains down and does not cross the cavity. Again, wall ties must be clean to prevent water crossing at the ties.

Another 'injected' insulation technique is to blow mineral wool fibres into the cavity through holes drilled in the outer leaf. The holes have to be much larger than those used for the injection of foam and so are more difficult to camouflage on completion.

A more worrying aspect in the early development of the technique was the habit of the fibres to slump to the bottom of the cavity after a period of time. This was quite a natural process caused by the excess air between the fibres being replaced with the fibres above. This effect could be exaggerated if there was vibration caused by traffic or the occupants, and was made even worse if the fibres became wet due to rain penetrating the outer leaf. Both were overcome by including a waterproof adhesive with the wool during injection. Once the adhesive set, the wool maintained its **loft**, and its open texture allowed water to drain down rather than pass across the cavity.

Placing the insulation on the inside face of the wall can be done in a number of ways. Traditionally, thin timber or **strapping** is fixed vertically to the inside face of the wall at centres. Further timbers are fixed horizontally at the top and bottom of the wall, as well as short lengths of strapping (**noggings**) in two or three rows between the verticals. Among

this framework, a quilt or batt of glass or mineral wool can be placed and the whole covered with a suitable sheet material – generally plasterboard. To prevent hot moist air from the room penetrating to the cold wall face, condensing and causing rot etc., a layer of polythene sheet is placed between the plasterboard and the strapping. The layer is known as a vapour barrier, vapour check or vapour control layer. Vapour barrier is too definite a term – it is impossible under construction site conditions to put up a vapour barrier. At best one can control or check – as in delay or reduce – the flow of moisture through the layer. Check is indefinite in meaning, so vapour control layer would be the best option. Figure 3.10 illustrates the wall strapping technique. Vapour control layers will be the subject of a more detailed discussion as they occur in various other parts of the construction process.

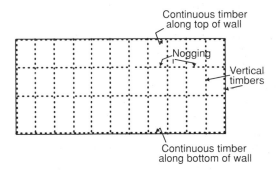

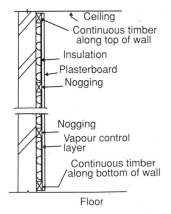

Fig. 3.10 Wall strapping on inner face of wall with insulation and boarding.

More modern practice uses a laminate of a plastic foam sheet onto a finishing type board, such as plasterboard, plywood, fibre board, etc. This material is applied direct to the inside face of the wall and fixed using a gap filling adhesive. The use of a vapour barrier under the finishing board is not usually necessary as the foam plastic sheet is fairly well moistureproof. Simple precautions with a sealing mastic can be used at joints.

If a very positive moisture seal is required, it is possible to have the laminate made up with a polythene or foil layer between the foam and the board. The joints of the boards can be made up with a rebate and the two layers of film or foil can be sealed together. This is illustrated in Figure 3.11. Adhesives used are generally based on a gypsum plaster and may be modified by the addition of a bonding agent. The sheets have the insulation bonded to the plasterboard so that the insulation projects on two adjacent edges and the board projects on the other two adjacent edges. The vapour control layer covers all exposed faces in the joint area and so can be sealed together.

Because of their friable nature, urethane foam boards are not suitable for laminating to a sheet finish or adhesive fastening to a sub-base. For this reason polystyrene foam is more usual.

Timber frame construction

Traditional timber frame

The definition of 'traditional' depends on how far back in time one wishes to go. Early construction forms used the natural shape of trunks and limbs, which might have been hewn with an adze into a roughly rectangular cross-section – all as recently discovered on reality television!

For the purposes of our explanation, we will consider timber frames built *on site* using only sawn timbers for the frame. Within our definition there are two distinct forms of timber frame:

☐ Balloon
☐ Platform.

Both are built off masonry/concrete walls and foundations.

Choice of ground floors is:

☐ Hung timber
☐ Solid concrete

For the hung timber floor shown in Figure 3.12:

☐ Both forms start with a 'ring beam' built on a wall plate, both beam and wall plate being fixed to the masonry base with **ragbolts**.
☐ The ring beam could be either a single timber or two timbers nailed together.

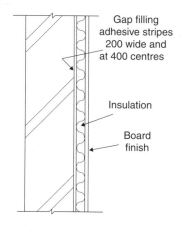

Gap filling adhesive stripes 200 wide and at 400 centres

Insulation

Board finish

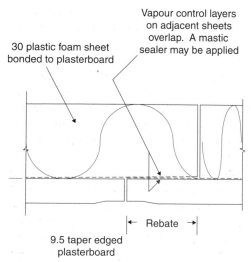

Vapour control layers on adjacent sheets overlap. A mastic sealer may be applied

30 plastic foam sheet bonded to plasterboard

Rebate

9.5 taper edged plasterboard

Fig. 3.11 Plasterboard/insulation laminate bonded to wall with adhesive stripes.

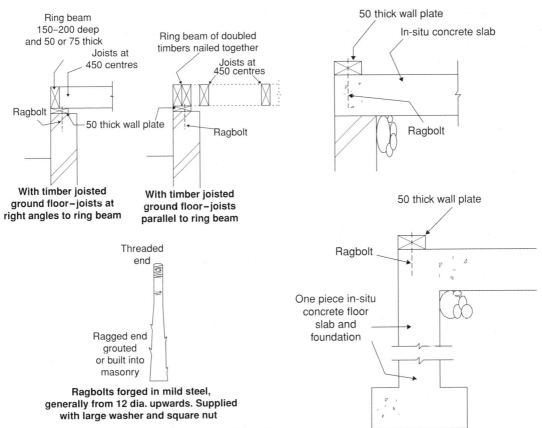

Fig. 3.12 Timber joisted floor as base for traditional timber frame building.

Fig. 3.13 Starting a timber frame building off a concrete floor slab.

For the solid concrete floor, two possible foundation details are shown in Figure 3.13:

☐ A masonry wall and a separate floor slab.
☐ A poured concrete foundation, wall and floor slab all in one. This is a common form of construction in North America where the poured concrete wall can form a basement or semi-basement.

In either case there is no need for a timber ring beam, and the wall plate is ragbolted to the edge of the concrete floor slab.

On the timber ring beam construction it was normal to fix a sole plate made up of two timbers on their flat. With a wall plate on a concrete floor edge, a single layer sole plate was used. In either case vertical studs rose, their height depending on whether the construction form was to be balloon or platform.

Platform construction means having the studs one storey high. On top of these studs the joists for a floor or platform are set; then another sole plate is set down and studs for another storey height are set up. In **balloon** construction, the stud height is set for two storeys. It is not generally possible to get timbers long enough for more. The intermediate floor is joined to the middle of the studs. Figures 3.14 and 3.15 illustrate both forms of construction.

In both instances there is a certain amount of prefabrication but this is done **on site** and is usually done by taking one layer of the sole plate and fastening studs to it, all for one wall plane, capping the studs off with a single runner. All this is done on the ground or the floor slab. Bracing of thin wide timber is let into recesses cut in the sole plate, studs and runners,

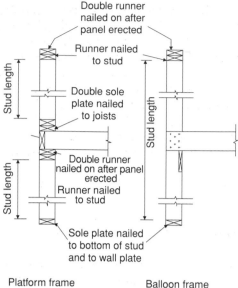

Platform frame
construction

Balloon frame
construction

Fig. 3.14 Schematic showing platform and balloon frame construction.

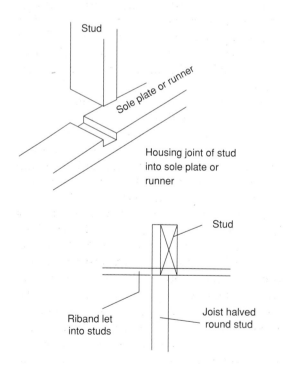

Fig. 3.15 Some joints in traditional frame construction.

and dwanging is added in the height of the studs at about 750 centres. Finally, the frame is erected on the wall plate or other half of the sole plate and the two are nailed together; temporary bracing keeps the panel vertical. A double runner is added on top of the panels.

With platform frame, the double runner on top of the studs is used as a bearing for the upper floor joists. Once these are in position, another single sole plate is nailed to the top of the joists, to which a further storey height set of panels is added; and so on. An additional runner is added to the top set of panels and it is to this double set of runners that the roof timbers are fixed.

In this form, and in balloon construction, it is important that all timbers – joists, studs and roof timbers – are fixed at the same centres so that loadings from roof and upper floors pass direct to studs, direct through joist ends, etc.

In balloon frame construction the floor joining halfway up the stud length is supported partly by a thin, wide timber called a **riband** let into the inner face of the stud, and partly by a halving joint of the joist to the stud and nailing to the stud. The top of the framed panel is made with a single runner and if a roof has to be refastened at that level, the runner is doubled up the same as in platform frame.

Typical sizes of timbers used are:

- ☐ Studs, runners, sole and wall plates, dwangs or noggings – from 150×50 to 200×75
- ☐ Bracing and ribands generally 32 or 38 thick and 200 to 250 wide
- ☐ Joists from 150×50 to 300×75
- ☐ Diagonal boarding 20 or 25 thick, 150 or 200 wide
- ☐ Clap boarding and weather boarding 15 or 20 mm thick
- ☐ Matchboarding 15 mm thick.

Once the frame is erected, the outer face will be covered with plain edged sawn boarding about 150–200 wide and 20–25 thick, laid diagonally; the edges are butted tightly together and nailed twice through the face to every timber – studs, sole plates, runners and dwangs.

This diagonal boarding has two purpose:

☐ Being diagonally laid, it contributes to the racking resistance of the frame
☐ Nailing twice to every underlying structural timber holds the frame tightly together and contributes significantly to its strength.

Once the diagonal boarding is fixed, a weathering surface is fixed in place. This can be of a variety of materials but is generally of the **impervious layer** type, where rain is shed rather than absorbed and allowed to evaporate. Materials commonly used are:

☐ Timber weather boarding, shingles, clay tiles and slate
☐ A layer of bituminous felt placed over the diagonal boarding. Very early examples used a 'tarred paper'.

Modern materials used in refurbishment work include plastics weather boarding, concrete tiles, synthetic slate and tile, fibre board sheathings, and profiled metal claddings. Whatever is used it is important to allow the old frame to breath. It must not be sealed off in any way.

The inner face used to be finished either with timber match boarding and/or with timber lath and plaster. The void between the two layers was frequently left empty, which means a cold house. If the climate was very cold, vegetable fibre such as straw, grass, reed, heather, moss, etc. would be packed in to improve the insulation. If the weathering face broke down and leaked water into the construction, rapid decay took place as none of the timber was treated with preservative. No vapour control layer was placed behind the internal wall finish – they hadn't been invented.

With no vapour control layer, moisture from the inside was able to pass through the wall until it reached the tarred paper or bituminous felt. Here it condensed and the now 'solid' water wetted the boarding, causing rapid rotting of board and timbers. Without either tarred paper or vapour control layer, the wall could 'breathe' and any condensation on the back of the outer boarding could evaporate away, even if only slowly.

Once the mechanisms of moisture passing through the wall to condense behind a cold impervious layer are understood, it is easy to work out a solution: provide a gap between the tarred paper or bituminous felt and the diagonal boarding, and allow this gap to be ventilated, as shown in Figure 3.16. This can be done by simply applying vertical battens to the face of the diagonal boarding to which the paper or felt is fixed, followed by the weathering skin. Allow this space to remain open top and bottom and it ventilates naturally.

Modern timber frame construction

Modern timber frame is built on the platform frame model and employs many of the same ideas as the traditional forms but applies factory prefabrication, regularised stress graded timbers, and plywood or oriented strand board sheathing panels instead of diagonal boarding, so there is no need for bracings, the sheathing shares the loading supported by the wall panels, timbers are much smaller in cross-section, and cladding is generally a masonry skin 75–102.5 thick of brick or block, facing brick or with a roughcast finish. Weather boarding is also used, either of natural timber or of plastics, and this would be fixed to vertical treated timber straps nailed over the studs in the panels. The skin of sheet boarding is such an integral part of the prefabricated panels that the construction form should really be termed a **stressed skin panel**.

Once the panels are delivered to site, they are erected on a sole plate or wall plate. They arrive covered with a layer of **building paper**. The building paper is stapled on through a heavy plastic tape which prevents the paper tearing at the staples. The tape is similar to that used to by suppliers and merchants to bind bundles of their material. Building paper can be a bitumen bound fibre paper or, more commonly now, a synthetic fibre (usually polypropylene) felted membrane such as Corovin or Tyvek.

Once the panels are erected and braced upright, the roof is put on. When weathertight –

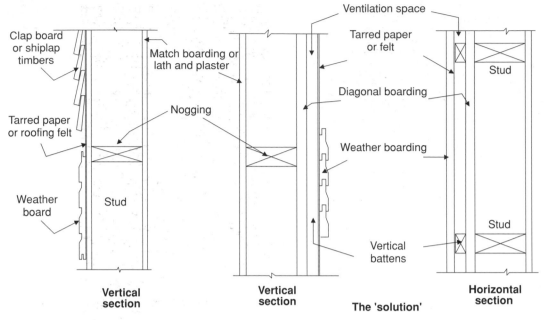

Fig. 3.16 Traditional timber cladding and improved methods of cladding over insulated walls.

at least the roofing felt in place and counter-battened – the panel frame is filled with a glass or mineral wool insulation, a vapour control layer *may* be fitted and a finish applied, usually plasterboard but also other sheet materials as well as timber panelling.

Figure 3.17 shows a timber frame construction at substructure level, Figure 3.18 shows

that construction at eaves level, and Figure 3.19 shows the junction of an upper floor with the timber panels. Note that a lot of the annotation has been omitted for the sake of clarity. On properly scaled detail drawings the annotation would be complete, either on the detail or in a notes column.

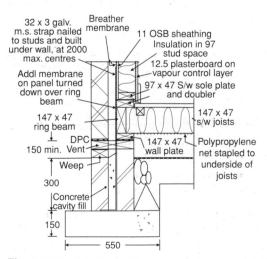

Fig. 3.17 Modern timber frame – substructure detail.

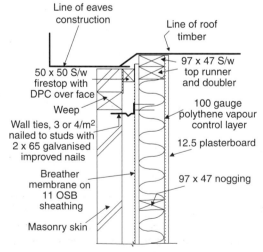

Fig. 3.18 Modern timber frame – eaves/wallhead detail.

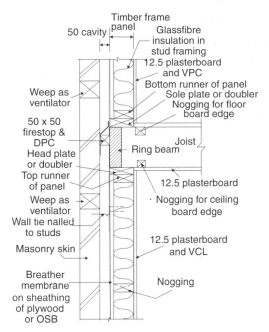

Timber frame panel
50 cavity
Glassfibre insulation in stud framing
12.5 plasterboard and VPC
Bottom runner of panel
Sole plate or doubler
Nogging for floor board edge
Weep as ventilator
50 x 50 firestop & DPC
Head plate or doubler
Joist
Ring beam
Top runner of panel
12.5 plasterboard
Weep as ventilator
Wall tie nailed to studs
Nogging for ceiling board edge
Masonry skin
12.5 plasterboard and VCL
Breather membrane on sheathing of plywood or OSB
Nogging

Note: The void in the depth of the floor immediately behind the ring beam is a potential cold spot and a roll of glassfibre insulation should be placed there

Fig. 3.19 Modern timber frame – intermediate floor junction detail.

It is important to note:

□ The hold down straps built in every 2000 mm or so and nailed to the studding – Figure 3.20. Note the building paper turned down to cover the joists and ring beam.
□ The weeps built in every fourth or fifth brick, top and bottom of the outer leaf of masonry to ventilate the cavity and keep the surface of the timber frame panels dry – Figure 3.21. Note how this weep projects beyond the face of the blockwork to accommodate the render to be applied.

On the outside, a layer of masonry is built off the foundation to give a 50 mm cavity next to the timber frame panel. Wall ties are nailed to the studs of the panel and built into the courses of the masonry – three to four per square metre with additional ties at quoins, intersections and openings. Nailing should really be with two nails per tie, and an improved nail would be best. Unfortunately few

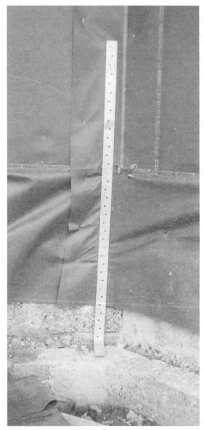

Fig. 3.20 Hold down strap from timber frame to top of foundation.

bricklayers are prepared to go to the length of putting in more than one nail and the use of improved nails is always a matter of what they are given!

Special ties are available which have holes for nails at one end and a 'grip' for bedding into a mortar joint. Figure 3.22 shows two samples of these ties.

The masonry can be of:

□ Half brick thick brickwork, either facing brick or commons, with a roughcast
□ Blockwork 75 or 100 thick and roughcast
□ 100 thick fair faced blockwork, natural, artificial or reconstructed stone.

Although the weeps have been previously described as 'ventilating' the cavity formed, the bottom weeps also act as cavity drains. The ventilation of underfloor voids must be sealed

Fig. 3.21 Weep as a ventilator at the wallhead.

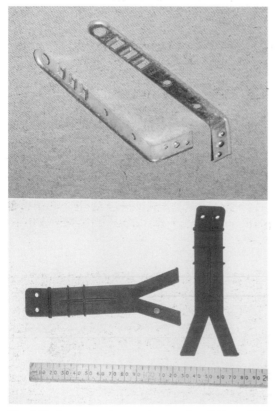

Fig. 3.22 Just two of many types of wall tie suitable for timber frame cladding.

off from the cavity by use of a liner to prevent the spread of fire from the underfloor void to cavity and vice versa. A fire stop of timber at least 50 thick (measured in the direction of the cavity) must be nailed to the top of the panels to prevent the spread of fire either to or from the intermediate floor or the roof void. Each 'separation' of the cavity must be ventilated with weeps top and bottom.

The details you don't see are often more important than the ones you do. The wall panels we have looked at earlier were all fire stopped horizontally to reduce the spread of fire up the cavity between panel and cladding. However, there is a need to fire stop vertically to reduce the spread of fire horizontally. This is done at corners; Figure 3.23 gives a detail while Figure 3.24 shows a photograph. What is not immediately apparent in the photograph is that the horizontal fire stop is there but shows only as a bump under the breather paper – just below the scaffold boards. Now why didn't the builder put up the corner ones as well, then put on the building paper? Well, the panels come with the paper already fixed in place and so when the paper from the upper panels was dropped, it covered the fire stop. Then they came along and fixed the corner fire stops.

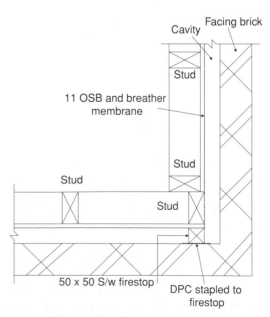

Fig. 3.23 Vertical fire stopping at corners of a building.

Fig. 3.24 Vertical and horizontal fire stopping.

beam should be fixed before the breather papers are fixed down so that the workmen can see where the studs are. On this site, they were fixed afterwards so that some of the straps were nailed only to the frame sheathing, the OSB (oriented stand board), which was only 11 thick, and they were bent to come over the horizontal fire stop. They were therefore not straight and tight across the junction. This also meant that they were pressing against the inside face of the masonry skin.

The rules, therefore, are:

(1) Fit straps across upper floor junctions before fixing the membranes down or fire stopping is fitted.
(2) Fix straps down into substructure before fixing the membranes down.
(3) Breather membranes should be completed over the whole of the timber frame panels before any other work is carried out.
(4) Fire stopping of **treated** timber should be fixed over the membrane.
(5) DPC material should be fixed over the fire stopping.
(6) No breather membrane between the fire stop and the masonry cladding.

Figure 3.25 is a photograph of a ground floor built on a block substructure and ready to have wall panels erected on it. The wall plate with the ring beam and the sole plate doubler are all there round the outside. Then there are sleeper joists with solid strutting across the mid span of the joists. What is not terribly clear is the two sleeper walls with wall plates supporting the joists at their mid third points.

But they got it all wrong. The breather paper, while repelling solid water, does so only when the surface of the membrane is open, free, and unobscured. If it is pressed between a fire stop and the masonry, it will draw in water like a sponge after a short period. What made things worse on this particular site was that the builder did not put any DPC on top of the fire stops, which were untreated timber – a recipe for rot and subsequent disaster.

But there was one mistake which was never photographed. The metal straps joining upper and lower panels across the upper floor ring

Fig. 3.25 A ring beam and joisted base ready for panel erection.

Fig. 3.26 Panels erected on a base.

Figure 3.26 shows panels erected on such a base[4]. However, one can identify all the timbers. Notice how the vent liner has not been properly built in.

Following chapters will deal with openings (Chapter 5), doors (Chapter 8) and windows (Chapter 9), and more detail of the wall techniques will be revealed then.

Figure 3.27 shows a photograph of the upper floor wall panel junction. The wall panel built in here incorporates a window opening and we will return to this photograph when we look at openings in walls in the next chapter. For now, note the top runner right across the opening and the ring beam and upper floor joists, with the nogging for the edge of the floor boarding.

Loadbearing and non-loadbearing internal partitions

Partitions can be built from a variety of materials. We will concentrate on:

[4] The reader will be aware by now that the photographs have been taken on different sites. This was not a good site; the last photograph, Figure 3.25, was a good site. Basically the difference was that good sites had the contractor's own workforce, who knew what was expected and had been trained in or inducted into the techniques required. Bad sites used labour hired by the squad, and the contractor knew little about and had no interest in their abilities or basic knowledge.

Fig. 3.27 Intermediate floor/panel junction over a window opening.

- ☐ Simple half brick thick brick and narrow block partitions, 50, 75 and 100 thick – these can be loadbearing or non-loadbearing.
- ☐ Stud partitions clad with sheet material both sides – these can be loadbearing or non-loadbearing.

Whether a partition rises from the floor or sub-floor surface or an independent foundation depends on:

- ☐ Whether it is load or non-loadbearing
- ☐ The floor structure
- ☐ The mass of the material from which the partition is made.

Brick partitions in domestic construction must generally be built off an independent foundation, whether load or non-loadbearing, as the self weight is high:

- ☐ Foundation minimum width 450
- ☐ Minimum thickness 150.

However, with a concrete sub-floor (*not* over site concrete) it would be possible, depending

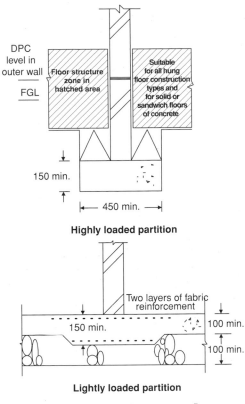

Highly loaded partition

Lightly loaded partition

Only suitable for single layer concrete floors with a slab thickness of at least 100 mm. NOT suitable for use with minimal sandwich floors

Fig. 3.28 Partition foundations (1).

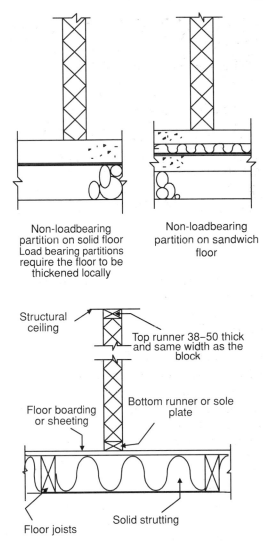

Non-loadbearing partition on solid floor
Load bearing partitions require the floor to be thickened locally

Non-loadbearing partition on sandwich floor

Block partition running parallel to the joists supported by solid strutting or a pair of joists nailed together. If running across the joists, a simple nogging would be adequate to prevent the boarding from sagging.

Fig. 3.29 Partition foundations (2).

on loadings, to build off the sub-floor – but see Figure 3.28 for the options available.

Dense concrete block partitions must be treated like brick partitions. They have similar or greater self-weight. Medium and light weight block partitions would generally need an independent foundation if they were highly loadbearing, or could be built off a sub-floor if lightly loaded, as in the brick partition detail in Figure 3.28.

Non-loadbearing block partitions are built off the floor surface if it is of concrete, solid or hung, or a timber runner if it is of timber. A timber runner is also fixed to the ceiling and the blocks are built between the two runners. These are shown in Figure 3.29. Note that the floor structure must include either an additional joist or noggings or, preferably, solid strutting between two joists under the runner.

A lightly loaded block partition built off a sandwich floor must be built off the lower layer of concrete, which in turn should be thickened locally as well as reinforced locally – see Figure 3.30.

Stud partitions are made from sawn or regularised timber starting with a runner or sole plate fixed to a floor surface; a runner is fixed to the ceiling surface and there is an infill of

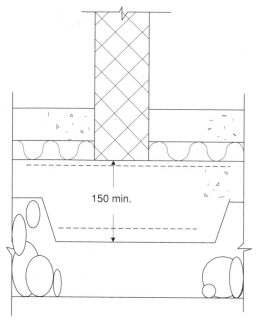

Fig. 3.30 Partition foundation (3).

vertical timbers or 'studs'. Generally, dwangs are fitted between the studs at a maximum of 800 intervals vertically to prevent buckling of the studs. This is particularly important if the partition is loadbearing. See Figure 3.31 for the range of options for junctions etc.

The frame is generally clad with plasterboard 9.8 or 12.5 thick, the 12.5 thickness being used to give fire protection on the side 'at risk' from fire damage. The plasterboard cannot contribute to the loadbearing characteristics of the partition. Timber sizes used are generally the same in any one partition and, in domestic construction, can range from 50 × 38 to 100 × 50 depending on the requirements placed on the finished partition.

Brick and block partitions, because of their mass, have good sound insulation qualities, whereas stud partitions tend to act like 'drums' and transmit sound easily; indeed they seem to magnify certain frequencies. Attenuation is possible by filling between the studs with fairly dense rockwool or fibre glass quilts, and even doubling up thicker layers of plasterboard – at least 12.5 thick – while fixing adjacent layers broken jointed.

A further technique shown in Figure 3.32 is the provision of a separate set of studs to support the boarding on each side of the partition, and the fibrous quilt is wound between the studs. Again heavy plasterboard – 12.5 or 19 thick – is used to give mass.

Stud partitions can be loadbearing, particularly in lightweight domestic construction.

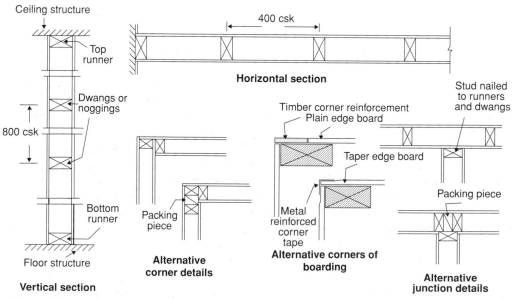

Fig. 3.31 General partition details.

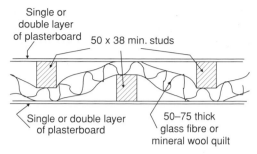

Fig. 3.32 Acoustic partition.

They are always supported by the floor structure which might require bolstering under the line of the partition. This is perfectly feasible, as the details show. A structural partition will require fire protection on both sides even if it is within a building of 'single occupancy', i.e. a single house.

In the mass local authority housing of the 1960s and 1970s, it was common practice to build 2400 high loadbearing partitions with 50 × 38 studs at 400 centres, 50 × 38 sole plates and two rows of 50 × 38 dwangs. The partition was clad with 12.5 plasterboard both sides. Such partitions were set on the ground floor of hung timber at right angles to the sleeper joists and within a quarter joist span of a dwarf or sleeper wall. They were designed to carry a share of the load from the first floor and sometimes a further partition above – but *not* carrying any of the roof load. Both partitions had to be directly above one another.

Racking resistance is generally not required as ends are usually fixed to outer walls, and top and bottom runners are fixed to ceiling and floor structure respectively.

Non-loadbearing stud partitions can be built off the upper surface of any concrete floor. Loadbearing stud partitions would normally be lightly loaded and so could be built off the top of a single layer floor without thickening it. With a sandwich floor, it would be better to build off the lower layer, as shown in Figure 3.33, purely because the upper layer is relatively much thinner and cannot be thickened locally without losing the insulation layer.

The sequence of construction of such a partition is as follows:

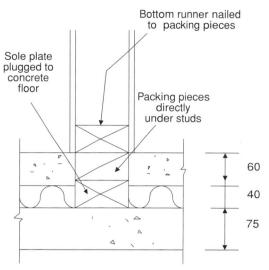

Fig. 3.33 Partition base in sandwich floor construction.

☐ Sole plate is plugged to lower layer of concrete, and short packing pieces or a continuous piece nailed to it. The top of the packing piece is set to finish at the same level as the upper layer of concrete. Short packing pieces must coincide with the ends of studs.
☐ Lay underfloor insulation.
☐ Lay upper layer of concrete.
☐ Nail bottom runner of partition to packing pieces and complete partition in usual way.

Expansion joints

External walls can be very long when applied to terraced housing, and when combined with materials with a high rate of moisture or thermal expansion and contraction there can be problems with walls cracking etc. in awkward places. Apart from being unsightly, random cracking can cause structural instability, largely due to the fact that the two parts of the wall are not tied together in any way and can move independently of each other under other influences – thrust from floors or roof, wind loadings, etc.

The effect of this moisture and thermal movement is most keenly felt in concrete block masonry. The simple solution is to break long stretches of wall into short lengths and at each length provide an **expansion joint**. There are two ways of doing this:

(1) Physically stop building the wall, leave a gap filled with some flexible material and then begin building the wall again, only joining the two separate walls with some flexible jointing material.
(2) Build in flexible jointing material in a length of the wall and then cut a slot in the wall with a power saw to about half its thickness.

The first is a conventional expansion joint, the second more of a stress relief joint or induced crack joint.

First of all, how far apart must these joints be placed? There is a lot of conflicting advice but it does appear to have some correlation with just how much the material in the wall is expected to move under different moisture content conditions. A two-storey terrace of six flats not too far from the author has an expansion joint in the centre of each gable and others back and front on the party wall line between the vertical pairs of flats – about 6 m apart. So in approximately 50 m of wall there are six expansion joints. Another building, a supermarket, has many fewer joints, one of which is shown in Figure 3.36 and which draws some adverse comment for the way it has been done – and yet there is no sign of any adverse effects of differential movement on the building.

So the best we can do here is look at how expansion joints and induced crack joints are built. Figure 3.34 illustrates both styles of joint and Figures 3.35 and 3.36 are photographs of two different joints.

In Figure 3.34, detail A is an expansion joint in a masonry cavity wall. The points to note are:

☐ The mastic pointing over the joint on the external leaf.

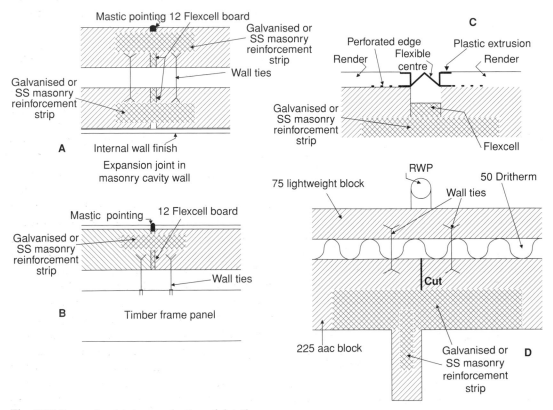

Fig. 3.34 Expansion joints – a selection of details.

Fig. 3.35 A proprietary joint of plastic in a rendered finish block wall.

Fig. 3.36 A mastic sealed joint in a complex wall of reconstructed stone and rendered blockwork.

☐ The two parts of both leaves are separated by a compressible bitumen impregnated fibre board – Flexcell.

☐ Both parts of each leaf are joined across the Flexcell with a strip of expanded metal lath such as is used for reinforcing mansonry. This is flexible but its position prevents the four parts of the wall from drifting apart in a horizontal plane and yet accommodates any movement apart or together.

☐ The internal wall finish simply passes over the joint, although many details will show a timber cover plate fixed to one side only of the point where the finish might crack.

☐ Most importantly are the additional wall ties built in here. These are over and above the 3–4/m² in the rest of the wall. They keep the four parts of the wall from drifting apart as well.

In the same figure, detail B is an expansion joint in the masonry skin of a timber frame house. All the things from the previous detail apply here, but only to the cladding of masonry. There is no allowance for movement of the timber frame.

Top right in the figure, detail C, is an enlarged view of the detail of the joint from Figure 3.36 showing the rendered finish on blockwork and the proprietary expansion joint cover of plastic bedded over the joint into the render on each side. For this purpose the plastic joint has perforated wings which key into the render.

Finally, detail D is an induced crack joint. The inner leaf of the wall was of 225 aac block (blocks built on their flat), and as this was being built the brick reinforcement strips were built in opposite the partition, as well as other strips being left sticking out to tie into the partition blockwork, which was also aac block but only 125 thick. Additional wall ties were also left protruding either side of where the block was intended to crack. Finally, the cut was made with a power saw. The outer leaf of lightweight blockwork and the insulation were added later. This blockwork was only 75 thick, and if the wall as a whole had moved, it would have cracked down a similar line to the main wall. To disguise any possible crack

Fig. 3.37 Faulty joint being built in the block cladding of a timber frame house.

Fig. 3.38 Proprietary expansion joint cover.

panels of plain blockwork which have been rendered with a wet dash. There is no sign of movement on the building but there is a lot of unsightly staining where there has been rainwater run-off concentrated at the mastic filling of the expansion joints. This is also a relatively new building.

Finally, Figure 3.37 is a photograph of a recently discovered joint being built in a timber frame housing development. The Flexcell is there against the wall, which is complete, but there is no expanded metal lath built across between the different parts of the complete wall. What the photograph doesn't show is the lack of additional wall ties. The whole wall is only one storey and therefore at most 3 m high. It is only a cladding and not structural but that does not make it correct.

Figure 3.38 is the expansion joint cover employed in Figure 3.35. It is made from two pieces of render stop material – the galvanished strip with the expanded metal edges – and a central PVC extrusion. Each side of the extrusion has a groove to take the edge of the render stop and these edges are hard plastic. They are joined by an inverted V-shaped piece of PVC which is much more flexible and allows the joint to open when shrinkage occurs.

in the outer leaf, a RWP (rainwater pipe) was erected over that line. That detail was executed twice on each of two 14 metre wallls approximately 18 years ago and there is still no movement to tell if the detail has been successful or not. The engineer who suggested it is confident it will perform as predicted when its time comes.

Figure 3.35 is the joint which featured at C in Figure 3.34. The building is relatively new and shows no sign of movement.

Figure 3.36 is a joint on a supermarket. The wall is about 6 m high with a lot of artificial stone facings, string courses, etc. with infill

4 Timber Upper Floors

Upper floor joists	103
Linear and point loadings on upper floors	112
Openings in upper floors	113
For pipes	113
For flues	114
For stairs	116
Alternative materials for joisting	118
Sound proofing	120
Modern sound and fire proofing	121
Support of masonry walls	123
Floor finishes	124
Ceiling finishes	124

By the term 'upper floor' we mean any floor not accessed at ground level or below ground level. Such floors have been developed using a variety of construction methods, materials and techniques. In this text we will confine our discussion to the most common in domestic construction – the timber joisted upper floor.

The functions performed by an upper floor may be summarised as:

☐ Providing a safe platform for the occupants of the house
☐ Providing support for partitions and equipment at that level
☐ Providing a space to accommodate services within the floor, serving the space above and below the floor level
☐ In buildings of more than one occupancy the floor may have to provide a barrier to fire, sound and even weather where part of it is built over a pedestrian or vehicular access.

Timber upper floors comprise:

☐ Joists which can span the narrower dimension of the building
☐ Floor boarding or sheet material
☐ A ceiling finish on the underside.

Plus as occasion demands, additional material and techniques to:

☐ Attenuate the transmission of sound through the floor
☐ Provide fire protection between upper and lower storeys
☐ Modify the effects of weather in pedestrian or vehicular accesses.

Upper floor joists

Upper floor joists made from a variety of materials and alternatives to the most common material – natural wood – will be discussed later in this chapter.

With regard to wood joists, the timber is used either as sawn or may be regularised, and considering the small additional cost involved should be treated with preservative at the saw mill against insect and fungal attack. The Building Regulations require that, as this is structural timber, it should be stress graded, the grade depending on the span required to be bridged and the expected loads of people and furniture; also the self weight including any sound/fire proofing on the floor as a whole or on individual joists.

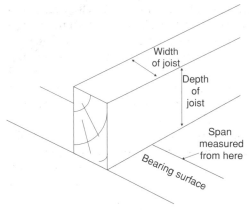

Fig. 4.1 Alignment of joists, span and wall rest.

It may seem obvious, especially after looking at ground floor construction in timber, but it has to be stated that *all* joists are built in with the long dimension of the cross-section in the vertical plane, as shown in Figure 4.1. The reader can test the effect of using a supporting timber on 'edge' or on 'flat' by placing a ruler across the gap between two books and pressing down at the middle of the span. The ruler is much stiffer when it is on edge but is prone to buckle sideways. This is due to the very small width/span ratio. Joists of course are restrained at each end and in short spans do not buckle because the ratio is much higher. In long spans we can employ techniques such as **strutting** to prevent buckling. The various alternative forms of strutting are discussed later in this chapter.

The ends of upper floor joists, unlike ground floor joists, are generally supported by building them into the wall, the use of cavity wall construction protecting the joist ends from dampness and decay. Before the general adoption of preservative treatment of structural timbers, protection was provided by one of the following:

☐ Brushing or dipping the ends in creosote
☐ Wrapping with DPC
☐ Building the joist end into a terracotta shoe.

Brushing or dipping in creosote was done to the last 300–450 mm of the joists. The process gave only a surface film of preservative and in order to allow the creosote to penetrate, many old specifications required the timber to be dipped and left for a number of hours or even days. Provided the timber is dry, the cut end will absorb the preservative but general penetration is poor in comparison to modern treatment methods.

The DPC used was invariably a hessian-based bituminous DPC, which if used in conjunction with creosoted timber could be softened by the solvents in the creosote, resulting in failure. Even if failure occurred, the likelihood of damage to the timber was slight.

Building the joist into a terracotta shoe was more commonly used where the supporting wall was shared with an adjacent property, the shoe being required to give fire protection to the end of the joist. It is very effective but the shoes are now difficult to find.

Alternatively, a timber runner can be fixed to the face of the wall and the joist ends 'housed' into the runner. Fixing of the runner can be done by masonry bolts or by building steel brackets into the wall and resting the bearer on these. (Note that steel brackets are not suitable for a shared wall. The brackets could become hot if there was a fire in the adjacent property and could set fire to the property being served by the supports.) Either way the centres of the fixings would be from two to three joist spacings, depending on the thickness of the bearer, the floor loading and the strength of the fixing.

Figure 4.2 shows methods of supporting upper floor joists other than building into the inner leaf of a brick or blockwork cavity wall. From top left, anti-clockwise:

☐ A timber runner Rawlbolted to the masonry and joists fixed to it using pressed steel joist hangers. These hangers are of thin galvanised steel and require multiple nailing to runner and joist to make a secure fixing.
☐ A timber runner against the wall, this time supported by a mild steel bracket and the joists housed into the runner and cheek nailed.

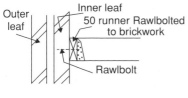

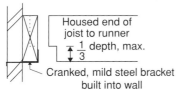

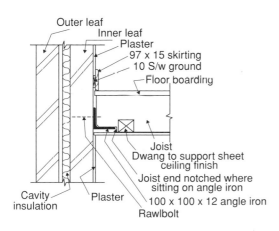

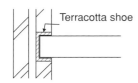

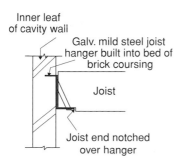

Note: Notching the joist over the angle or hanger allows the ceiling finish to run through flush

Fig. 4.2 Alternative methods of joist support.

☐ A mild steel angle iron is Rawlbolted to the masonry and the joists are set on the flange of the angle iron with a shallow notch to allow the ceiling finish to pass under the angle iron. It is not normal to fix joist and angle iron together, but the angle could be drilled and a nail or screw driven into the bottom of the joist.

☐ Using a galvanised steel joist bracket built into the masonry, one bracket per joist end. These must be of 3–4.5 mm thick galvanised steel, folded and welded; thin pressed steel folded hangers are *not* suitable.

☐ Finally, a terracotta shoe is built into the masonry and the joist ends are set into the shoe. Not a general duty method but can be used to allow building in of joists in cavity walls shared between two properties where fire regulations will not allow joists to be simply built into the wall lest fire spread across the cavity.

Figure 4.3 shows a joisted construction using heavy gauge (3–4.5 thick steel) hangers built into the wall to support both a trimmer joist (the very thick joist with bolt) and common joists. The picture shows dwangs/noggings to support the edge of a sheet flooring material. Note the trimmer joist built up from a number of layers of common joist bolted together with 12 mm coach bolts, nuts and washers. The wall is a shared wall, commonly called a **party wall**, a **mutual wall** or, where it rises to the top of a pitched roof, a **mutual gable**.

In modern timber frame construction, the upper floor joists are supported by the wall

Fig. 4.3 Heavy gauge joist hangers supporting upper floor joists.

panels, and a 'ring beam' is included for the full depth of the joists. The ring beam serves:

☐ As a support and fixing point for the joist ends
☐ As a 'stop' to prevent the spread of fire from the cavity into the floor void.

Figure 4.4 shows upper floor joists in a timber frame construction and is a repeat of Figure 3.19, but the annotation and dimensions have been stripped off to make the actual junction of the joists with the timber frame more clear.

Figure 4.5 is the closest it was possible to get to Figure 4.4 for comparative purposes. In the

Fig. 4.5 Upper floor joists in timber frame construction (2).

photograph the joists are seen resting on top of the timber frame panels; the ring beam can be clearly seen, as can the dwangs or noggings for the edge of the sheet flooring material. The timber frame panel shows a double top runner at the left-hand side but uses a steel channel section to provide support over an opening under the remaining joists. See Chapter 5, Openings in Masonry Walls, where the use of metal sections in timber frame construction is discussed in more detail.

The size of the floor joists is determined by:

☐ The loads and span to be carried and the need to limit deflection to $\frac{1}{380}$th of the span
☐ The spacing or 'centres' of the joists
☐ The grade of timber selected
☐ And, not so important in domestic structures, the vertical space available or required for other purposes such as hidden services – ducts and large pipes etc.

While the sizing of upper floor joists follows the same rules as for ground floor joists, note that spans will generally be greater. Provision of heavier joists round large openings is covered later in the chapter.

Domestic loads are not high in comparison with other structures but spans can range up to 7 or 8 m on the shortest distance across a building. Such spans could well require joist depths in excess of 250 mm. A depth of 200 mm marks an important economic boundary for timber, so spanning such distances with single lengths of timber can be quite uneconomical. It is not uncommon to span from 3.50 to 4.00 m from support to support with single joists.

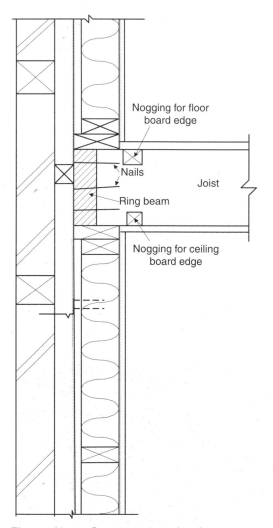

Nogging for floor board edge

Nails

Joist

Ring beam

Nogging for ceiling board edge

Fig. 4.4 Upper floor joists in timber frame construction (1).

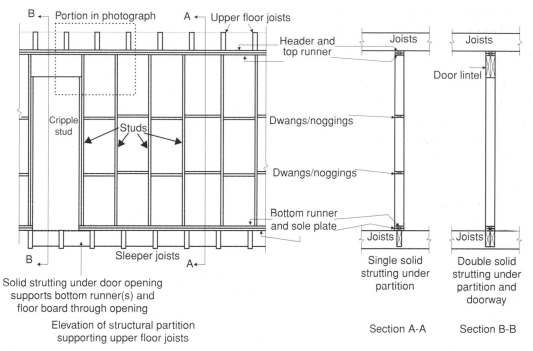

Fig. 4.6 Upper floor joists supported by a partition (1).

When spans start to exceed this range it is time to think of alternative ways to support the joists – in other words to break the span into two or more manageable proportions. The best options are to have a loadbearing partition under the joists or provide a beam of some kind. This can be done by making use of partitions on the lower storey as loadbearing partitions, as shown in Figure 4.7. The partition has a door opening spanned by a timber lintel supported on cripple studs with doubled top runners supporting the first floor joists. Note the joists do not fall directly above the studs as these timbers are at different centres, but the top runner of the partition is a double layer of timber.

Figure 4.6 shows a detail of upper floor joists supported by a partition, while Figure 4.7 is a photograph of the same situation. In Figure 4.7 the partition is shown on the right-hand two thirds of the picture, and a door opening with a heavy timber lintel, not a beam, is shown on the left-hand side. The prominent vertical timber in the foreground is a temporary strut.

Breaking the span of joists can also be achieved by:

☐ The use of beam(s) spanning from wall to wall at right angles to the joists
☐ A combination of the above with a partition.

Floors constructed over beams are commonly referred to as **double floors**. When beams only are used, it is common to have the beam span the shorter distance across the building, thus dividing the longer distance into two even

Fig. 4.7 Upper floor joists supported by a partition (2).

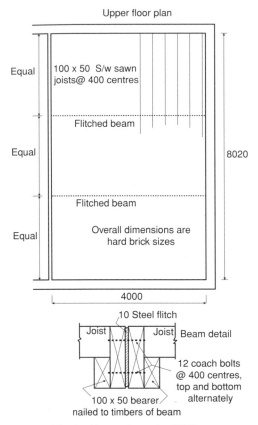

Fig. 4.8 Flitched beam from the 1970s.

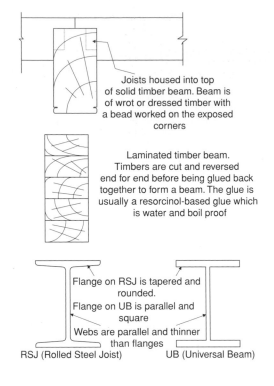

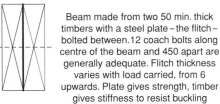

Fig. 4.9 Alternative beam constructions.

shorter spans for the joists. Two or more beams can of course be used.

An example from a house built in the late 1970s is shown in Figure 4.8. The only difference if this floor was being constructed now is the need to have all the timber stress graded; C3 would be perfectly adequate. It is interesting to note that when the calculations for the beam were carried out, the two 100 × 50 bearers for the joists were not taken into account except for the self weight of the beam, and yet the design was perfectly adequate in all respects.

Any supporting partitions can be built using brickwork, blockwork or timber stud. Refer to Chapter 3, Walls, for details of partition construction.

Beams can be of timber, steel or a combination of timber and steel. Figure 4.9 shows some alternative beam constructions and Figure 4.10 shows a flitched beam elevation, illustrating the extent of the steel plate. Note

that the steel plate was not the same thickness end to end. The middle part was calculated to require 10 thick plate, and the two end eighths 6 thick plate. The whole was welded together and packers used to ensure that the beam ended up a uniform thickness for its entire length.

Supporting joists at flitched beams have already been shown – a bearer nailed to each side (Figure 4.8). Alternatively, joists could be housed into the timbers, pressed steel hangers used or the joists passed over the beam. The large amount of notching for the housings would have to be considered when calculating the size of the timbers to be used. In practice, the next larger timber size available would be used. Similar means of support could be used with laminated and solid timber beams.

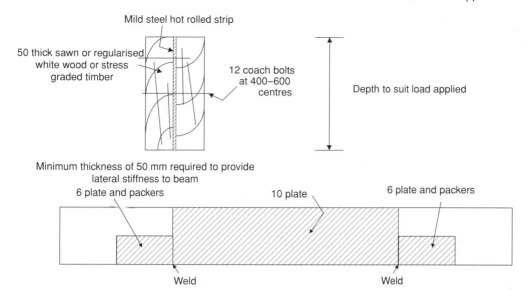

Mild steel hot rolled strip

50 thick sawn or regularised white wood or stress graded timber

12 coach bolts at 400–600 centres

Depth to suit load applied

Minimum thickness of 50 mm required to provide lateral stiffness to beam

6 plate and packers

10 plate

6 plate and packers

Weld

Weld

Elevation of a beam with 'steel saving'.
The hatched shape shows three pieces of steel welded together and the clear area represents packing of plywood or timber strip. The precise extent of the steel can be easily calculated.

Fig. 4.10 Flitched beam detail with minimal steel content.

Occasionally reinforced concrete is used but its high self weight makes it expensive to put into place. With steel beams, it is not possible to notch into the beam or nail hangers to it, so joists can only be set over the top of the beams or into the **bosom** of the beam. Figure 4.11 shows typical details of a steel beam supporting joists.

Combining beams and partitions can be successfully done using brick or block partitions with steel or flitched beams, although modern examples of timber frame construction are increasingly showing steel beams resting on cripple studs. Where timber stud partitions are built, timber or flitched beams can be used. To keep beam sizes down, columns can be provided in the lower floor. These are generally of masonry, one brick square.

The spacing of joists is dictated to some degree by the loadings involved and there are three standard centres which can be used – 400, 450 and 600. There is nothing, frankly, to stop the joists being placed at other centres, the only snag being that the standard load span tables don't cater for centres other than the standard ones already mentioned. However, spacing out the joists is dictated more by

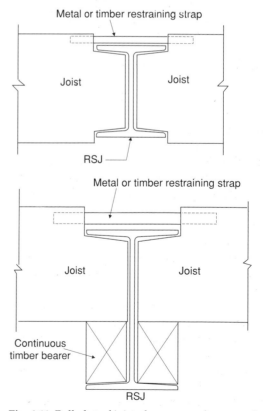

Metal or timber restraining strap

Joist

Joist

RSJ

Metal or timber restraining strap

Joist

Joist

Continuous timber bearer

RSJ

Fig. 4.11 Rolled steel joists, beams or columns as floor support beams.

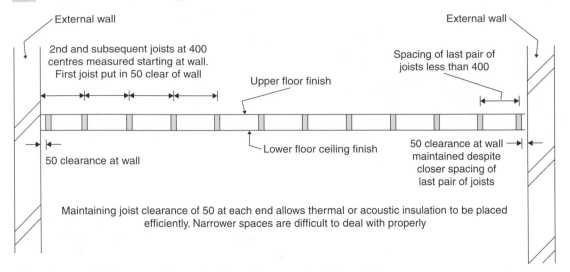

Fig. 4.12 Centres of joists in relation to wall and 'last space' and last wall.

the need to support a sheet material for the lower floor ceiling finish and occasionally by the floor finish as well.

Modern construction more often than not uses plasterboard, having a sheet size of 1200×2400. Joist centres therefore must be at 400 or 600 mm centres. Joist layout follows the same principles as for ground floors; the first joist is set 50 mm clear of the wall and the remainder fixed at centres, finishing, irrespective of the gap left, with a joist 50 mm clear of the opposite wall. Figure 4.12 shows a section across floor joists to illustrate the regular centres of fixing and the odd space at the end.

Should economy or other requirement dictate that the joist centres be some other dimension, then light timbers can be fixed at 400 centres and at right angles to the underside of the joists, and the sheet finish nailed to these. These timbers are known as **branders** and the process as **brandering**. It can be a useful technique in refurbishment work for bringing an old uneven ceiling down to a better 'level', as well as sorting out joists at odd or uneven centres. Figure 4.13 illustrates how this can be done.

Figure 4.14 illustrates an alternative to brandering using **cheek pieces** to support a ceiling. This can only be used if the centre of

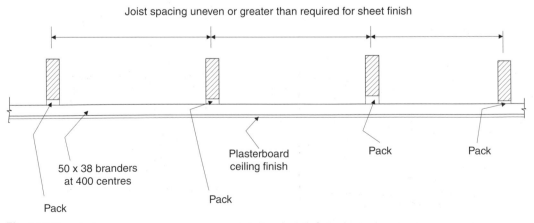

Fig. 4.13 Brandering.

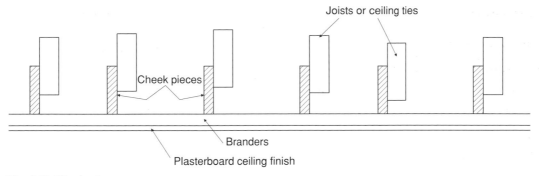

Joists or ceiling ties

Cheek pieces

Branders

Plasterboard ceiling finish

Fig. 4.14 Cheek pieces.

these timbers is not critical – unless branders are nailed to the cheek pieces.

Having the joists or branders at 400 or 600 centres gives support to only two edges of a standard plasterboard sheet. The edges at right angles cannot remain completely unsupported so additional timbers must be fitted to provide that support. These timbers are lighter in section than the joists (38 × 38 to 50 × 50) or the same section as the branders. These timbers, dwangs or noggings, are fixed at a maximum of 800 centres but most commonly at 600.

The spans of upper floor joists are inevitably longer than ground floor joists, and, with the greater depth of joist, twisting or buckling under load is a possibility. To prevent this, one of two forms of **strutting** is built in. This must not be confused with the dwangs or noggings already mentioned. Indeed, because this strutting is never flush with the underside of the joists, it cannot be used for edge support. It is done once at half the span or twice at a third span, depending on the span. The object should be to reduce the free length of joist to a maximum of around 2.5 m.

Figure 4.15 is a detail of solid and herringbone strutting in steel and timber, and Figure 4.16 is a photograph of a mock-up of the galvanised steel strutting. Note that the joists

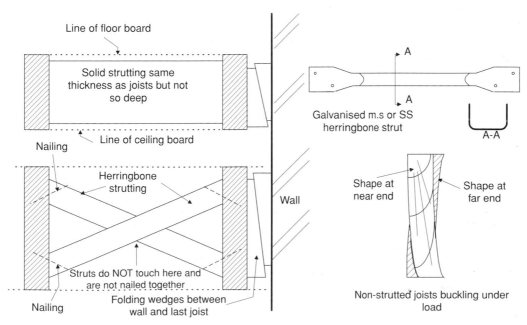

Line of floor board

Solid strutting same thickness as joists but not so deep

Nailing

Line of ceiling board

Herringbone strutting

Struts do NOT touch here and are not nailed together

Nailing

Folding wedges between wall and last joist

Wall

A

A

Galvanised m.s or SS herringbone strut

A-A

Shape at near end

Shape at far end

Non-strutted joists buckling under load

Fig. 4.15 Solid strutting and steel and timber herringbone strutting.

Fig. 4.16 Galvanised steel herringbone strutting.

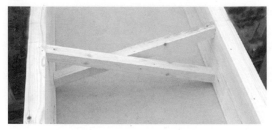

Fig. 4.18 Timber herringbone strutting.

are 200 deep and the strutting is only 150 deep and fixed flush with the upper surface of the joist. This leaves 50 mm immediately above the ceiling board for running services such as water pipes and electricity cables.

Figure 4.17 shows timber solid strutting and Figure 4.18 shows a photograph of a mock-up of timber herringbone strutting.

Linear and point loadings on upper floors

The loads referred to here are loads caused by partitions built across the floors, or items such as hot water storage cylinders, cold water tanks, baths, etc. The use of load span tables will not allow these items to be taken into account and their effect must be calculated separately. Generally, partitions running parallel to the run of the joists are supported by placing a joist immediately under the partition. The joist would be at least double the thickness of the ordinary joists; indeed, it may be two ordinary joists laid close together and joined by a pattern of nails.

Point loads can be caused by partitions which run at an angle to the run of the joists, or by bath support legs or cradles or by water storage systems. The increase in joist width or the introduction of additional joists are both solutions, but they usually require individual calculation. An illustration of doubling up ordinary joists – even triple and quadruple thicknesses – is shown in Figure 4.19 and the joists are identified in the key in Figure 4.20:

- □ The joists numbered 1 are common joists 50 mm thick.
- □ The joists numbered 2 comprise two 50 thick joist timbers nailed together. The increased thickness is required to carry a partition load as well as the share of floor load.
- □ The joist numbered 3 is a special joist – a trimmer or trimming joist; it comprises two 50 thick joist timbers nailed together and is required to support all the previous joists.
- □ The joist numbered 4 is another special joist – a trimmer joist; it supports one end

Fig. 4.17 Solid timber strutting.

Fig. 4.19 Common joists, trimmers and trimming joists.

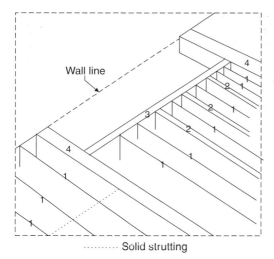

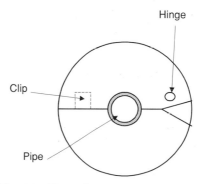

Fig. 4.20 Key to Figure 4.19.

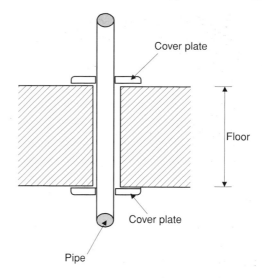

Fig. 4.21 Floor plates.

of the trimmer or trimming joist as well as a partition load and share of the floor load. It comprises four 50 mm thick joist timbers bolted together with 12 mm coach bolts, nuts and washers.

Note the solid strutting in the centre foreground.

Openings in upper floors

Openings in domestic upper floors occur for a number of reasons:

- Passage of services – water pipes, disposal systems, small ductwork, cables, etc.
- Flues for boilers, fireplaces, etc.
- Stairs – the opening shown in Figure 4.19 is for a stair.

For pipes

In the case of small pipes for hot and cold water, or even disposal piping up to around 50 mm bore, passage through the floor is simply done by piercing the floor board and ceiling finish and making good round the pipe.

Where the hole is exposed to view, a polished metal, plastic or wooden cover plate, often termed a 'floor plate', is sometimes fitted to mask the edges of the hole, especially where

the pipe might damage the underlying finish due to expansion/contraction, or to cover a roughly cut hole. Figure 4.21 shows a sketch of this.

Holes need not be vertical to the floor plane for small bore pipes and conduits. They can be in the joists themselves and it is here that special rules apply, and the Regulations have something to say on the subject. Holes such as these can be formed in two ways:

(1) Drilling a circular section hole completely in the joist
(2) Cutting into the edge of the joist forming a notch.

The hole should only ever be drilled through the neutral axis of the joist. The neutral axis of a rectilinear joist is always on the centre line of the vertical face of the joist. Holes should only be drilled from a quarter length of joist to 0.4 length of the joist from either end, and individual hole diameters should not exceed a quarter of the depth nor be closer than three diameters.

Notching should be avoided if at all possible but is often the preferred method when accommodating services in old floor structures and floor boards have to be lifted to gain access to the void. Confine notching to the first and last fifths of the length of the joist and keep the start of the notching not less than 0.7 of the span from the end of the joist. Overall depth of the notching should be no greater than one eighth of the depth of the joist.

Any variation from the above cannot be allowed unless the loadbearing capacity of the joist has been recalculated to take account of the hole(s) or notch(es). Any pipes or cables set into notches should be protected where they pass over the joist by fixing a metal strap across the notch.

For larger piping, such as soil and vent piping of 80–150 mm diameter, it is usual to 'frame' up the joisting. It may seem strange to require framing of the joisting to accommodate a pipe of some 100 mm bore, but the alignment of these pipes is critical. Only two bends are allowed from the junction with the uppermost appliance to the open vent at the top, and the pipe must generally be fixed back to a wall or into a corner. In the either case, the joist could be in the way.

Figure 4.22 shows how bridling of joists round a large bore pipe might be carried out. In the case of the opening shown, the floor loading only falls on two sides of the opening and only *one* joist is shortened by the bridle. Because of this, the bridle would be made from the same size of timbers as the general joisting.

If more than two sides were supported or more than one joist cut, the bridle and the joist supporting the end(s) of the bridle would have to be thicker. The increase would be in the order of 25–50%, so in a floor joisted with 200×50 joists, the bridle and the supporting joist would be 200×63 or 75 mm thick, depending on availability of the timber size.

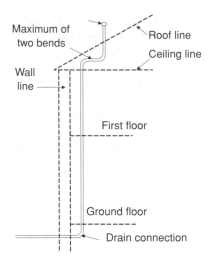

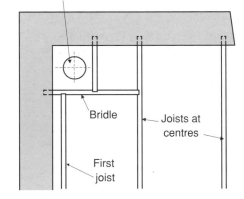

Fig. 4.22 Bridling joists around large pipes.

For flues

Flues are built into buildings to carry away the gaseous and fine particulate products of combustion from open fires and stoves of various types. Fuels range over gas – natural and LPG – various grades of oil, wood, peat, coal and derived fuels, straw and other 'waste' products. A variety of flue types is available to accommodate the widely differing conditions of temperature and products in the flue gas.

Extremes of temperature can be met with different fuels burned in quite different appliances. For example, a light fuel oil in a central heating boiler can produce flue temperatures of around $500°C$, requiring either a lined masonry flue or a double metal wall with ceramic insulating lining. Solid fuel appliances can produce even higher flue temperatures.

A fully condensing, gas fired, sealed boiler with a flue temperature of 65°C or less can be flued with a plastic pipe, usually of ABS (acrylonitryl-butadiene styrene). We will concentrate on flues with higher temperatures.

Flues carrying high temperature gases and passing through floors introduce the possibility of setting fire to the joists, floor and ceiling boarding, etc. by:

- Hot gases and/or flames escaping
- Combustible material touching the hot flue
- The flue radiating heat towards combustible material.

Precautions are necessary to prevent this and depend on:

- **Integrity** of the flue structure itself
- **Isolation** of the flue from any combustible materials using an air gap
- **Insulation** of the flue from the combustible material.

Integrity of the flue material

Provided the flue does not allow flame or hot gas to escape, it fulfils its function in terms of preventing fire. Flues can be of a variety of materials but basically domestic construction will use:

- Masonry flues – built with bricks or blocks with a clay liner. A clay liner is mandatory, the masonry on its own being considered insufficient to stop flame or hot gases reaching the rest of the structure
- Metal flues – a pipe of either single or multiple wall construction with sealed joints. These are proprietary systems. The days of sticking any old piece of cast iron or asbestos cement drainpipe in are far behind us now.

Isolation and insulation of the flue

Isolation of the floor structure can be carried out by keeping the flue completely clear of any combustible material, with an air gap 50–75 across and all round. The flue material itself may afford a degree of insulation, e.g. brickwork, blockwork – especially lightweight or aac blocks, or in-situ concrete – and this in many domestic situations is perfectly adequate.

Insulation of the structure from the hot flue is generally carried out using mineral wool, which has a high melting point. Glass fibre should not be used as it could simply melt and set fire to some other part of the structure. Other materials which can be used are vermiculite or mica, both best used as the aggregate in a refractory concrete rather than as a loose fill, when retaining the granules can be problematical.

It is difficult to separate the ideas of isolation and insulation as the solutions for both are almost identical. Depending on flue temperature, a simple gap may be enough to prevent radiant heat igniting the combustible components of the structure, as well as separating flue and materials. However, isolating sleeves or a combination of gap and sleeve may be required when flue temperatures are higher, especially where the space for a very large gap is not available.

Figure 4.23 gives the solutions for a combustible floor, in diagrammatic form. Full details are given of how to deal with two types of flue for very specific fuel types in specific appliances:

- Solid fuel open fire burning either coal, coal derivatives or wood
- Gas fired closed appliance with an all metal flue.

There is a large range of proprietary flues available for most types of appliance and fuel combination. Please refer to manufacturers' literature.

The masonry flue for an open fireplace will generally start off at the lower floor levels as a mass of masonry approximately 1000 wide and 450–600 deep on plan. This is known as a **chimney breast**. This mass of masonry accommodates the open fireplace and is built to very specific minimum dimensions. At ceiling level the dimensions of the chimney breast reduce to accommodate the flue, its liner and the

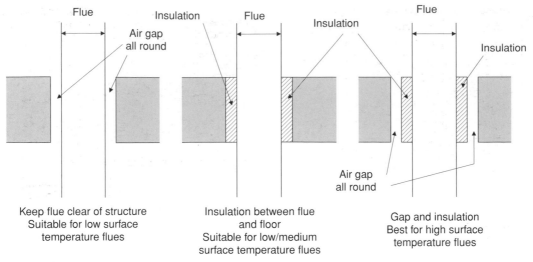

Fig. 4.23 Schematic of flues passing through floor.

surrounding layer of masonry. In brick-built flues this would give an overall plan dimension of 440 × 440.

Masonry flues *must* now include a **liner** which is generally made of fired clay; a typical example is shown in Figure 4.24. The liner is made in lengths of around 450 mm and the joints are **joggled**. Note that the joggle can be built upside down, which would allow any tars from the fuel being burned which condense on the cooler wall of the liner to run out of the joint, thus staining the surrounding surfaces. Figure 4.25 shows a flue through a floor and top of chimney breast, with the joint in the liner laid correctly.

Fig. 4.24 Clay flue liner.

Where the flue passes through a floor which is entirely within one dwelling, i.e. the upper floor of a two-storey house, there is no need for the void between the flue and the joists to be fire-stopped. Only support is required. If the upper floor separates two different dwellings, i.e. flats, then there must not be a gap around the flue, so only a mineral wool or a concrete fill can be used for isolation and insulation of flue and structure, as it must also perform as a separation between the dwellings.

If the upper floor separates two different dwellings i.e. flats, then the plates above and below the floor must give support and fire resistance. Proprietary flue systems have both types of 'plate' as standard fittings. This is illustrated in Figure 4.26 which shows a flue through the floor of a two-storey house, support only.

For stairs

Larger openings generally occur to allow access by stairs. Depending on the run of the joists in relation to the long dimension of the opening, this can mean cutting a large number of joists, supporting the ends with a heavy timber(s) whose ends are in turn supported on the remaining joists. These supporting

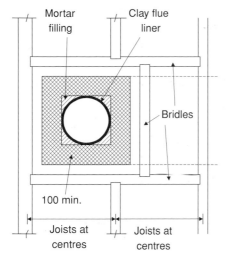

Plan of flue

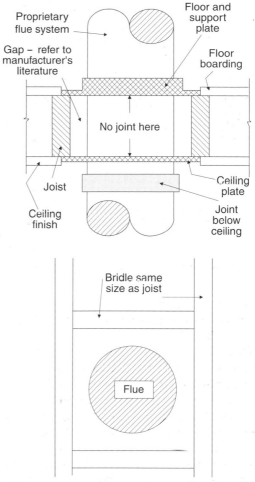

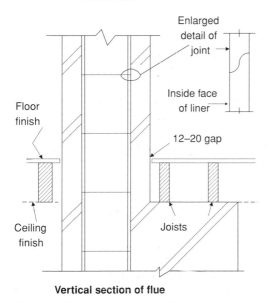

Vertical section of flue

Fig. 4.25 Flue passing through floor at top of chimney breast.

Fig. 4.26 Proprietary flue system support and sealing plates.

timbers must be larger than the general joisting, and joints must be made which provide proper support. Readers should refer back to Figure 4.19 which shows a stairwell opening in a timber joisted floor.

When framing up openings in floors with larger joists, the trimmer and trimmer joist joint is usually a tusk tenoned joint (see Appendix C). Housed joints are used for joists to trimmer junctions. Modern domestic construction frequently uses pressed steel hangers for all joints. Figure 4.27 shows joists fixed to a trimmer with pressed steel hangers. Note that these hangers are of thin galvanised steel sheet; that many small galvanised nails are used to fix them; and that the top of the hanger is bent flat over the top of the joist and nailed down. The latter can be seen on some of the earlier photographs.

Figure 4.28 shows a heavy gauge hanger for fixing to masonry and Figure 4.29 shows a lightweight pressed steel hanger suitable for joining joists etc.

Chapter 10 on Stairs shows how the edges of the opening are finished off.

Fig. 4.27 Floor joists fixed to trimmer with pressed steel hangers.

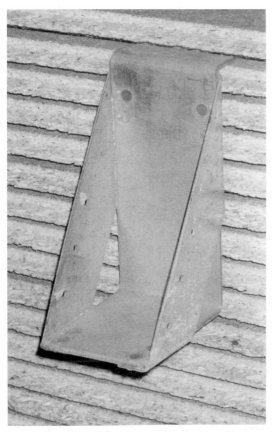

Fig. 4.28 Heavy gauge hanger for fixing joists to masonry.

Fig. 4.29 Light pressed steel hanger for joining joists.

Alternative materials for joisting

From an economic point of view there is no material better than solid timber for joists of average length. However, when solid timber cannot provide carrying capacity over a particular span, or where services of large cross-sectional area require to be contained within a floor structure, then alternatives must be sought. As already described, long spans can be broken up by using beams or partition walls under the joists, but these may not be desirable or may not exist.

Proprietary joists are available in a variety of material combinations and configurations and a few typical alternatives are described here. This is by no means an exhaustive list but serves to illustrate the types available:

☐ Combinations of timber and steel
☐ Combinations of natural timber and 'reconstituted' timber

□ Combinations of natural timber and plywood or fibre board.

The first, a steel/timber combination, is the Posijoist by Mitek Industries. The company supply only the pressed galvanised steel lattices in pairs, each pair having three sets of teeth similar to gang nailing plates. Manufacturers of Posijoists then supply stress graded timber for the upper and lower chords and press the pairs of lattice sets into the chords with a hydraulic press. The system can obviously be designed to meet a variety of loads by altering the cross-section of the timbers used. The spacing of the chords is fixed, being a function of the size of the lattices. The author is grateful for permission from Mitek Industries to publish the photograph in Figure 4.30 which shows a floor structure made using Posijoists. It amply illustrates the versatility of a lattice system joist structure.

For more information try the website http://www.mii.com/unitedkingdom/

Another lattice style joist is made by a company called Metsec. The strength is from the all steel construction, the top and bottom chords being made from pressed steel channels and the lattice from a continuous length of steel tube bent and welded between the channels. To make the joist 'joiner friendly', timber inserts are crimped into the steel channels, providing a fixing for both floor boarding and ceiling boarding.

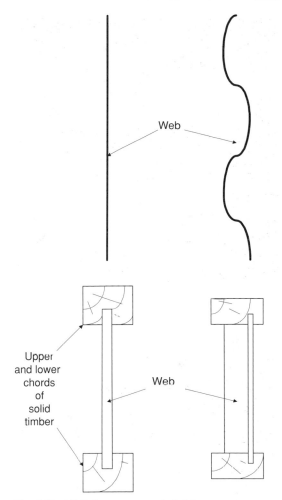

Fig. 4.31 All timber, man-made joists.

For more information try the website http://www.metsec.co.uk/metsec/en

'All timber' manufactured joists are made, and employ stress graded timber chords each with a groove machined into them. A web of man-made board is then glued into the grooves in the timber chords. Their construction is illustrated in Figure 4.31 and the straight web style is also shown in Figure 4.32, supporting an upper floor. Note in the photograph how trimmers and trimmer joists are formed by doubling (or more) the number of joists fixed together (the large timber running from left up to right across the whole picture). Note also the use of light pressed steel hangers to fix the joists to that doubled

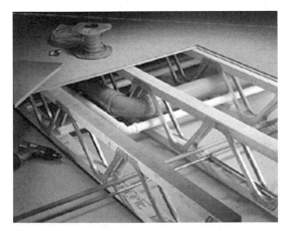

Fig. 4.30 Posijoists.

Fig. 4.32 Upper floor supported with Finnjoists.

joist. The straight webbed joist can use a heavier material for the web and it is frequently a type known as OSB – oriented strand board – in thicknesses from 11 upwards. There is a proprietary system marketed under the name Finnjoist which uses OSB for the web, but instead of solid timber for the flanges, it uses laminated timber.

For more information try the website http://www.finforest.co.uk

Figure 4.33 shows the surface of a sheet of OSB with a metric rule to give some idea of the size of the particles or 'strands'.

Because the wavy webbed joist requires a more flexible web material, plywood or even hardboard is used. If you doubt the strength of hardboard you really must revise your thinking. As part of any stressed panel construction even 3 thick hardboard can be incredibly

Fig. 4.33 OSB sheet.

strong – and of course it is available in other, thicker sections.

Sound proofing

Traditional methods of sound proofing relied entirely on increasing the mass of the floor structure – placing a layer of clinker or ashes in the joist void and sealing between joists and joists and walls, by covering the ashes with a layer of lime putty[1]. Figure 4.34 shows this traditional sound proofing. The detail shown in Figure 4.34 could only be carried out using large section joists; 250 × 50 was a minimum and this was frequently increased to 300 × 75. These sizes were necessary to carry the large dead load of the ash layer, the supporting boards and the lime putty.

The heavy timbers also gave great rigidity to the floor, and so even with the floor boarding nailed direct to the joists, attenuation of sound caused by impact was relatively effective. The thick flooring boards themselves, spanning only a clear 400 mm, were also very rigid and this reduced 'drumming' of that 'skin'.

Surprisingly, the construction as detailed proved to be a good fire barrier due to the thick layer of lime plaster on the ceiling and the lime putty over the ash layer, which not only gave protection against flame but also gave a seal to resist the passage of hot gases, and had the layer of ash to provide insulation.

The author had experience of a fire in a late nineteenth century house which had been converted into offices; it illustrated how well these old floors could resist fire. The floor structure was 20 mm oak parquetry on 32 mm softwood tongued and groove flooring on 300 × 75 joists, with ash deafening and lime putty on 25 boards and a ceiling finish to the room below of 20 mm lime plaster on split softwood lath. The fire was started by a radiant electric fire, plugged in and switched on to help dry the parquetry floor which had been

[1] Slaked lime, which can be mixed to a variety of consistencies before being used. Each can be referred to as lime putty. In the instance quoted above the consistency should be 'creamy' to allow it to be poured over the beaten ash base.

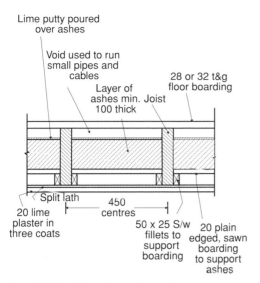

Fig. 4.34 Traditional sound proofing.

mopped during the evening. The fire had been propped up in an unstable way. It was forgotten about and during the night fell over, face down, onto the parquetry. During the night it slowly burned its way through the floor and the ceiling, showering the room below with hot embers. The room below was an office with desks littered with paper, which caught fire, and this room was gutted.

After the fire brigade had been called and the fire extinguished, the damage to the ceiling was found to be minimal. Some had come down and joists and support boards for the ash layer were charred but were still structurally sound. The only damage to the room where the electric fire had been was due to hot smoke coming up through the hole burned by the electric fire. Papers on desks were charred and fragile but nothing had actually ignited. Smoke had discoloured everything and was even 'burned' into the surface of a glass tumbler sitting on one of the desks. The electric fire was still plugged into the wall socket and was dangling through the hole suspended by its flexible cord. Of course the insulation to the cord was totally destroyed.

Modern sound and fire proofing

Modern construction has been forced to adopt smaller sections of timber etc. due to ever

escalating costs. Thus we cannot rely entirely on mass with the need for protection of large joist sections nor on thick and rigid layers of boarding. Mass does still play a part but resilient materials are introduced to reduce noise caused by impact and vibration.

A detail common in flatted dwellings of the 1960s used smaller joists, mineral wool 'pugging' to give mass (up to 100 kg/m^2) and a glass fibre mat as a resilient layer between relatively thin flooring boards and the joists. The pugging was formed by making the mineral wool into small pellets which could be poured into the void between the joists; 12.7 ($^1/_2$ in), 20 gauge galvanised chicken wire was nailed to the underside of the joists to provide support for the pugging in the event of the plasterboard ceiling finish failing in a fire. Two layers of 12.7 mm plasterboard on the ceiling gave the requisite fire protection and additional mass. Plasterboard sheet joints were staggered in adjacent layers to prevent the passage of smoke and hot gases.

The floor boarding was isolated from the joists by a layer of fibreglass quilt draped over the joists. The floorboards were nailed to a light batten running between the joists. The batten was not fastened to, nor did it touch, any part of the structure. A roll of fibreglass quilt was placed between the first and last joists and the wall, and the quilt draped over the joists was turned up against the wall at the edges of the floor-boarded area.

Figure 4.35 shows a detail from the 1960s – a floor with sound and fire proofing as used

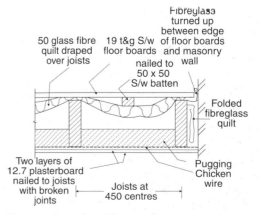

Fig. 4.35 Sound and fireproofing 1960s style.

between low rise flatted housing. Note the following:

☐ The floorboards are *not* nailed to the joists, only to the free batten between them.
☐ There must be a triple thickness of fibreglass quilt stuffed between the first/last joists and the walls.
☐ There must be a single layer of fibreglass between the edge of the floor boards and the wall.
☐ The pugging is a layer of mineral wool pellets laid at a rate of 90 kg/m².

To keep the joist size down and still maintain good rigidity, it was common to break the span of the joists with partitions in the lower floor, or use an RSJ. Both partitions and RSJ had to be protected from the effects of fire. Plastered partitions of half brick thick brickwork or 75 blockwork were used as these were inherently fireproof and with their mass attenuated the noise transmitted. The most common type was **stud partitioning** of 50 × 38 sawn S/w studs, runners and a minimum of two rows of noggings/dwanging. Cladding was 12.7 plasterboard both sides to protect the wooden framing from fire. This gave a protection period of half an hour.

RSJs were frequently used in association with one or more one brick square columns, which were plastered. The columns broke up the span of the RSJs. Steel distorts badly in a fire so the RSJs had to be protected on each vertical face and on the bottom. This was done by framing round the RSJ with timber and applying two layers of 12.7 plasterboard as shown in Figure 4.36.

Fire proofing an RSJ spine beam is detailed in Figure 4.36. Points to note are:

☐ There are four layers of plasterboard between the bottom of the RSJ and the lower compartment – this provides insulation.
☐ The two layers immediately below the RSJ are temporarily taped into position and finally secured by the chicken wire layer.
☐ Plasterboard round the RSJ has joints 'broken' to give a better 'gas' seal.

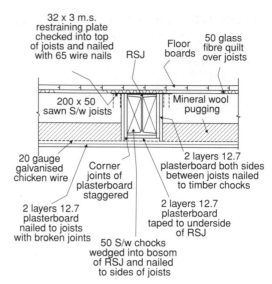

Fig. 4.36 Fireproofing an RSJ spine beam.

☐ The metal restraining strap is fixed across the top of the joists where it is least vulnerable to fire. Many authorities insisted that it was screwed down rather than nailed.
☐ The fibreglass layer is continuous over the RSJ.

A more modern form of fire and sound proofing is shown in Figure 4.37. It should be noted that the pugging can be of a variety of

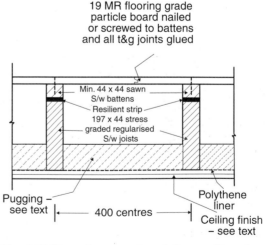

Fig. 4.37 Up-to-date sound and fireproofing of an upper floor.

materials laid to a density of 80 kg/m². Fulfilling these requirements would be:

- Ash, 75 thick
- Dry sand, 50 thick
- 2–10 limestone chippings, 60 thick
- 2–10 whinstone chippings, 60 thick.

The polythene lining keeps any dust from penetrating the ceiling, especially while the pugging is being poured. For the resilient strip on top of the joists 25 mineral wool, density 80–140 kg/m³ is recommended. The ceiling finish can be:

- 6 plywood nailed to the joists, then two layers of plasterboard with broken joints to give a total thickness of 25 minimum, *or*
- Metal lath with 19 of dense plaster.

Support of masonry walls

All thin or slender vertical structures are prone to buckling when loaded excessively. Masonry walls are no exception, and to break down the height of the wall into smaller unsupported heights it is a requirement of the Building Regulations that walls of more than 3.00 m in length are tied to the intermediate floors. Where the joists are built into the walls with a bearing of a minimum of 90 mm, nothing further need be done, but on adjacent walls there is no connection between wall and floor so metal restraining straps must be built in. The straps are of galvanised steel and must have a minimum cross-section of 30 × 3 mm. They must extend across three joists with dwangs between the joists, and packing or folding wedges between the first joist and the wall. **Solid strutting** was frequently used instead of dwangs/noggings when this regulation was first promulgated. Clarification in more recent editions of the regulations required the dwang/nogging to be at least **half the depth** of the joists.

How the work is done is illustrated in Figures 4.38–4.40. Figure 4.38 is the detail, Figure 4.39 is a photograph of the metal strapping and Figure 4.40 shows the metal strap

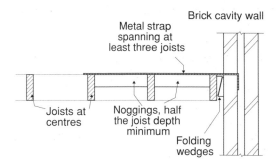

Note that the thickness of the strap has been exaggerated and that much of the construction detail annotation etc. has been omitted for clarity

Fig. 4.38 Detail of support from floor of masonry walls (1).

extending into the wall and nailed over not just three joists but four. Figure 4.41 shows the view from the underside of that strap, and one can see that the square section noggings under the strap would appear to be undersized if it was meant to comply with the Building Regulations. The end of the strap can just be seen in the top right hand corner of the photograph.

The joists in this particular building are regularised out of 250 sawn timbers so the dwang should be at least 125 deep. It shows around 50 × 50. What is not shown in the photograph is the packing between the wall and the first joist. This was a plain square section timber. Wedging would have been better. This was

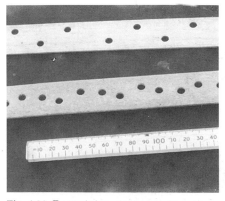

Fig. 4.39 Restraining straps.

Fig. 4.40 Detail of support from floor of masonry walls (2).

not a 'good' site. The support straps should be placed no further than 2.00 m from each corner and a maximum of 2.00 m apart. The site where these particular photographs were taken showed a lack of understanding of the need for these straps; not nearly enough of them were used.

Floor finishes

As with the hung timber ground floor construction, floor finishes on joisted upper floors use either solid timber boards or manufactured sheeting. All that is said in Appendix C about this applies equally to upper floors.

Ceiling finishes

Ceiling finishes have been mentioned in several of the earlier sections of this chapter, without a great deal of explanation, because they:

☐ Influence decisions regarding the spacing or centres of joists or the need to use brandering

Fig. 4.41 Detail of support from floor of masonry walls (3).

t&g and V jointed boarding

t&g and beaded boarding

t&g and rebated boarding

Fig. 4.42 Match board sections.

☐ Contribute to fire and sound insulation of the completed floor.

In every instance the finish mentioned has been plasterboard. Plasterboard is not just available in one form, but in a variety of sizes, thicknesses, configurations, compositions, finishes and a host of sensible combinations of these. Much of that information is summarised in Appendix H, Gypsum wall board, or can be readily obtained from the websites quoted there.

Fig. 4.43 Match board fixing clip.

As a reminder, here is an extract from Appendix H:

> At its most basic, this sheet material is a sandwich of two layers of special paper with a layer of gypsum (calcium sulphate) between which can be nailed or screwed to a framed timber or light steel section background. Thus it can be used to cover the underside of upper floor joists to provide a ceiling, or applied to walls and partitions to provide a smooth wall surface.

Other materials can of course be used to finish a ceiling, and solid timber is probably the most common after plasterboard. The section used is usually a decorative one but incorporating a **t&g** joint on the long edge, and end joints of board are always made at a joist. Like floor boarding, the benefit here is that, unlike sheet material, the centres of the joists are less important in reducing waste. Boarding like this usually finishes around 12–15 thick and is machined out of no more than 75 wide sawn stock

or less, so the cover width at a maximum is around 60–65. All the boards in one batch are the same width. Boarding like this is known as **match boarding**.

Figure 4.42 illustrates some of those sections. The top and middle sections in the figure have been used for panelling of walls for nigh on a century and a half. They can both be secret nailed in the same manner as floor boards (see Appendix C). The lower section in the figure came into vogue in the early 1950s and while some of it was nailed to the background framing, that tended to leave nail heads exposed, which had to be punched down and filled over. If the timber was varnished, the stopping showed up as lighter coloured spots. The result was the invention of the **match board clip** shown in Figure 4.43. The clips are nailed to the background timbers and the match board pushed against the spiked end of the clip. For reference the nails shown are 20 long.

This style of boarding can also be used as an alternative wall covering.

5

Openings in Masonry Walls

For small pipes and cables	126	Threshold arrangements	137
For larger pipes and ventilators	127	Partitions of masonry	139
Large openings in masonry walls	127	Openings in timber frame walls	141
Alternative sill arrangements	136		

As we did with floors, we can classify openings in walls roughly by the size of opening required.

So we will look at:

□ Openings for small pipes and cables <50 mm diameter
□ Openings for pipes and ventilators >50 mm diameter but <225 mm
□ Openings in excess of 225 mm – large openings.

For small pipes and cables

Openings for small pipes and cables can now be drilled with a power tool fitted with an appropriate drill bit – solid for holes up to about 50 mm and a core bit for larger holes up to about 150 mm diameter. There are even some specialised forms of bit which can cut large diameter holes but they generally require water cooling. All that is required is a powerful enough tool, core bits requiring the greater power.

Because of the extensive use of cavities, sleeves must be built in across the full width of the wall to prevent the leakage of pipe contents into the cavity and to prevent vermin getting into the cavity. Typical examples of drill bits and a sleeve in a masonry cavity wall are shown in Figure 5.1.

Sleeves should be built in with a slope down and towards the outer face of the wall to prevent water running through the sleeve. This does not need to be excessive – the sketches exaggerate the slope.

Sleeves can be made of any solid drawn pipe material with a bore which will allow an easy fit for the service pipe or cable passing through. Typically, copper piping or plastics piping can be used. Ferrous material should be avoided, as should aluminium which corrodes in contact with a strong alkali. Plastics pipe does seem to be the favoured choice on building sites one visits, but it is uncertain whether or not compatibility with cable insulation etc. has been considered in the choice of pipe material.

Joints of the service pipes and cables are not allowed in the sleeve. With soft-walled pipes and cables, the ends of the sleeve should be rounded or be fitted with some form of cushioning to prevent wear on the pipe or cable. Chapter 4, Timber Upper Floors, gives a description of 'cover plates' where small pipes etc. pass through a floor. These can be used with holes in walls.

If the space between the sleeve and the pipe or cable is to be sealed, this should be done with a non-setting mastic or sealer. Take care to ensure that it is compatible with the material of both sleeve and service pipe or cable. The space should be left large enough to make

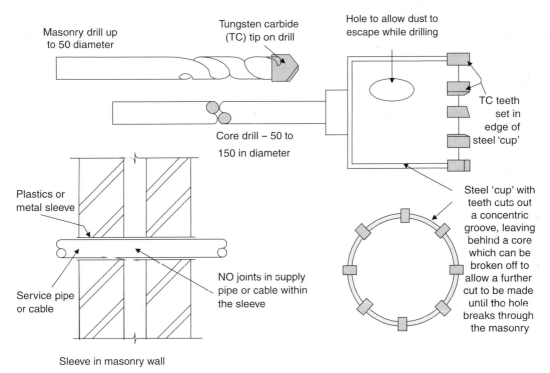

Fig. 5.1 Drills for drilling through walls.

it easy to inject the mastic. The whole space from end to end is never sealed, only about 25 mm at each end.

For larger pipes and ventilators

The only large pipes to pass through domestic walls are generally soil pipes or flue pipes – occasionally air extract ducts. In both cases the norm is to cut a hole in existing walls or build in the pipe or flue with mortar, fitting the masonry units around them. The cutting is done with a hammer and chisel or a core drill in a power tool. The latter causes much less damage to the new masonry.

In the case of a flue, the heat generated may be sufficient to damage cavity insulation. In such a case an approved sleeve (metallic or asbestos cement pipe) with proper clearances must be built into the wall. Manufacturers' literature should give the necessary advice on the type of material to use for a sleeve and the clearances required, together with the expected temperatures with various fuels for which the flue is suitable.

For pipes, ducts and flues up to 150 mm diameter, holes would be formed or drilled to coincide with a junction of two bricks or blocks so that these units can be **saved over** rather than use a separate lintel, as shown in Figure 5.2. Also shown is a ventilator (*not* an airbrick) which generally tends to be rectangular in section and made in brick coordinating sizes. For ventilators up to one brick wide therefore, there is no need to do anything other than ensure that two bricks in every half brick thickness are evenly distributed over the ventilator opening. Saving over relies on the self arching effect of masonry.

Large openings in masonry walls

Large openings are made for doors, windows, hatches and the like. What differentiates them from previous openings is the need to provide extra support over the opening as the self

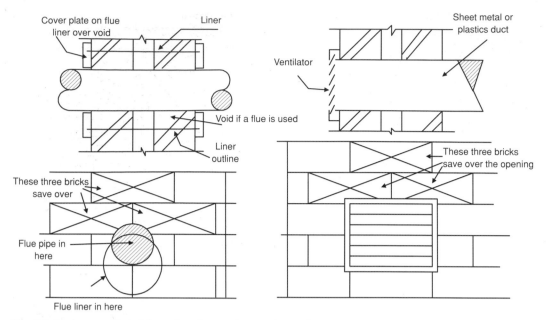

Fig. 5.2 Saving over in brickwork at ducts, pipes and ventilators.

arching effect can only be partially successful. Support is provided by **lintels**.

Lintels can be made from a wide variety of materials and a variety of shapes. Materials include stone, reinforced precast concrete, prestressed concrete, cold rolled mild or stainless steel sections, and hot rolled mild steel sections. Old buildings may have timber lintels or 'safes' but these are no longer used and should always be replaced during any refurbishment work. Shape will be discussed as we look at each application.

There is some terminology to learn. Any large opening has:

- ☐ A bottom – the sill for windows or threshold for doors
- ☐ A top – the head which has a soffit or soffite
- ☐ Two sides – the jambs which have reveals.

Once a frame is put into the opening, the soffit is divided into an inner and outer soffit, the reveals are divided into inner and outer ingoes, and the sill into inner and outer sills. These are all illustrated in Figure 5.3.

Cavity walls have the cavity closed at the head, jambs and sill or threshold. A further feature is the positioning of the door or window frame within the opening. Where there is a risk of wind-driven rain, the frame is positioned back in the wall thickness, thus giving shelter to the joint between frame and wall. This situation arises generally in Scotland and in northern and western and southwestern England, Wales and Northern Ireland. In more sheltered areas, such as the Home Counties, Norfolk and Lincolnshire, the frame was traditionally placed flush with the outside face of the wall, and the majority of textbooks showed frames in this position. The situation has changed somewhat and the textbooks now tend to show the frames set back behind the face of the wall. Setting frames flush with the outer face of the wall had implications for making a weatherproof joint between frame and wall which did not rely on some form of mastic. Like flattery, mastic is the last resort of the rogue builder in situations like that. Mastic is advocated in the details which follow but it is not relied on entirely; the detail itself has to be correct first.

Another area for concern was the closing of the cavity at head, jambs and sill. With the frame so far forward of the cavity, closure was haphazard in practice and the positioning of

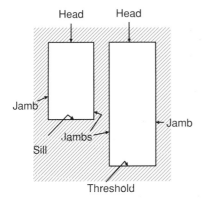

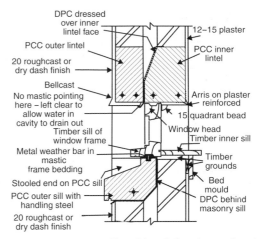

Fig. 5.4 Pre-1970s sill and head details, rendered external finish.

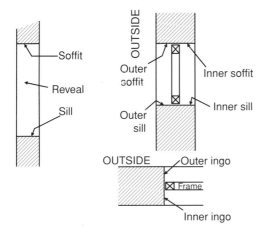

Fig. 5.3 Terminology associated with openings in walls.

the DPC at jambs and head was never quite satisfactory.

This chapter will deal with frames set back from the face of the wall – indeed the outer face of the frame should align with the inner face of the outer leaf of masonry in a cavity or timber frame wall construction. The details shown all work, and work well. They function in the most extreme weather conditions in the UK.

Closure of cavities was traditionally, pre 1970, done by:

☐ Making inner lintels wide enough at the head with a DPC at the closure
☐ Making sills or thresholds wide enough, with a DPC at the closure

☐ Returning the inner leaf of masonry against a DPC layer on the outer leaf at the jambs.

These details were sound and have stood the test of time since the early 1900s with only minor adjustment. The principles should not be abandoned for the sake of fashion or whim. They are illustrated in Figures 5.4 and 5.5 in a cavity wall which has a roughcast finish.

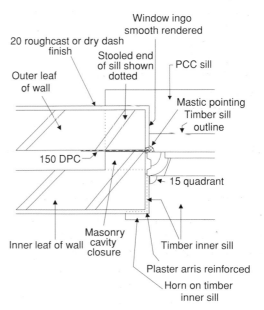

Fig. 5.5 Pre-1970s jamb detail, rendered external finish.

A few important points to note about the details:

☐ PCC (pre-cast concrete) lintels of that period were frequently only reinforced at the bottom. Top and bottom reinforcement with links is now much more common.

☐ The roughcast finish stops short of the head and jambs of the timber frame and under the PCC sill. At both head and sill this allows water to drain out. This should *not* be filled with mastic. At the jambs, the space is used to seal frame to wall to DPC with mastic.

☐ Joints of plaster to timber frame are covered with a timber quarter round beading to hide the joint which would inevitably show a crack as the timber and plaster shrank away from each other.

☐ Good joinery practice would have the bed mould tongued into a groove in the timber inner sill. Economics now dictates otherwise.

☐ Had the wall been finished with facing brick, many architects would have left the brickwork in exactly the same position as the common brick shown, relying in that case on the mastic alone to give a weather seal instead of having the roughcast provide a check or rebate. It would be better to have this small rebate built into the wall as shown in the masonry only outlines of Figure 5.6.

Figure 5.7 is a photograph of the sill and part of the jamb of an opening in a blockwork wall. The details shown are the 200 thick outer leaf of masonry; the sill with the upstand with weather bar groove; the weathered surface of the sill; the stool on the end of the sill; the lack of protection to the finished surfaces of the sill and how dirty they are – not a good site.

Figure 5.8 is a photograph of the same window opening which shows both sill and lintel and masonry mullion. The mullion is bedded between sill and lintel but is also restrained in position by metal dowels both top and bottom. Figure 5.9 shows how this would be done. Note how the sill has a sloping surface generally to make water run to the front edge, and how the underside of the projection has a

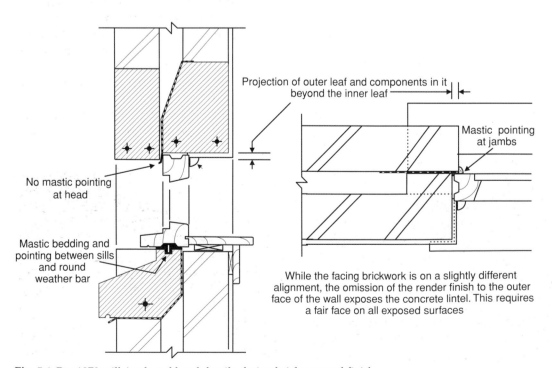

No mastic pointing at head

Mastic bedding and pointing between sills and round weather bar

Projection of outer leaf and components in it beyond the inner leaf

Mastic pointing at jambs

While the facing brickwork is on a slightly different alignment, the omission of the render finish to the outer face of the wall exposes the concrete lintel. This requires a fair face on all exposed surfaces

Fig. 5.6 Pre-1970s sill, jamb and head details, facing brick external finish.

Fig. 5.7 Photograph of 1990s sill and jamb.

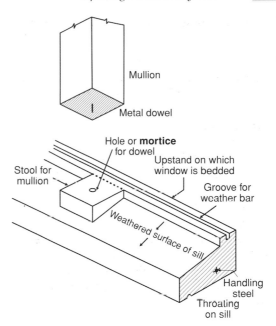

Fig. 5.9 Detail of mullion–sill junction.

groove – the **throating** – to prevent water running back onto the face of the wall. Because the upper surface slopes, the mullion sits on a **stool**. This device is also used at the ends of sills which are built into walls, to allow the masonry to be bedded on a horizontal surface.

Figure 5.10 is a photograph of the inside of another window on the same site, which shows a galvanised steel lintel holding up the inner masonry leaf over the opening. Steel lintels are sketched a little later in the text.

It is important to note that frames for windows or doors are only fastened to the walls through the jambs and that in the traditional methods this fastening went into the masonry, closing the cavity.

Figure 5.11 is a photograph of a cavity closure in blockwork. Note how a cut block closes

Fig. 5.8 Photograph of 1990s head, sill and mullion in artificial stone.

Fig. 5.10 Steel lintel at inner leaf of opening.

Fig. 5.11 Detail of checked jamb.

the cavity and bonds into the inner leaf in alternate courses, and that a cut of blockwork is used for the intermediate courses, all bedded up solid in mortar and with a vertical strip of DPC between closure and the outer leaf. Compare these photographs with Figure 5.4, the general sections.

For nailing to plugs, proprietary framing anchors or **cramps** are the most common methods, although some window frames are still only wedged into place. More of this when we examine windows in Chapter 9.

Note that in Scotland, traditional practice is to form the opening and then fit the frame into the opening, whereas the practice in the rest of the UK is almost entirely to set the frame up on the sill, rack or brace it vertical, and then build up the rest of the wall round it.

The window sections shown in all the drawings so far have been based on standard single glazed casement windows. A range of sections will be studied in Chapter 9, Windows.

The practice of building in frames as the wall is built up means that fixings are generally

galvanised or stainless steel 'cramps' fixed to the back of the frame and bedded into the mortar joints. Screws should be used with timber frames, and machine or self tapping screws with metal or plastic frames. A typical fixing is shown in Figure 5.12.

Since 1970 there have been several editions of the Building Regulations, and the stricter requirements regarding U-values of walls and the need to avoid **cold bridging** have meant that the traditional details have had to be updated. This is shown in Figures 5.13 and 5.14.

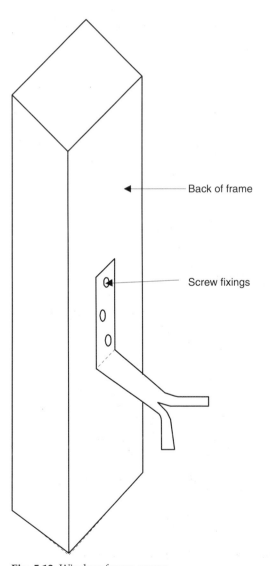

Back of frame

Screw fixings

Fig. 5.12 Window frame cramp.

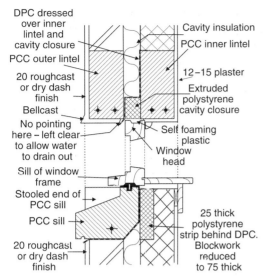

Fig. 5.13 Eliminating cold bridging at sill and head.

Variations on finishing

The figures which follow show variations on the finishing of various parts of openings incorporating different materials and

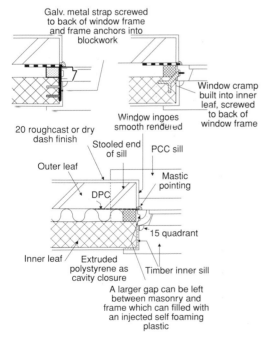

Fig. 5.14 Alternative jamb arrangements and cavity closure to eliminate cold bridging.

components but always including wall insulation and avoiding cold bridging in terms of the current Building Regulations. The good practice in relation to the positioning of DPCs and frames, and mastic pointing in appropriate places, have all been maintained from the pre-1970 details.

Textbooks will show weeps formed or built into perpends of the outer leaf of brickwork over window and door openings. This cannot be done when lintelling with PCC and/or a roughcast finish. How do you build a weep through a solid concrete lintel? There were two possible routes for water in the cavity to use to escape. The first was between DPC and outer lintel, a space which was never pointed with mastic, as has been noted on the previous figures. The second was off the ends of the DPC into the cavity itself, but at the back of the outer leaf. This was made easier by the simple expedient of having the DPC project about 100 mm beyond the end of each lintel. With no insulation filling the cavity, water running off the low end of the DPC did so against the inner face of the outer leaf and was in no danger of crossing back across the cavity.

In recent years in the construction press, much has been made of **cavity trays**. The DPC over inner lintels is such a tray and in the bad old days they formed up from hessian-based bituminous DPC where appropriate. There is a widely held opinion that the ends of *all* cavity trays should be 'boxed' to prevent water running off the end into the cavity. In many instances this is correct but is not really necessary at ordinary PCC lintels unless the cavity is filled with insulation. There are three solutions to the problem: only part fill the cavity and let the water run off the ends of the DPC in the usual manner, or put the insulation somewhere else, completely inside or outside the wall, or redesign the lintelling when we have a facing brick finish.

Alternative lintelling arrangements are shown in Figures 5.15, 5.16 and 5.17. Figure 5.15 shows a steel lintel made up from three pressed sections spot welded together, zinc plated, passivated and painted in the factory. A separate DPC is shown but

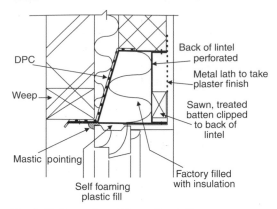

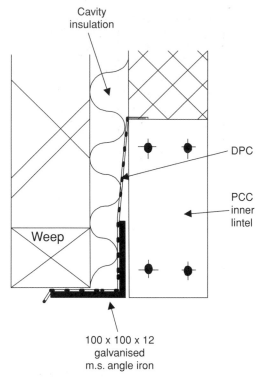

Fig. 5.15 Pressed steel inner lintel (1).

some manufacturers do not recommend one. Figure 5.16 shows an IG lintel, available in galvanised steel and stainless steel. Sealing the window frame and foam filling etc. is all the same as the earlier lintels. Figure 5.17 shows a very old form of lintelling when using a facing brick outer leaf: the mild steel angle iron. Galvanised angles are more readily available and are to be preferred to plain steel. Note that the

Fig. 5.17 Angle iron lintel.

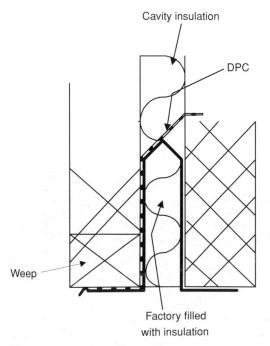

Fig. 5.16 Pressed steel lintel (2).

inner lintel here is rectangular and that a DPC is dressed over it and down over the angle to project beyond the angle iron. DPC projection is necessary in all the details where it is shown to provide a good drip and prevent water being drawn under the lintel and so coming in contact with the frame. Mastic pointing *is* used at the head of the frame.

The illustrations so far show an outer leaf of facing brick. All of these lintel alternatives can be used with a roughcast or dry dash finish on common brick or blockwork. The general detail does not alter significantly. The weeps can be blocked when the roughcast is applied and, if possible, the use of weeps with a removable strip or a strict regime of cleaning up after rendering has to be adopted. Figure 5.18 shows a possible solution using the IG lintel and without using weeps, instead using the natural porosity of blockwork to allow water to escape. The detail would be the same for any steel lintel.

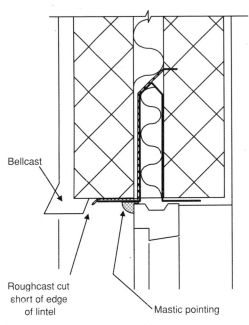

Bellcast

Roughcast cut
short of edge
of lintel

Mastic pointing

Fig. 5.18 Pressed steel lintel (3).

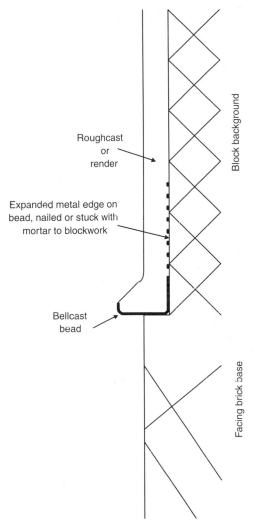

Roughcast
or
render

Block background

Expanded metal edge on
bead, nailed or stuck with
mortar to blockwork

Bellcast
bead

Facing brick base

Fig. 5.19 Bellcast bead.

Before leaving lintels the term 'bellcast' requires expansion. It is obvious how this projection of a roughcast or rendered finish got its name. It is not formed directly by applying the mortar in that shape; that would not be possible. Instead, early craftsmen fixed a temporary batten to the masonry background and worked to the thickness of that batten. Modern practice uses either a galvanised steel bead or a plastic bead, which is fixed to the wall and the render brought to that. Figure 5.19 shows a typical section and its application, while Figure 5.20 is a photograph of three galvanised beads, from the left:

☐ The bellcast bead.
☐ A render stop bead used to finish off plaster at an open edge of a wall or where separating plaster or render finishes either side of an expansion joist etc.
☐ Angle bead for protecting arrises in plasterwork. It features in most of the sketches in this book on openings in masonry walls.

Note that all the beads feature a mesh-like finish to the edges which are bedded into the plaster or render. This mesh is **expanded**

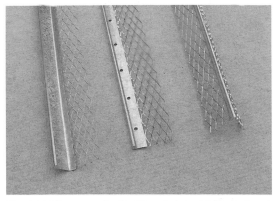

Fig. 5.20 Photograph of bellcast bead and corner bead.

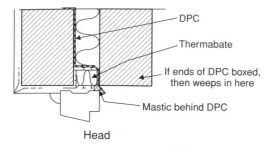

Head

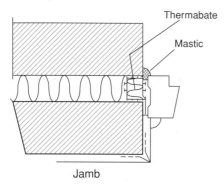

Jamb

Fig. 5.21 Proprietary cavity closure at head and jambs.

metal and is made by slitting the metal sheet and then pulling it apart.

The only alternative to the use of masonry or extruded polystyrene strip to close cavities at jambs is the use of a proprietary closure. The component is an extruded uPVC section filled with expanded plastic foam. The outside has grooves or ridges to provide a fixing point for fittings and to direct water flow. Because the PVC outer is waterproof, there is generally held to be no need to place a separate DPC against the outer leaf. This is *not* necessarily the case when the frame is set back in the opening, as is common in much contemporary construction. Figures 5.21 and 5.22 show how this can be dealt with.

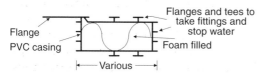

Fig. 5.22 Schematic of cavity closure.

Alternative sill arrangements

Figure 5.23 shows a PCC sill with a proprietary closure immediately under the sill of the window frame. The flange of the closure should be over the PCC sill, bedded in mastic and trimmed back if necessary. The projection of the timber window sill may have to be larger than usual. Note the use of and position of the DPC. Provided the under sill DPC starts on the inside face of the closure, there is no need to place DPC between the closure and the outer leaf at the jambs detail.

An alternative shape for a PCC sill is shown in Figure 5.24. This shape can be cast one course high, although it makes a very slender sill if there is a wide window opening.

Tiles, both plain roofing of clay or concrete and quarry floor tiles, make good sills as they are made to prevent the entry of water. The detail is illustrated in Figure 5.25. They must, however, be laid in a minimum of two courses with the bond broken and in a strong mortar, mix 1:5 or even 1:4. Any DPC must be brought down at least one, preferably two, brick courses below.

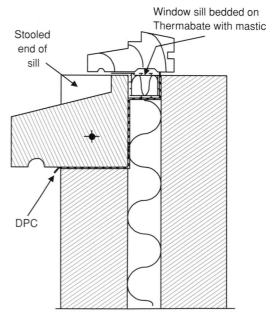

Fig. 5.23 Proprietary cavity closure at sill.

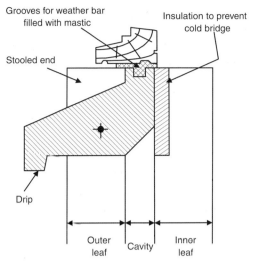

Fig. 5.24 Alternative sill in PCC.

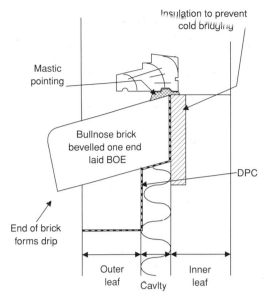

Fig. 5.26 Bullnose brick sill.

Bullnose bricks can be laid on edge to provide a sill. The plain header end is cut to a bevel against the inner leaf and the bricks are canted up at the inside end to form a weathering. The sharp arris at the bullnose end forms a satisfactory drip. A strong mortar is required to provide a reasonably weather resistant joint. The detail is illustrated in

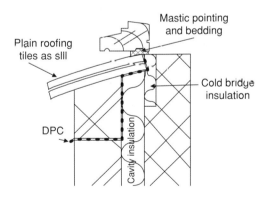

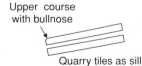

Fig. 5.25 Tile sills.

Figure 5.26. DPCs are best taken one full course below the bullnose sill.

Note that it is only on the PCC sill details that stooled ends are shown because only PCC (or stone) sills are built into the walls at the jambs. Tile and brick sills are built only between the jambs. PCC (and stone) sills are also made without stooled ends, to be built into the opening after the walls are complete and either just before or after the window frames are fixed. Such sills are termed 'slip sills'.

Threshold arrangements

What a sill is to a window, the threshold is to a door. The same requirements apply – the need to make this opening joint wind and watertight. Many variations are seen, but with frames recessed into openings, the ideal situation is to provide a step up to the floor level with the face of the step in the same plane as the face of the door. Figure 5.27 shows this detail and points out one possible weakness – the fixing of the weather bar. Figure 5.28 shows a solution to that problem which does not jeopardise the integrity of the detail but provides a secure anchorage for the weather bar. The

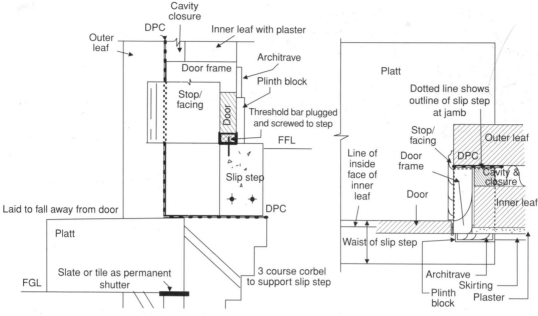

Fig. 5.27 Threshold (1).

detail is shown in Figure 5.29 and a photograph in Figure 5.30.

Note that the first detail, Figure 5.27, shows the traditional full width frame in the checked reveal while the second detail, Figure 5.29, shows a more modern small section frame with the reveal finished in the same way as the wall. Note that the second arrangement reduces the amount of timber used for the door frame and the stop facing; there are no plinth blocks to make and fix and no architrave; there are $\frac{1}{4}$ round beads; and plaster and skirting

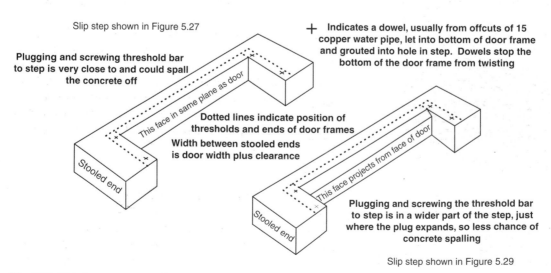

Fig. 5.28 Slip steps.

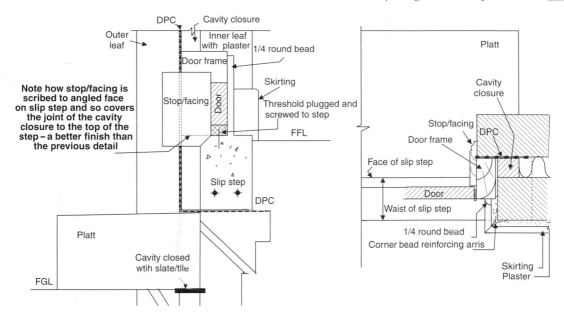

Vertical section through platt, slip step and bottom of door

Horizontal section above slip step

Fig. 5.29 Threshold (2).

have to return into the reveal from the wall face. This detail is more economical than the first arrangement.

Note that in both details the type of floor has not been shown. This is irrelevant to the threshold construction, although for some

Fig. 5.30 Photograph of threshold (2).

floors the DPC under the slip step should be turned up between step and floor – very important if the floor is of hung timber or concrete construction. For solid concrete floors, the DPC should be joined to the DPM.

It is usual to keep the top of the slip step slightly higher than the finished floor level of the building. This allows for any thickness of floor finish put down by the occupiers such as carpet, vinyl sheet, etc.

Threshold bars come in a bewildering variety of types, sizes, materials and finishes. Some have moving parts (to be avoided, as they inevitably break down); others have fixed sections. Some have only one section fixed to the slip step; others have a number of sections fixed to both slip step and door.

Partitions of masonry

The fitting of door frames into the openings in partitions can be done using full width or part width door frames. The effect is similar to that seen with the external doors shown earlier.

Figure 5.31 shows lintelling arrangements. Figure 5.32 shows two typical examples – a part width frame with a PCC lintel and a full width frame with a pressed steel lintel. The choice is deliberate as it is difficult to finish the soffit at a part width frame where there is a pressed steel lintel of this type – how is the finish attached to the corrugated shape of the lintel? It might be possible to introduce a strip of metal lath, but how can one conveniently fasten that to the lintel?

Frames for doors in partitions will normally include a timber threshold plate (not to be

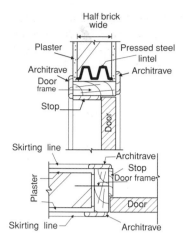

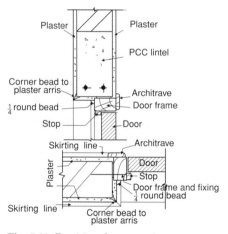

Fig. 5.32 Partition door openings.

confused with a threshold bar), a plain timber section, usually hardwood. This may be only the width of the door (irrespective of the actual frame width) or the width of the frame. From a purely practical point of view, a full width plate is best as any floor covering fitted to the threshold plate does not incur excessive waste. Indeed, the occupant may be forced to purchase the next width up in floor covering because of the door 'recess'.

For fastening frames, see Appendix G, particularly nailing to dooks and proprietary framing anchors. Note in the partial width frame situation how close the frame fastening is to the edge of the frame, or if it is set in the middle how close it would be to the edge of the wall.

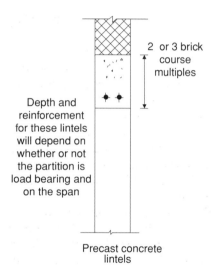

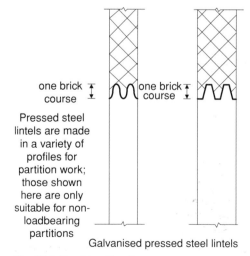

Fig. 5.31 Partition lintels.

In all the door frames fastened to masonry, 3 or 4 frame anchors or plugs should be used in internal walls and 4 or 5 for external doors, depending on the weight of the door leaf.

Openings in timber frame walls

Openings in timber frame walls share only three things in common with masonry walls:

☐ The need to provide support for the structure built over the opening
☐ The need to provide a finish to the structure below the opening
☐ The need to close cavities and maintain a waterproof interface with whatever the openings will accommodate.

Additional features include:

☐ The need to include fire stopping round openings
☐ Proper weather proofing where the outer skin is of the rain screen variety (this will be explained as we go along)
☐ Provision for differential movement at openings in structures over three storeys high between the frame and the facing material, particularly a facing of masonry. This will not be included in this text.

We begin by looking at suitable details for one- and two-storey buildings with domestic or light commercial loadings, in each case taking different cladding systems/techniques into account. As we have done before, we can classify the openings according to size:

☐ For small pipes and cables < 50 mm diameter
☐ For pipes and ventilators >50 mm diameter but <225 mm
☐ In excess of 225 mm – large openings.

Openings for pipes and cables up to 50 mm diameter can be treated in the same way, no matter what the storey height.

When cutting through a timber frame wall for a pipe or cable, the elements shown in Figure 5.33 will all be pierced and the cavity bridged; if not treated properly this will be

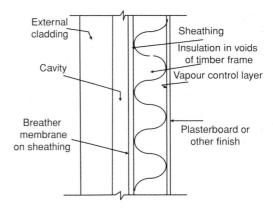

Fig. 5.33 Timber frame wall schematic.

a source of potential trouble as the building ages.

Starting at the inner face, the hole in the plasterboard can be disguised by using a floor plate. Immediately behind is the vapour control layer (VCL) which if badly torn will allow water vapour to collect in the insulation – a possible source of condensation and rot on the back of the sheathing. Cutting through the sheathing causes no problems but piercing the breather membrane could allow solid water to pass to the sheathing.

The breather membrane should be cut with an upside down T-shaped cut which forces any solid water to run round the cut and down the face of the membrane. This is illustrated in Figure 5.34. Bridging the cavity must be done

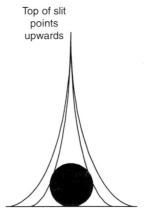

Fig. 5.34 Cutting waterproof fabric for pipe passing through.

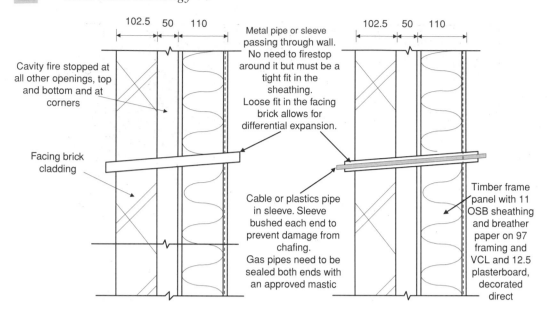

Metal pipe or sleeve ≤ 50 diameter

Fig. 5.35 Pipes and ducts through timber frame wall (1).

with any material sloping down towards the cladding, thus preventing solid water reaching the breather membrane. Cutting through the outer cladding must be done with care, not only from the point of view of appearance but also to keep the hole to the minimum required, so preventing vermin and larger insects getting into the cavity. Larger holes will also allow weather, particularly rain and wind-driven snow to penetrate. This is more important where the external cladding is of the rain screen variety.

Commencing with a masonry layer, Figures 5.35 and 5.36 show some typical examples; the slope to the outside is, of course, exaggerated. In Figure 5.36 a duct is illustrated which could be circular or rectangular, plastic or metal. Whichever of these combinations it is, there is every need for fire stopping around it. Sealing of all the ducts, sleeves and pipes to the masonry cladding could be done, but rather than using the ubiquitous silicone sealer, one based on polysulphide might prove to be more long lasting. There is nothing wrong with silicone sealers but frequently the incorrect grade is used and often in the wrong places for a silicone-based product.

The examples shown have only illustrated a cladding of masonry which to some degree or other will absorb rain driven onto its face. Penetration to the cavity is not a problem as measures are built into the general detailing to cope with this, particularly the provision of a DPC layer between fire stopping and cladding.

Rain screen cladding

Mention was made earlier of rain screen cladding. These types of cladding do not absorb water to any great extent and some not at all:

☐ Slightly absorbent claddings include slate and tile, both clay and concrete, timber such as shiplap or weather boarding, and boards made from timber, asbestos cement or silica fibre cement compounds.
☐ Non-absorbent claddings include metals and glass.

Metals and glass are seldom used in domestic structures in the UK but are common in North America, as are boards of timber fibre and asbestos cement. This leaves tile or slate

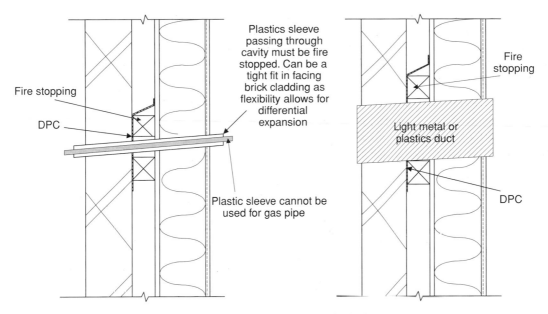

Plastic sleeve < 50 diameter and light metal or plastic duct > 50

Fig. 5.36 Pipes and ducts through timber frame wall (2).

cladding and timber weather board or shiplap cladding. All are quite common.

Vertical slating and tile hanging is usually done on horizontal battens nailed to vertical battens which in turn are fixed through the breather paper and sheathing to the studding of the timber frame. The vertical battens allow a space to be formed from top to bottom of the wall cladding, and this should be finished off top and bottom to allow a free air flow, thus allowing any water getting past the tiles onto the breather membrane to dry out.

Timber boarding is normally fixed horizontally to vertical battens which in turn are fixed through the breather paper and sheathing to the studding of the timber frame. Shiplap boarding is always fixed horizontally but weatherboarding can be fixed diagonally, although this is only done for decorative effect. Any attempt to have timber boarding fixed vertically should be avoided as it defeats the rain screening properties of the boards.

Figures 5.37 and 5.38 show the same pipe, sleeve and duct situations already examined for a masonry cladding. Here the cladding has changed to tile, weatherboarding and shiplap boarding.

Permutations of various finishes showing small pipe, sleeves or ducts through all the various combinations of cladding for timber frame construction would be excessively repetitive here. The details already covered can be extrapolated to cover most situations.

Lintels

Large openings are those which require support provided by lintelling over doors and windows, and here techniques have similarities to masonry construction:

☐ The opening in the timber frame is spanned by a lintel – in this case made from timber or a timber/steel combination.
☐ The jambs provide support for the lintel by the insertion of additional studs called **cripple studs**.
☐ There can be multiple cripple studs in one jamb, some supporting the lintelling arrangements and others supporting a sill.
☐ The bottom of the opening is finished off with a sill or a threshold

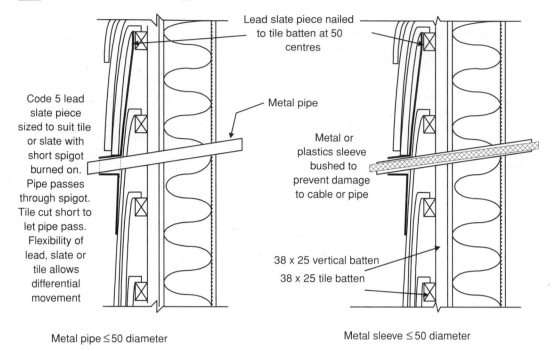

Fig. 5.37 Pipes and ducts through timber frame wall (3).

☐ Sills require support from cripple studs.
☐ Masonry claddings require support over the opening and this can be done with conventional lintels of concrete or steel or by special steel lintels restrained against the timber frame
☐ Jambs are treated conventionally externally and internally but the structure within the

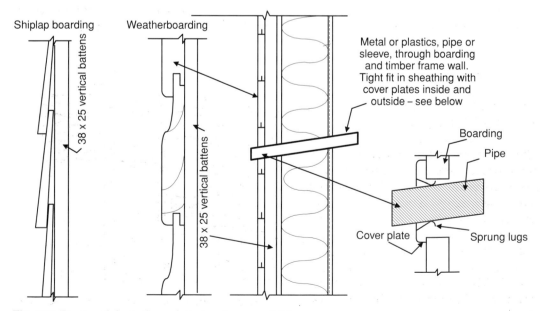

Fig. 5.38 Pipes and ducts through timber frame wall (4).

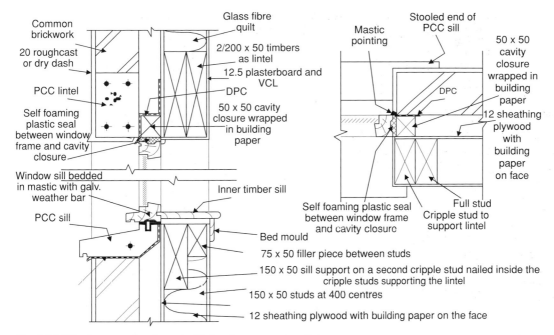

Fig. 5.39 Window opening in early timber frame constructioin.

jamb is made up of multiple studs and cripple studs in the timber frame.

☐ Masonry sills can be treated conventionally in all the accepted ways but the actual sill butts back against the timber frame panel with the sill DPC between. The inner sill is treated in a conventional manner with packers rather than grounds for fixing down.

Figure 5.39 shows a version of a window opening in an early timber frame wall. Note that all the structural timbers were sawn and not stress graded. These details are fairly typical of early timber frame construction, which tended to mimic the masonry cavity wall in the order in which it was built – first the frames were erected, then the masonry skin was built and then the openings were filled with windows or doors. Now the trend is to erect the frame, fix the cavity closure to the window or door frames, fix these back to the timber frames and then build the skin against all the timber work. This does not leave a gap between frame and timber panel to be filled with a sealant.

Figure 5.40 shows the inside of the timber framed panel once it is erected on a wall plate, the doubled top runner put in place and the upper floor joists in place. Note the use of crippled studs – studs cut short to provide support at the ends of horizontal members.

Figure 5.41 shows a more modern approach to timber frame panels. This is a photograph of

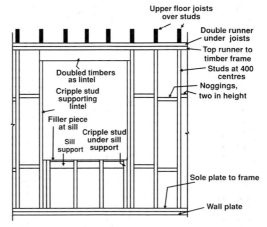

Fig. 5.40 Elevation of early timber panel at window opening.

Fig. 5.41 Modern timber panel at window opening.

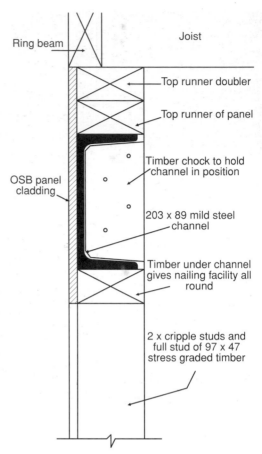

Fig. 5.43 Detail of rolled steel channel as lintel in timber panel.

the inside of a timber frame panel but note the OSB skin, the regularised and stress graded timbers, the timber lintel supported on double cripple studs and the narrow sill support on cripple studs. Now compare with Figure 5.40. The studs, runners and dwangs are regularised and 97 × 47 but the studs are put in

at 600 centres and so the joists are put in at 600 centres and have to be much deeper. A further complication is the need to keep ceiling heights down. If wooden lintelling is used for wide openings, the top of the glass in the window would be much too low. This can be

Fig. 5.42 Rolled steel channel as lintel in timber panel.

Fig. 5.44 Flitched beam over large opening.

Fig. 5.45 Flitched portal frame round opening in timber panel.

solved in one of two ways depending on the loadings involved.

The first way is shown in Figure 5.42 and uses a combination of a mild steel channel sandwiched between two layers of the 97×47 timbers, with blocks at each end set into the channel and nailed to the studs. These keep the channel in position. Note that there are doubled cripple studs supporting each end of the channel/timber lintel. Channel irons are made in a range of sizes but any in the range 76×38 to 305×89 would be suitable. Figure 5.43 shows a detail at the head of the frame.

The second way is to 'flitch' the frame members – both studs and lintels–to make a 'portal' round a large opening such as a patio door. Figure 5.44 is a photograph of a flitched beam spanning a wide opening and is followed by a photograph of a flitched portal frame in Figure 5.45. The steel strip in the portal frame was welded at the corners before sandwiching it and bolting up between the timbers. Note that the steel does not extend to the bottom of the jambs but stops short. A plywood packer is put between the timbers.

6

Roof Structure

Roof classifications	148	Flat roofs in timber	162
Prefabrication	149	Insulation, vapour control layers	
Trussed roofs	150	and voids and ventilation	164
The trussed rafter	153	Traditional roofs	167
Verges meet eaves	159	Roof insulation	169
Roof bracing	160		

Before looking at structure, coverings, etc. for roofs it is important to understand the general terminology of roofs. The sketch in Figure 6.1 covers much of what we will need to know for this chapter and for Chapter 7, Roof Coverings. The remainder we will pick up as we go along.

Roof classifications

Roofs can be classified in a number of ways and these are a few of them:

- By roof pitch. We talk about **pitched roofs** and **flat roofs**. Pitch is the angle that the general waterproof layer makes with the horizontal. Up to and including an angle of 10° the roof is a flat roof. With an angle over 10°, the roof is pitched.
- Flat roofs are never described as pitched but as **laid to fall** or, if there is more than one slope, laid to falls, and the fall is given not as an angle but as the fall in millimetres over a horizontal distance in millimetres.
- Pitched roofs may be monopitch, symmetrical or asymmetrical.

The next classification we will make is the roof form:

- Flat roof

- Lean-to roof
- Monopitch roof
- Symmetrical pitched roof
- Asymmetrical pitched roof
- Gabled roof
- Hipped roof
- Mansard roof

Not an exhaustive list but enough for our purposes here. Figure 6.2 shows examples of these in outline form. As the reader becomes more familiar with construction, other forms will become apparent; many are permutations of the above list.

Pitched roofs can be further sub-divided by construction type or form into **stick built** or **prefabricated.**

Stick built roofs are not often seen now, for the term means that timber is brought onto site in lengths to suit the roof, but not exact lengths. Then a squad of carpenters set up a **jig** on the site and cut all the timbers to length, forming all the notchings and halvings in them before assembling the roof on the top of the walls, stick by stick. The term was even extended to roofs with trusses or purlins etc. as the trusses were built on the ground in a jig and then hoisted into position. But we are getting ahead of ourselves.

Each stick in a roof has a unique name, so before going into prefabrication we need to

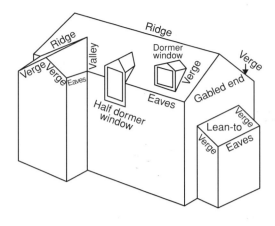

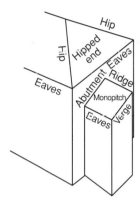

Fig. 6.1 Roof terminology.

know what these names are. Figure 6.3 identifies a few of them.

Prefabrication

Prefabricated roofs are very common now but early attempts were seen by the author in the 1950s. The roofs of the no-fines houses he saw were supported on **trusses**, and these and the **rafters, purlins** and **ceiling binders** were all assembled on the ground and fixed to the **wall plates**. Some additional bracing was built in to ensure that the whole frame stayed plumb and straight, and then it was lifted onto the top of the walls by a large crane, where it was set down on a layer of wet bedding mortar. The technique worked extremely well but was confined to the few very large sites where a covered work area could be built and the jigs

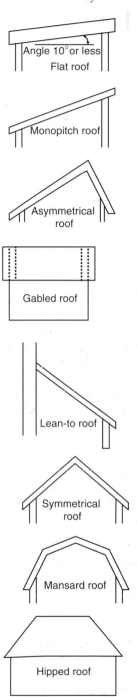

Fig. 6.2 Roof form outlines.

for the trusses and the whole roofs could be set up inside. Also, small sites could not support the large squad of carpenters required, nor employ their joinery skills later when it came to finishings etc. Power tools for cutting

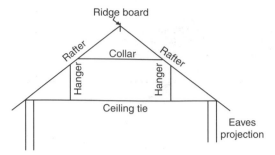

Fig. 6.3 Stick built roof timbers.

the timber were installed, and as soon as the site was operational roofs were being prefabricated and stockpiled until the concrete walls had been poured and the shuttering stripped. The same cranes used for all that lifted the roof structure into place.

Prefabrication now has come about with the introduction of the **trussed rafter** roof – not a complete roof but a set of units which are laid along the line of the roof and give the pitched form required. There are no **purlins** and no **ridge**; **binders** are sometimes used. The **eaves** and **verges** can all be prefabricated and easily lifted into place. The main point of this type of roof is that the components are a maximum two-man lift for all but a very few roofs – no need for a crane.

Trussed roofs

Before the trussed rafter, timbers (1) for roofs had to be of sufficient strength to span quite considerable distances using nails and simple jointing techniques. If the timber couldn't make it on its own, then strong beams (2) could be used to break up the spans of these timbers, and if that was not enough extra strong frames (3) were made up and put in to break the span of the beams. Thus we had:

(1) These timbers would be the **rafters, ties, collars** and **hangers** of an ordinary roof.
(2) The beams used would be **purlins** to support the centre of the rafters, and **binders** to support the ties, generally at their third points.

(3) The frames are called **trusses** and could range from king and queen post and other wonderful trusses from medieval times, to the TRADA trusses of the latter half of the twentieth century.

Figure 6.4 shows some layouts of these timbers in (1) in the above list for various forms of roof. For the same timber sizes, roof C is a larger span version of A, while B, the collared roof, is common on old properties and farm buildings but is not as strong as roof A. The provision of a tie in roofs A and C prevents the rafter ends at the **wall head** pushing outwards and pushing the wall over. While a collar can also do this, the lower half of the rafters are put into bending as well as compression and so much stronger timbers have

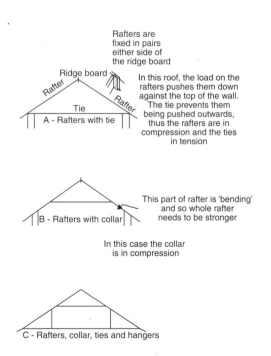

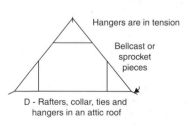

Fig. 6.4 Pitched roof shapes and types.

to be employed for the same span of roof. D is the form of C with a steeper pitch and the possibility of having attic rooms. With the steeper pitch, lighter section rafters, collars and hangers can be used but the ties now double up as floor joists and so must be heavier and possibly even supported at mid span by a beam or partition. Roof D also shows a device from the early roofs, the sprocket or bell-cast piece. These were separate short lengths of timber attached at the eaves and forming part of the eaves projection. The pitch angle on them was usually less than the main roof, which slowed the run-off of water from the slate or tile into the gutter.

Figure 6.5 shows, in step-by-step build up format, a roof using purlins and binders spanning from gable wall to gable wall and supporting rafters and ties.

Finally, Figures 6.6 and 6.7 show how a trussed roof would be built up step by step. The truss shape is a king post truss and it would be made from heavy natural timbers each in one piece, with the centre strut, the king post, shaped to take the struts and

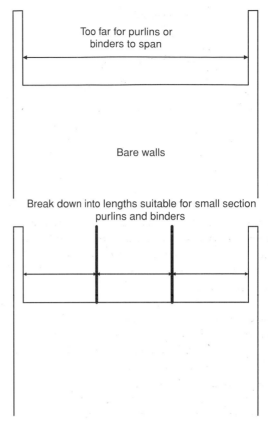

Fig. 6.6 Trusses and purlins.

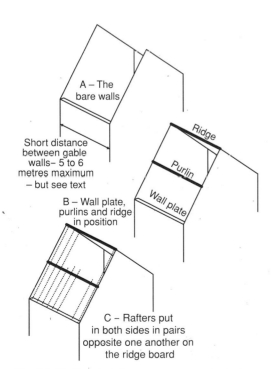

Fig. 6.5 Purlined roof.

the truss rafters. Note that the rafters which actually support the roofing material are called common rafters and are laid out across their supports, the purlins, without taking any notice of the trusses. The trusses are only there to support the purlins.

It is possible to do without trusses if the length between the gables is not too long, by using **trussed purlins** or **I beam purlins** or **box beam purlins**. Figure 6.8 shows sections through these, together with the position between the gables that they would take. Note that there will still be a wall plate on top of the wall for all the rafter feet, and a ridge where the pairs of rafters join. A plywood I beam with a depth of approximately 900 and with four 100 × 50 SC3 timbers as flanges, was capable of supporting a tiled roof over a span of 4000. These beams can all be made on site, the fastening being either bolts and connectors

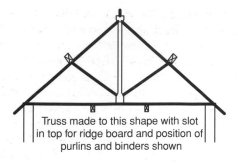

Truss made to this shape with slot in top for ridge board and position of purlins and binders shown

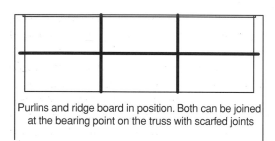

Purlins and ridge board in position. Both can be joined at the bearing point on the truss with scarfed joints

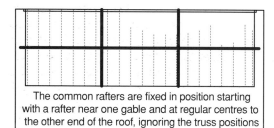

The common rafters are fixed in position starting with a rafter near one gable and at regular centres to the other end of the roof, ignoring the truss positions

Fig. 6.7 Trusses within the roof outline.

or **pattern nailing**. Figure 6.8 shows what is meant by pattern nailing.

Other truss shapes have been made from solid timbers and Figures 6.9 and 6.10 show a selection of these. The queen post truss will span much further than the king post truss and the hammer beam truss further still. Variations on the latter have been used for many public and religious buildings and that is probably the best place to see them. From its shape, it is obvious that the queen post truss is very suitable for containing attic rooms so many of these fine trusses are hidden from view, as is the mansard truss. The mansard truss is a combination of a modified queen post truss with a fairly low pitch king post

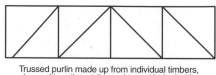

Trussed purlin made up from individual timbers, layered together at the nodes and bolted with Bulldog connectors between the timbers

I and Box beams–I beams have a plywood web and four solid timber flanges. Box beams have two plywood faces and a solid timber chord top and bottom. In addition there are solid timbers across the ply to prevent buckling of the ply layer(s)

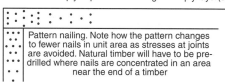

Pattern nailing. Note how the pattern changes to fewer nails in unit area as stresses at joints are avoided. Natural timber will have to be pre-drilled where nails are concentrated in an area near the end of a timber

Fig. 6.8 Trussed, I beam and box beam purlins.

truss on top. In every case each truss has rafters and the purlins they support in turn support common rafters. The Belfast truss is, by the standards of the others, relatively modern – late nineteenth, early twentieth

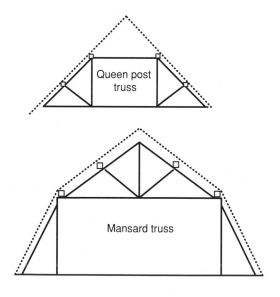

—— Truss timber
······ Common rafter line
□ Purlin position

Fig. 6.9 Queen post and mansard trusses.

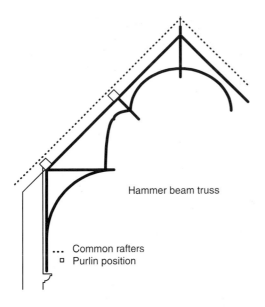

Hammer beam truss

... Common rafters
□ Purlin position

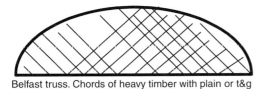

Belfast truss. Chords of heavy timber with plain or t&g
boarding nailed each side but in different directions

Fig. 6.10 Hammer beam and Belfast trusses.

century – when it was widely used for industrial buildings.

Figure 6.8 brought up the possibility of layering timbers at the nodes of the trussed purlin. Indeed, this was the tactic adopted by the Timber Research and Development Association (TRADA) in their series of trusses for domestic and light commercial roofs. Figure 6.11 illustrates this type of truss. Appendices C and G give details of the timber connectors used.

The outline shows the general arrangement of the timbers. The pair of rafters are each made up of two pieces of timber, as is the tie. There is extensive use of packing timbers or spacers which are used to connect the pairs of timbers together. The purlin struts are broad and have a check in the top to support the purlin. The long struts are bolted to the ridge support at the top and, although slightly out of alignment, are bolted in between the tie

timbers. Most of the **node**[1] details are given in Figure 6.11. Purlins are generally from 150 × 50 to 200 × 65 as the trusses are set at 1800 centres with common rafters between them at 450 centres, so there are three common rafters then a truss rafter, three more common rafters, a truss rafter, and so on right down the roof. These trusses have rafters which take a share of the roof covering load along with the common rafters. Using trusses this way in the late 1940s and early 1950s was the first time that it had been done.

The author built one set of trusses in 1964, following the TRADA designs, which had double timbers for the ties and triple timbers for the rafters. The purlin struts were double but the long struts were single. The timber was Douglas Fir – stress grading was unknown at the time. The pitch was 22° and the roof was covered with aluminium on felt on wood wool slabs. The roof is still there with nothing other than routine maintenance. About a year after the roof was completed and the house was occupied, the author had occasion to go into the roof space and while there decided to have a closer look at the joints of the trusses. To his dismay he found that every one of the nuts on the bolted connections was only fingertight at best and in many instances was actually loose. This was not caused by vibration but was solely due to shrinkage on the multiple layers of timber. With packers and splice plates some of the joints were some five layers thick, which represented something like 250 of timber. The nuts were all retightened and while the author was in occupation they were checked regularly but never came slack again. What was holding the trusses together? Only the Bulldog connectors' teeth embedded into the timbers.

The trussed rafter

The point of including the Belfast truss in Figure 6.10 was to show that the main timbers of the truss were joined together and the

---

[1] A node is the point on any frame where timbers, bars, lines, etc. meet and join.

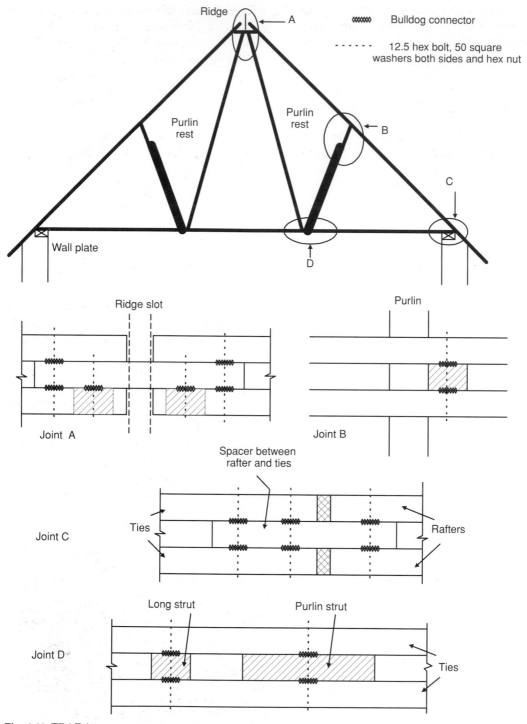

Fig. 6.11 TRADA trusses.

truss stiffened with thin boarding nailed each side – very effective. So it is not a surprise to learn that the idea of putting wooden 'plates' on each side of a node is a good, cheap way to make up trusses. With cheap plywood available, this can also be used and often with much less waste; certainly the plywood can be a lot thinner and so reduces the bulk at the node points. Pattern nailing is required – thin nails and lots of them and long enough to penetrate almost through to the opposite side of the timbers. Of course we are only talking about single layers of timber, not the multiple layers used in the TRADA trusses. The processes are simple but labour intensive, and before the advent of stress graded timber and the variability of the workmanship at each node, there was no way of making any great saving in timber. Everything had to be overdesigned to give an excessive margin of safety.

This changed rapidly with the advent of both stress grading and the invention of the gang nailing plate. Stress grading allowed engineers to treat timber in much the same way as metals had been treated; the performance of timber could now be predicted with some accuracy. This meant that smaller sections of timber could be used. The gang nail plate as a method of jointing had a number of advantages over nails and bolts:

☐ Only two pieces of metal are involved in any one joint
☐ There are no holes to drill
☐ There is nothing to tighten up
☐ There is no manual labour – other than placing components – involved in making the joints
☐ The results of using the plates are highly predictable
☐ There are no notches etc. to cut on timbers – only square or splayed cuts
☐ Timber is cut in a jig and assembled in a jig
☐ There are no glues, heat or other components involved.

Figure 6.12 is a sketch of a single gang nail plate and Figure 6.13 is a photograph of a node on a roof truss. The plates are pressed in each side simultaneously using a hydraulic press. The result is that the little spikes on the plate

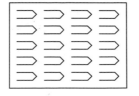
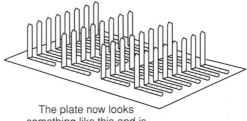

A plate of steel is punched with a pattern and then the pointed pieces are pushed through until they stand at right angles to the general surface

The plate now looks something like this and is hot dipped galvanised

Fig. 6.12 Gang nail plate schematic.

all penetrate at the same time, under an even pressure. Different sizes of plate are available to suit the conditions which will prevail at any node in a structure.

The manufacturers of the plates provide a complete service to the industry. They don't make trusses themselves but they supply all the machinery necessary:

☐ The jigs in which to cut as well as set out the timbers and the joining plates
☐ The hydraulic presses
☐ Computer software to design the trusses
☐ The plates.

Fig. 6.13 Gang nail plate.

Fig. 6.14 Trussed rafter roof.

What is very different about these trusses is the fact that they don't support purlins which support common rafters etc. These trusses support a section of roof which extends halfway to the next truss either side. So we are putting up a whole series of lightweight trusses and the roof covering is laid immediately on these.

Figure 6.14 is a photograph of some newly erected trussed rafters on a two-storey house. They are spaced out at 600 centres and are set on and fixed to a wall plate bedded and fastened down to the inner leaf at the wall head. Figure 6.15 shows the wall plate bedded down and two of the straps which hold it down can be seen on top of the plate. In Figure 6.14 the ends of these straps can be seen

Fig. 6.15 Wall plate secured to wall head.

Fig. 6.16 Trussed rafters secured to wall plate.

running down the wall at the far side. They are masonry-nailed to the blockwork.

Figure 6.16 shows the detail at the eaves of the roof. The ends of the trussed rafters sit on the wall plate and a fastener can be seen nailed to the wall plate and to the trussed rafter – every trussed rafter. The long board fixed to the ends of the trussed rafters is the **eaves fascia board.** It is simply nailed to the ends of the projecting rafters and finishes off that vertical face. Supports for the guttering are fixed to it. Figure 6.17 is a photograph taken down through the centre of the trussed rafters. Notice that the ties of the trussed rafters are in two pieces and are joined with a gang nail plate.

With these four photographs we have covered a lot of construction and probably raised a number of questions in the reader's mind. A little further down we will attempt to answer what some of these questions might be. Before that, there is one important thing to know about the trussed rafters. They can be made up in a variety of shapes using varying sizes of timber according to the loads being imposed.

The shape and layout of the trussed rafters in the photographs is known as a Fink truss and is easily recognised because of the distinctive W shape of the internal strutting.

Fig. 6.17 View down centre of trussed rafters.

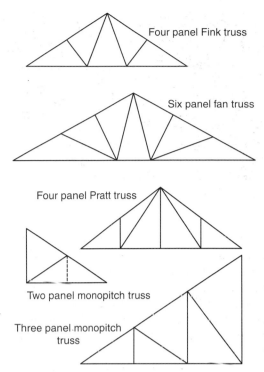

Fig. 6.18 Various truss shapes (1).

Figure 6.18 illustrates the Fink layout and a few other common arrangements. More are shown in Figure 6.19, including the mansard truss and the attic truss. The term **panel** in relation to these truss shapes refers to the unsupported length of rafter in each. The first, Fink truss is of four panels – each rafter is supported by the struts at the centre of its length, dividing each rafter into two, therefore four panels.

A book, which is really meant for the timber engineer but which is well worth referring to, is the *Timber Designers' Manual* by Ozelton and Baird, now in its third edition and published by Blackwell Publishing. There is a wealth of information, over and above the design calculations and methodology, which is not available from other sources.

Once a roof structure has been started with the erection of the trussed rafters, there is still the need to tie masonry walls to it at both the tie level and the rafter level. This is done in exactly the same way as floors are tied to masonry walls, but nowhere in my travels have I been able to find a correctly carried out example to photograph. Figures 6.20 and 6.21 show badly carried out examples. Figure 6.20

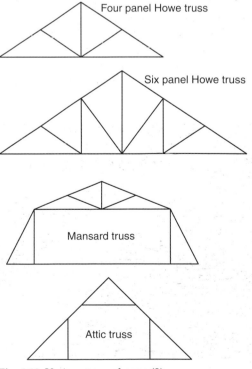

Fig. 6.19 Various truss shapes (2).

Fig. 6.20 Wall to roof stability (1).

shows the correct style of metal strap but the blockwork has been hacked down to accommodate it, and it is nailed to a thin strap of wood laid and nailed over the top of the ties. The hole hacked in the blockwork breaches the integrity of a mutual wall, which is now no longer up to fire resistant standard. The nail hold for the strap into the thin timbers must be questioned. There are no noggings between the ties and no solid packing or wedges between the last tie and the wall.

Figure 6.21 shows a correct style of strap, this time fixed to a fairly heavy horizontal bearer which in turn is fixed along the face of the struts of the trussed rafters just under the rafters. So far so good, but there is no wall in sight; it is obviously a case of get the tie in now and we'll build the wall to it later. That would be all right, but on this particular site there must be grave doubt that the masonry even came close. The scope of the photograph in Figure 6.21 has been deliberately kept fairly wide to show up one or two other points. At the very bottom left it is possible to see the diagonally opposite corner of this roof with another strap projecting out waiting for masonry. Also, note how small the gang nail plates are at the top of the row of struts from the centre and running down to the right of the photograph. These struts are in compression and so only need restraining in position while being erected etc.

The first part of the prefabrication of the roof was done when the trussed rafters were made. They have been erected now, and fixed down securely to a wall plate which in turn is fixed to the masonry wall below it. In Figure 6.21 horizontal and diagonal bracing can be seen and we have already criticised the arrangements for wall restraint.

The next part of the prefabrication is done for the formation of the verges. A **verge ladder** is made, so called from its appearance not because it is used for climbing. Each verge has a ladder and Figure 6.22 is a good example. This is on a timber frame house but the rest of the roof is not made from trussed rafters; it uses three trussed purlins.

Figure 6.23 has drawings of a verge ladder and how it is fitted. Note that the fully prefabricated version has the finish to the edge and to the underside. These panels are frequently

Fig. 6.21 Wall to roof stability (2).

Fig. 6.22 Roof ladders.

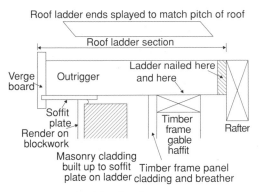

Fig. 6.23 Roof ladder details.

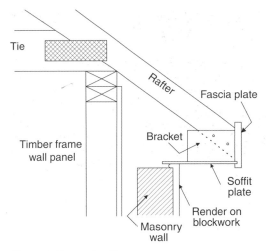

Fig. 6.24 Eaves detail.

painted up to undercoat stage and many are made with PVC verge boards and soffit plates. The sketches explain the terminology. In traditionally constructed stick built roofs these timbers – and some more – were all cut and fitted individually from the scaffold.

The verge ladder has four components:

(1) One side of the ladder is the **verge board**.
(2) The other side of the ladder is a simple timber plate but more often a length of 9 or 12 plywood.
(3) Between (1) and (2) are nailed a number of short lengths of timber, the **outriggers**.
(4) To finish the underside of the projection of the verges there is a **soffit plate** which is let into a groove on the back of the verge board and nailed to the outriggers.

To erect the verge ladder is simple. Hoist it into place and nail the plywood 'side' of the ladder to the last trussed rafter at the gable end. Side nail the outriggers to the runner in the top of the timber frame panel. Job done. The ends of the ladder are cut to a splay which matches the pitch of the roof so that there will be an easy join to the eaves finish, which is put on next.

The eaves finish can be prefabricated as well. First of all the ends of the rafters of the trussed rafters must be cut to the correct angle and length. The ends must be in a straight line. The easiest way to achieve this is to stretch a chalk line across the top of the rafter projections the correct distance from the wall face, pull the line tight and lift and release – snap the line – against the rafter ends. A chalk line

is made on a rafter – which will be in a straight line with all the others. Cut at the correct angle and the work is ready to receive the eaves boxing. Figure 6.24 shows one way of doing this.

Once the rafters have all been cut, the fascia and soffit plates can be put in place. Prefabricated, they come with the soffit plate glued into the groove on the fascia and with a few rectangular off-cuts of timber pinned and glued inside to keep the soffit and fascia at right angles. The whole unit is offered up to the rafter ends and nailed into the ends of the rafters. A more traditional approach would be to fix the brackets to the rafters, fix the fascia in place on the rafter ends, then put the soffit in the groove and nail it to the brackets.

Verges meet eaves

The tricky bit for someone trying to draw the detail is what happens where the verge detail meets the eaves detail?

Figure 6.25 shows a solution which has long been used on traditional roofs. In it the verge soffit is left short of the join between the verge board and the fascia plate. The verge board is joined to the back of the fascia plate. The eaves soffit is cut to match the width of the verge soffit, and the eke piece, a triangular piece of timber matching the verge board, is grooved

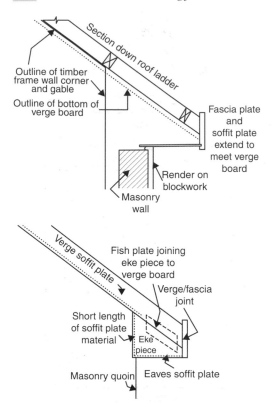

Fig. 6.25 Eaves and verge junction.

on one edge to support the eaves soffit. The eke piece is also grooved on the adjacent edge to take a short length of soffit plate material with a width matching the verge soffit. It is made a tight fit between the verge soffit and the eaves soffit and is wedged in the other direction between the eke piece groove and the masonry. To hold the eke piece in place, a plate of plywood is screwed to the back and allowed to project. This allows the joiner to secure the eke piece to the verge board.

Roof bracing

The roof structure is not complete without being properly braced, and gable walls have to be attached to the roof structure. The latter was dealt with when we looked at some badly executed strapping work next to blockwork gables in Figures 6.20 and 6.21 and also in Chapter 4, the section 'Support of masonry

walls', where similar principles apply. Properly done, these would be more than adequate. What we have not dealt with is how to connect timber frame panel gables to the roof structure. Figure 6.21 gives a clue. While the photograph was taken on a site with all masonry walls, note that there is a long continuous timber fixed to the top of the trussed rafter struts and it has a galvanised strap fixed to it. That timber is part of the roof bracing and in that figure was showing that it could also be used as an anchor point for the tie-in to the gables. That is what happens in timber frame construction. The brace extends into the depth of the panel and is secured either directly to the studs of the panel or to a dwang or nogging secured between the pair of nearest studs.

The author has seen many direct connections and few if any have overcome the minor problem of the different angles of the brace and the studs, the latter vertical and the former tilted over with the struts in the trussed rafter. What usually happens is the carpenter takes an axe to the end of the brace and chops it roughly into line with the stud, and then hammers a large nail or two through the thinned end – which splits! If noggings are fitted they are done badly and not properly secured to the studs.

Figure 6.26 shows a method of dealing with both situations which might take a little longer to execute but would improve the final result beyond all measure.

The other braces within a trussed rafter roof are:

- □ Diagonal braces across the underside of all the rafters in each roof plane
- □ Horizontal braces on the outside, both sides of the peaks of the trusses
- □ Horizontal brace along the upper edge of the ties.

Figure 6.27 illustrates these.

The last part of the structure is whether or not the rafters are covered with boarding of timbers or sheet material such as plywood, OSB or some other fibre-based board. Tradition varies across the UK. With the old bituminous felts as underlays below the slates

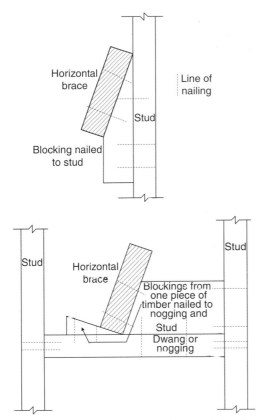

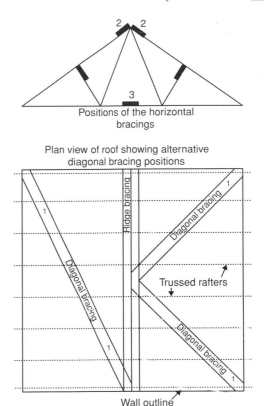

Fig. 6.26 Horizontal brace to gable panel detail.

Fig. 6.27 Roof bracings schematic.

or tiles and the manner of fixing slates and tiles regionally, there were arguments for and against boarding. With the advent of alternatives to the bituminous felts, it is largely a matter of how the slates or tiles are fixed and whether the boarding plays some structural function.

The traditional boarding used is a sawn and now treated, plain edged board 150 × 12.5, 150 × 15 or 150 × 20. This boarding is called **sarking**. It is nailed to each rafter with two 65 galvanised nails, pressing the edges close together as nailing is done. If there are to be battens fixed to the roof, then 12.5 or 15 thick boards are adequate depending on the span from rafter to rafter. If there are slates to be fixed direct to the sarking, at least 20 thick boards should be used.

Where no sarking is to be fixed, then an underlay which can be stretched across the rafters without tearing must be used. The old bituminous felts had a reinforced variety, the reinforcing being a mesh of hessian yarn. It was sometimes known as **sarking felt**. These felts, while being impervious to solid water, did not allow air to pass through and this is an important feature of modern roof construction. Roof ventilation will be covered in Chapter 7, Roof Coverings. The modern equivalent of the old felts is generally a plastic sheet; one is made that has micro-perforations at about 50 centres each way. This is marketed under the brand name Tilene by Visqueen Building Products. Figure 6.28 shows how it works and how it is impervious to solid water and yet allows air to pass each way. Others are made from felted polypropylene fibres, such as Tyvek. The felted polypropylene fibre coverings are impervious to solid water but are breathable.

Many of the roofs examined recently are boarded with 12 sheathing plywood, then a

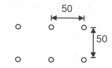

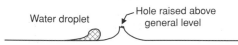

Water droplet — Hole raised above general level

Plastic sheet is perforated with a needle which first raises a blister and then pushes through. This work hardens the plastic which retains the 'volcano' shape. It is this as well as the extremely small hole which repels solid water

Fig. 6.28 Breathable plastic underslating felts.

breathable membrane such as Tilene is laid and battens are laid to which the tiles are fixed. Slated roofs are boarded with timber sarking or thicker sheathing plywood and covered with one of the polypropylene fibre felts. The slating in the author's home area is direct onto the sarking boards.

Natural timber sarking imbued a certain amount of bracing or stiffening of the roof structure to which it was fixed. This is also true of the sheet materials used in its place and these now become a necessary part of the structure in tightly designed roofs in timber frame construction.

Our discussion so far has followed a progression which has ended in the concentration on trussed rafter roofs as if they were the only thing built today. While this is not the case, trussed rafters account for the majority of roofs erected now.

Flat roofs in timber

Flat roofs deserve some attention for, although in a minority in new work, they are still popular for extensions.

There are few of these roofs which are truly flat, i.e. the surface is horizontal. The majority have a fall or falls to take the rainwater off to one side or to a collection point. We will discuss a truly horizontal roof a little later.

What do we have to do to introduce a fall(s) in a flat roof? Before we start, the timbers which provide the main structural support are properly termed **roof joists** although we will end up calling them simply joists – in the current context this is fine.

There are various ways of giving a fall to a flat roof built in timber:

(1) The roof joists which support it are laid to a slope, which in itself can be done in two ways:
    (a) the joists slope end for end
    (b) the joists are horizontal but each is set at a different level to the adjacent joists
(2) The joists are laid horizontally and additional taper cut timbers are placed:
    (a) on top of them to introduce a slope – **firring pieces**
    (b) across them to introduce a slope – **declivity pieces**
(3) The upper surface of the joists is cut to a slope
(4) The structure is built horizontally but the insulation applied to the bearing surface varies in thickness to give a fall(s).

Figure 6.29 shows how these techniques can be used for roofs with a single slope to one edge of the roof. Only an indication is shown – these are not detail drawings.

When a roof requires more than one slope – it is laid to falls – then the joists are usually laid horizontally and the added timbers are arranged to give the multiple slopes required. Modern requirements for insulation of these roofs has brought in the idea that the roof shown in (4) of Figure 6.29 is a good way to achieve multiple slopes. The insulation is designed using a CAD[2] system and the

---

[2] A CAD (computer aided design) system will allow the designer to prepare a plan of the roof with all the key dimensions (sometimes referred to as CAAD – computer aided architectural design.) This can be input to another dedicated design system which calculates the shape of a whole series of blocks of insulation required to cover the roof and which will give multiple falls on the roof. Each block is cut by a machine which takes its data from the dedicated design system. Each block is numbered and a drawing is prepared which shows the laying sequence.

1a Joists slope end for end

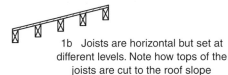

1b Joists are horizontal but set at different levels. Note how tops of the joists are cut to the roof slope

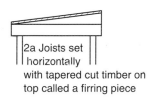

2a Joists set horizontally
with tapered cut timber on top called a firring piece

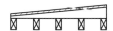

2b Joists set horizontally but taper cut timbers set at right angles now known as declivity pieces. Note thin end still has some thickness to take load from deck

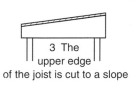

3 The upper edge of the joist is cut to a slope

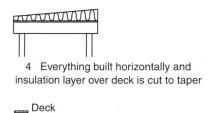

4 Everything built horizontally and insulation layer over deck is cut to taper

━━ Deck

Fig. 6.29 Flat roof – options for structure.

Table 6.1 Roof angles and rise in horizontal distance in millimetres.

Angle in whole degrees	Natural tangent	Roof slope in mm/1000
1	0.01746	17
2	0.03492	35
3	0.05241	52
4	0.06993	70
5	0.08749	87
6	0.10510	105
7	0.12278	123
8	0.14054	141
9	0.15838	158
10	0.17633	176

horizontally. That ratio of vertical height to horizontal distance is better known as the natural tangent of the included angle, i.e. between the slope and the horizontal. The natural tangents for the whole degree angles from 1 to 10 are listed in Table 6.1, together with the flat roof slope to give the reader some instant appreciation of what these slopes are really like.

There are a lot of different ideas about how a flat roof might be constructed to provide a fall(s). The illustrations in Figure 6.29 show a **deck**, and in an all-timber roof this has to be of timber as well. Most traditional material was a tongued and grooved boarding – usually flooring boards simply nailed to the roof joists. Depending on the centres of the joists, 20–22 thick boarding was usually adequate. On top of this deck a waterproof layer was built up by first nailing roofing felt to the timber and then sticking further layers onto the first with hot bitumen. These old felt roofs when laid by experts had a surprisingly long life but only if the pitch was kept as high as possible. Not flat roofs, but the author knows of several roofs laid in the 1960s which lasted well over 20 years with only minor repairs. In fact the company that laid them gave a 20 year warranty on their work. That is not done today despite the 'advances' made in bituminous felt technology – polymerised bitumens and all the rest.

The old roofs were not insulated, and all that was between the occupier and the great British

insulation is cut using numerically controlled machinery.

At the start of the chapter we defined a flat roof as one with a slope not exceeding 10° and that slope for flat roofs was usually given as the fall in millimetres per 1000 millimetres climate was the felt on the deck, then a gap where the joists ran, and a simple ceiling. As heating fuel rose in cost, insulation was given some thought – although not a lot, for many of us spent hours sorting out the problems caused by the insertion of insulation in new and existing roofs. The problem had a lot to do with the fact that putting a layer of insulation into a structure causes one side to become warmer and humid and the opposite side to be colder, and if the moisture gets through the insulation it also becomes wet because of condensation. High humidity levels cause rot and increased insect attack in timber, so flat roofs were rotting down all over the place.

Insulation, vapour control layers and voids and ventilation

Figure 6.30 shows a series of flat roof sections with insulation, a deck and waterproof layer and a ceiling. They are in different positions

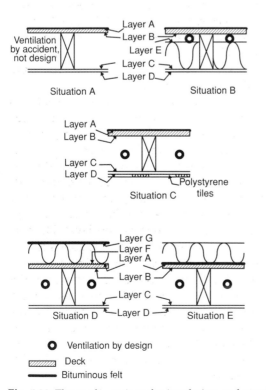

Fig. 6.30 Flat roof – options for insulation and ventilation.

relative to each other and the provision of a VCL may cure or kill the roof.

Taking each in turn, situation A shows what was the traditional approach to the flat roof. Any water vapour-laden air passing into the void between ceiling and deck had some chance of escaping along the joist void and out at the eaves or head of the roof, such was the looseness of the build. Ventilation was accidental, not planned and allowed for. Nevertheless, in times of cold weather and increased heating activity below the ceiling, a lot of the water vapour would condense out on the underside of the deck – Layer B. Layer C also became wet due to solid water dripping off Layer B. Moisture would pass through the deck but could not pass through the waterproof layer and so condensed within the deck structure or material at Layer A. With the copious amounts of heat being pushed through the ceiling, the ceiling itself would remain dry until the heating was turned off and it cooled rapidly, leading to condensation at Layers C and D. Being in kitchen and bathroom extensions, there was plenty of moisture around.

Situation B was the next step. Insulation was stuffed into the joist void by taking down the ceiling, insulating and replacing it. An alternative is shown at situation C. Here the keen DIY man has covered the entire ceiling with polystyrene tiles, only 10 or at the most 12.5 thick so as insulation almost ineffective and a waste of time. The adhesive used was a strong wallpaper paste and the tiles were of polystyrene beadboard. Neither would present any real resistance to the passage of water vapour. In both these situations there was occasionally some attempt to increase ventilation flow rates down the joist void. Unfortunately many people including tradesmen did not realise that each joist void was a sealed cell on its own and required individual ventilation. The effect was to further increase the temperature gradient across the roof, so not only was there condensation at layers A and B but also on top of the insulation at layer E. If the latter became serious enough the water could get back down to layer C and the insulation could be rendered ineffective – except as a sponge!

The polystyrene tile solution was no real solution; besides, there was a hazard using such tiles as, decorated or not, they presented a real fire hazard. The author spent some time on 'fire report' duty with a company that owned a number of houses. Monday morning was the busy day, with drunks setting fire to their kitchens with an unattended chip pan, and in the 1960s the whole thing was exacerbated by the use of polystyrene tiles on the ceilings, which melted, dripped and caught fire, so spreading the fire across the room. Situation B could have been salvaged if the tradesmen DIY experts had only realised that stopping the water vapour getting into the joist void was as important as putting in the insulation. The key element missing was the vapour control layer (VCL), a simple layer of polythene fixed to the underside of the joists before the plasterboard went on.

This layer could not be 100% vapourproof; there were too many things to go wrong for that. First, there were a few hundred staples or tacks holding the polythene in place before the plasterboard went up. Then lighting cables had to come through, perhaps a pipe or two, a ventilation stack for plumbing or an extract fan, and finally a few hundred nails held up the plasterboard. Many experts argue that driven nails and staples present no problem as the plastic stretches round the nail shank and seals to it. Certainly the cables, pipes and ducts do present a problem and there are simple ways to secure the plastic to these items with gaffer tape etc., which should reduce the flow. But, at the end of the day some water vapour is going to get through and we don't want to have it condensing above the insulation layer at Layers A and B.

This is where the ventilation comes into play. There has to be sufficient ventilation to have a flow of air which can carry any water vapour away – ideally before it condenses on Layer B, but that is not always a possibility. The flow should ensure that Layer B will dry out rapidly and so prevent a build up and condensation at Layer A. The felt layer(s) are left exposed to the weather and to ultra-violet radiation. UV does break bitumen down, and the improved bitumens now used have gone some way to improving the resistance to decay caused by UV. The weather also has its effect on the felts. The constant wetting and drying and heating and cooling cause the felt to expand and contract far beyond the scope of the material to continue to move for its expected life span. The result is blisters and cracks and splits in the layers which have to be patched, and there are more thicknesses over some of the roof giving rise to more inconsistent layering which makes the whole situation worse.

One novel approach which met with great acclaim for a short period was to make the waterproof layer absolutely horizontal and put high upstands round all the edge of the roof. Rainwater outlets were arranged at high level in these upstands, or rainwater disposal pipes were brought through the felt and terminated high above it. The result was that the roof would be permanently under a volume of water and a depth of about 300 was supposed to be ideal. If there was no rain and the ponding was evaporating away, the owners were supposed to fill it up with a hose! The water protected the felt from UV radiation – a few millimetres are enough. The mass of water would also help to regulate the temperature of the felt layers and keep it from fluctuating wildly. Problems solved!

The snags came immediately the roof was constructed – it was very expensive because the structure had to be super strong to hold up about 300 litres/m^2 of water and its own weight. Then there was the workmanship aspect. Some of the roofs leaked from day one. Others started to disintegrate at the one point that wasn't protected by the water – the upstand above the water run-off level. The water certainly kept the main felt cool and UV free, but the little strip round the edge was very vulnerable. The cost of pumping out the roof – no-one had thought about building in a drain down facility – on top of the repair cost at the upstand without any guarantees, put paid to the idea. It was to be at least 20 years before a better method of protecting the felt was found.

Situation B is commonly referred to as a **cold roof** construction – the insulation is at ceiling finish level. Situations D and E are warm roof constructions and were seen as the answer to a lot of problems, including one or two left by situation B and the ponding disaster.

Situation D came first and simply moved the insulation to the top of the deck. This meant that the deck stayed warm and so there was no condensation at this layer. The insulation used was either extruded polystyrene with a plywood or cork layer bonded on top, or a polyurethane layer with aluminium foil one side and kraft paper the other. The idea of the plywood or cork bonded to the polystyrene was to protect the polystyrene from the heat used to lay the felt. Polystyrene melts very easily at a low temperature. The extruded version has a very high resistance to the passage of water vapour. Layer F in that situation is a cold bituminous emulsion – a water-based mixture of bitumens – which is used to stick the polystyrene down. Solvent-based adhesives would dissolve the polystyrene. By now the reader will be wondering why we bother to use polystyrene. Well, the problems with polyurethane can be just as bad: incompatibility with adhesives, not so bad with heating but a terribly friable material even when placed between paper and foil layers. There is also a cost difference in favour of polystyrene, as well as its very much lower vapour transmission rate. Polyurethane must have a foil face.

So polystyrene it is. And on top of the plywood or cork, a layer(s) of felt can be bonded in hot bitumen. The bitumens were still improving but were being challenged by single skin materials of synthetic rubber and some plastics. But no matter what the covering, it was still exposed to weather and UV. It was now that the idea of forming falls in the insulation was introduced.

Then along came situation E – a simple idea but instead of using bituminous emulsion to stick down the polystyrene, use the whole waterproof layer, bed the polystyrene into the bitumens – cork or plywood down – and the felt layers are kept at a constant temperature under the insulation, and the UV radiation can only get at the insulation. High winds might be a problem – foamed plastics can be literally shredded by high winds. The slightest imperfection in the surface is enough for the wind to get a grip, and soon there is a bigger hole and then a bigger hole and soon lumps of plastic are scudding through the sky like Frostie the snowman on speed! The answer was again simple: weigh it down. Gravel was tried but on high level roofs could be blown off to shower people below with pebbles. There is no point in using a layer of concrete as it prevents inspection and maintenance of the insulation layer and ultimately the felt. The answer was to simply to make the stones bigger and/or to couple their use with laying paving slabs for access by maintenance workers and in areas with higher exposure ratings.

In both situations D and E, the bitumen layer under the insulation acts as a primary vapour control layer. The use of extruded polystyrene acts as a secondary VCL.

How these roofs are faring now the author has no idea, but he would like to hear of anyone's experience with them.

To return to the earlier mention of ventilation – a vexing question for the energy conservationist. Who needs a cold draught blowing down the joist voids between the ceiling and the insulation? Some conservationists have advocated ventilating these voids to the building itself, even the room covered by the roof, but if it were a bathroom or kitchen that would hardly be suitable. They cannot be ventilated into other voids such as wall cavities and so on because of the risk of fire spreading. So it would appear there is no alternative to providing holes or slots in the roof edge to allow air to move down these voids. But what is not wanted is a howling gale which will carry away all that expensive heat.

An old-fashioned Victorian idea comes to mind which is still found in extant buildings of that vintage, and most of them schoolrooms or lecture halls etc. – the **Tobin tube**. This is a hole part way into a wall with a ventilating grid on the outside; the hole turns up in the wall for about 1200 and then turns into the building with another adjustable grille set about 1400–1500 above floor level. Sometimes a flap valve is fitted near the bottom of the tube, with linkage to a lever on the grille inside the building. The effect of this device was to induce a flow of air without the flow getting out of hand even on quite windy days. It cannot be beyond the technology of our present civilisation to put this idea into practice fitted with some automatic valve operator such as one finds on greenhouse windows and so on.

Traditional roofs

The chapter would not be complete without a brief look at the stick built traditional roof. This is best illustrated with a few photographs, some of them very recent – the ones of the pole plate roof are only three years old and yet the technique was first shown to the author some 45 years ago and was very old then.

Figures 6.31, 6.32, 6.33 and 6.34 are of a fairly old roof (1961). It had a 54° pitch – very steep – and three rooms in the attic space formed. The ceiling ties were laid in and supported partly on brick partitions and partly on an RSJ as they formed the attic rooms' floor. A timber about 35 square was let into a groove along the point where the rafters would meet the ends of the ties. The ends of the rafters were splayed off at the correct angle and lengths, and a notch was cut in one end to go over the timber let into the ties. Then the rafters were erected in pairs at intervals along the roof with the ridge board.

Fig. 6.32 Roof with traditional sarking boards.

Collars were put in to support a ceiling in the attic rooms and hangers were finally added to break the span of the lower part of the rafters. They bore down on a plate nailed to the top of the ties. The ties were simply nailed to the wall plate, which was only bedded in mortar on top of the brick walls.

Temporary bracing was fixed inside the rafters, and sarking boards were nailed to the overall roof surface which was finally covered with underslating felt and concrete interlocking tiles. The photograph of the roof with the sarking shows some boards missing. These were put in place as the boarding went on but they were not nailed into place. Once two more boards had been nailed above them these boards were taken out and nailed temporarily to the boards below. This gap gave a foothold for the carpenters to step up and complete more of the boarding. Once the last

Fig. 6.31 Stick built roof (1).

Fig. 6.33 Stick built roof (2).

Fig. 6.34 Stick built roof (3).

Fig. 6.36 Pole plate (1).

board at the ridge had been fixed in position, the carpenters came down one step and fixed the previous boards into the step just left. And so on down the roof.

These steps you see in the photograph were left out so that the plumbers could go up and fix the lead round the dormer window. They had been used earlier by the carpenters who were making the dormer window framework and cladding. With the exception of the ties, the roof timbers were all 100 × 50 sawn, and timber was selected from a slight overorder in quantity in order to get straight grained timbers with not too many knots for all critical timbers.

Figure 6.35 is not about the roof so much. It is, after all, a trussed rafter roof, which has been shown earlier. It is more about the wall which sticks out at you in the centre of the

picture. This is the gable wall which runs away to the left of the picture. Notice how the designer has used a corbel 'stone' – a concrete projecting block – to carry the wall out beyond the front wall of the building and thus facilitate the joint of the verges to the eaves. A sketch of how this detail might have looked is given in Figure 6.38.

Figures 6.36 and 6.37 show the rafter feet to tie junction on a pole plate roof. The problem with rafters is that they tend to spread their feet out when under the load of tiles and wind and snow and so forth. Many roofs have the rafters joined to the ties by making a halving on the rafters and nailing this to the side of the tie. The joint is entirely reliant on the nailing. In a pole plate roof, the pole plate itself acts as

Fig. 6.35 Corbel at eaves verge junction.

Fig. 6.37 Pole plate (2).

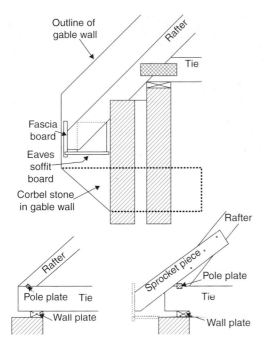

Fig. 6.38 Roof details – corbel and pole plates.

a key between the rafter feet and the ends of the tie and prevents the rafter feet sliding off.

Figure 6.38 includes a detail of how the critical junctions on both these pole plate roofs were done. Also included is how the sprocket piece was used to finish and support the eaves construction on the 1960s house.

Roof insulation

Insulation is even more important in roofs than ever before. Now that we have a better understanding of the inter-relationship between the need to insulate and the control of vapour passing through a structure, the possibilities for condensation and the need for ventilation, it is possible to design warm or cold roofs, whether flat or pitched, to suit every condition we could hope to meet in practice. One or two points merit mentioning here.

Ventilation of roofs is vital. While masonry walls can be built as breathable structures and there is now serious doubt about the need for vapour control layers (VCLs) in them, this is

not the case in a roof where the structure is predominantly of wood and wood products. The ideal must be to keep the wood products on the warm side of the insulation and to keep them protected from moisture coming through the ceiling finish by placement of a VCL immediately behind the ceiling finish. There has to be a void to carry ventilating air in and moisture laden air out, and whether the roof is a cold roof or a warm roof will determine where that void will be. Looking back at Figure 6.30, situation B is a cold roof so the void is to the outside of the insulating layer. Situations D and E show warm roofs and the void is to the inside of the insulating layer. No matter what else happens in the construction of these roofs, the Regulations require a minimum 50 deep gap between the insulation and the deck or ceiling.

Pitched roofs can also be built as warm or cold roofs. Figure 6.39 shows schematics for both warm and cold roofs. Cold roof A

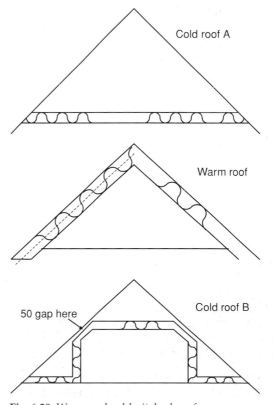

Fig. 6.39 Warm and cold pitched roofs.

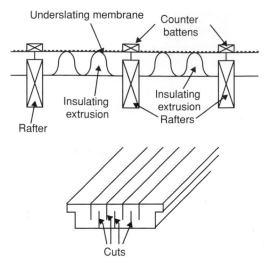

Fig. 6.40 Preformed plastic foam insulation for a warm roof.

requires ventilation at or just above the eaves level, and at the ridge. At eaves level it is usual to perforate or form a continuous slot in the eaves soffit and ensure that there is open passage for air over the top of the wall into the roof space. At the ridge, special tiles or roof ventilators can be fitted. The warm roof has insulation under the sarking and if not used

for forming rooms there is a need for some ventilation, with control over the amount but not the means to cut it off completely. The cold roof B is common in speculative housing, and the insulation of the framing forming the room in the roof space is often so spectacularly badly done that it is a waste of money putting it in. Note that there has to be a minimum 50 gap at the slope on the attic ceiling. The slope is called a **coomb**. Ventilation is required top and bottom of the roof.

Figure 6.40 shows a detail for a warm roof using a purpose-made foamed plastic extrusion. This product has a rebate on each edge which fits over the rafter. The slots either side of the extrusion run the complete length. As the extrusion is forced between the rafters these slots close, ensuring that the plastic is a tight fit between the timbers. An underslating membrane is placed over this and counter battens nailed through the foamed plastic into the rafters.

The plastic used is the ubiquitous polystyrene, this time in bead board form but a fairly dense grade. Tiling or slating can be carried out over the counter battens just as in a conventional roof.

7

Roof Coverings

Tile and slate materials	171	Bituminous shingles	176
Slates	175	Pan tiles	177
Plain tiles	175	Spanish and Roman tiles	177
Interlocking tiles	176	Edges and abutments	178
Timber shingles	176		

In the previous chapter on roof structure, we reached a stage in construction where there was a waterproof layer over the structure – the underslating[1] felt or membrane. This is not a long-term solution to keeping the building watertight but is frequently used short-term to allow internal work to proceed while outdoor work continues, dependent on weather conditions, at the same time. This makes good use of that scarce commodity, time. Eventually the underslating membrane will have to be covered over with something a little more permanent, and that is what this chapter is about – a short discussion on the materials which can be used, how to apply the more common types and how they are finished at edges, junctions and changes in direction. As a finale, complete details of basic roof edges will be given. We will not consider metal or plastics roof coverings in this text except for the necessary leadwork at abutments.

[1] Some in the industry always refer to this layer as under-tiling felt or membrane (even if it isn't felt); others always refer to it as an underslating felt or membrane. One could be pedantic and always apply the correct term in relation to the final materials used. Is it a bituminous felt or a plastic membrane? Is it slate or tiles which are being laid? It makes little difference to what is actually used. We will use underslating membrane irrespective.

Tile and slate materials

Mock-ups are used extensively in this chapter, with cardboard slate and tiles, to around one-fifth scale. They may not be as life-like but they have the benefit of using less manual effort to set up and allow annotation to be added with a marker pen. We hope they are useful.

Figure 7.1 is a photograph of several terraced houses with roofs at different levels. Note how the roofs are covered with a membrane, and that membrane is held down by timber strapping running up the roof over the rafters. These are the **counterbattens**. There are also at least two horizontal straps, again to help hold the membrane down. A roof light can be seen above the prominent doorway and the membrane is fitted to that to provide a watertight junction. This roof is ready to have the tiles or slates fitted.

Materials natural and man-made, sizes, colours, techniques for all sorts of situations – they all abound in this one area of construction. The easiest way to come to terms with it is to set out much of this information in a table, Table 7.1.

When laying any of these units it is obvious that they must overlap adjacent units all the way down the roof slope, forcing rainwater to flow from one waterproof surface to the next.

Fig. 7.1 Houses ready for roof tiles.

So we can group these units according to the type or number of laps they require.

Cutting across the material differences, the first grouping can be taken as those units which require to be laid with double laps and those with single laps to obtain a watertight covering. Slates, timber shingles and plain tiles require a double lap; the interlocking tiles

and bituminous shingles only require a single lap. The double lap is only necessary in the direction of the slope; laterally all units rely on a single lap. To help illustrate these laps a short series of photographs has been prepared where the units were modelled by rectangles of card, something which students can do for themselves to help their understanding and visualisation of double lap joints in particular.

Figure 7.2 illustrates the double lap fixing of plain tiles or slates. The work always starts at the eaves and works up the roof, generally from right to left, forming verges and abutments as course upon course of tile or slate is laid. We will refer to the tiles or slates of whatever type as units.

So the bottom of the mock units is the bottom of the work – the eaves. The first course,

Table 7.1 Tiles and slates – materials, fixing and laying method.

Material	Man-made/natural	Product	Fixed direct to sarking	Fixed to battens	Lap
Slate	Natural	Slates from Wales, Cornwall, Lake District, Scotland, Spain and China	Yes	Yes	Double
Artificial slate	Man-made of asbestos or silica fibre in OPC	Slates which are recognisable as artificial	Yes	Yes	Double
Reconstituted slate	Man-made of slate particles in resin	Slates which attempt to imitate the natural product	Yes	Yes	Double
Clay	Natural	Fired clay tiles; plain or interlocking. Can be obtained glazed	No	Yes	Plain tiles double; interlocking tiles single
Concrete	Man-made	Pressed concrete tiles; plain or interlocking. Colour and texture added	No	Yes	Plain tiles double; interlocking tiles single
Timber	Natural	Shingles	Yes	No	Double
Bituminous felt	Man-made	Shingles	Yes	No	Single
Clay	Man-made	Pan tiles	No	Yes	Single with special side lap arrangements
Clay	Man-made	Spanish and Roman tiles	Yes and can be bedded in mortar on a concrete slab	Yes	Single with special side lap arrangements

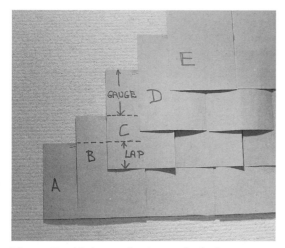

Fig. 7.2 Double lap roof unit – method of setting out.

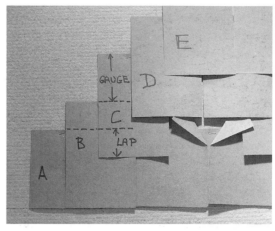

Fig. 7.3 Double lap roof unit – showing lap with alternate courses.

A, is a course of short units either specially cut, as in the case of natural materials, or specifically made for man-made units. If there were battens for fixing, the first batten would be under the bottom of this course and the second batten would be under the head of these units. The course is termed an **eaves course** or a **cut tile/slate course**.

The first full unit course is course B, and it must be observed that the units in this course are displaced horizontally by half the width of a unit. Run the eye up the stepped ends of the courses to be aware of this effect. The courses are laid to break bond just like laying bricks or blocks. The object here, however, is to keep water out of the house.

The third course, course C, overlaps the heads of the units in course A by a specified amount, **the lap**. It is because C overlaps both courses A and B, that the system is termed double lap. Lap varies according to the type of unit and the pitch of the roof – less pitch, more lap. The associated term **gauge** is the distance from the top of one course to the next immediate course and is therefore the centres at which any battens used for the units are fixed, or for marking out where no battens are used.

Without the double lap, rainwater running off the unit would simply go down between the two units below. With the double lap, that rainwater is caught and drained away by the middle unit in the group of three that make up the double lap. This is illustrated in Figure 7.3 where a unit has been removed in course C to expose the heads of two tiles in course B, which have been folded back to reveal the head of the cut unit in course A.

This all looks quite simple but there is one thing missing in the little mock-up – unit thickness. Thickness is not of great concern when using Welsh slate, which is thin, or even the modern Spanish and Chinese slate, which are of similar dimensions, but it is of concern when using thicker units such as clay or concrete tiles. Even slate from other UK sources tends to be slightly thick but just manageable without too much trouble.

Units are generally nailed at the head, two holes being provided. Not every tile has to have two nails. In sheltered areas it is common practice to nail with one nail entirely or even omit nails for tiles in the centres of large areas. The manufacturers are the best guide to nailing requirements – see also the comment on interlocking tiles later in this chapter.

Figure 7.4 uses some books to illustrate what happens when thick units are piled up three high at the overlap. The cut course is drooping and a gap shows between it and the next full course. This is a point of potential weakness. Eaves, like other edges of a roof, are very vulnerable in high winds, and a unit cocked up like that second course would be caught in a gust and ripped out. Apart from any weak-

Fig. 7.4 Eaves tile 'cocking up'.

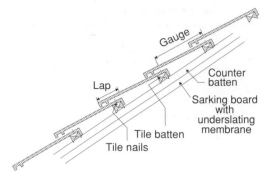

Fig. 7.5 Interlocking tiles on battens and counter battens.

ness, it is also unsightly and we will return to this aspect when we examine single lap fixing.

The answer to the problem is quite simple: cock the first course up. This is done by resting it on the top of the fascia board which is raised above the general level of fixing for the units, be that direct onto sarking boards or battens. This will be seen in the details in the last section of this chapter.

Slates, natural and manufactured, must all be nailed to the roof. Manufacturers of thick tiles as opposed to natural slates have overcome the problem simply by introducing a convex curve to the upper surface of the tile. This effectively lifts the centre of the tile up and away from the intervening head of the course immediately below and ensures that the tile sits down on the batten at the head. In addition, tiles may also be curved across their width, ensuring that all four corners sit down on the underlying surface. There is another theory put forward regarding the compound curvature. Perfectly flat tiles have a tendency to lift and fall in a high wind on an exposed roof. They make a noise, which has led to the term chattering being used to describe the phenomenon. The noise is one thing but the incessant hammering of tile upon tile starts to chip the edges and gradually larger pieces come away and so on. The double curvature is supposed to stop chatter.

Figure 7.5 illustrates single lap fixing, and Figure 7.6 is a photograph of the underside of a typical, concrete, interlocking tile.

In Figure 7.5 we show a schematic of interlocking tiles, the section being taken along the roof slope. The photograph in Figure 7.6

shows the tile top at the top. One feature can be seen that is common to all man-made tiles – nibs which hook over the tile batten and prevent the tile sliding down the roof. There are ridges running horizontally across the back of the tile. These are water stops and prevent water being blown up under the tile from the surface of the one below.

Figure 7.7 is a schematic view of the ridges and rebate on the side of these tiles. This is what stops the water passing down between the tiles onto the surface below.

On a roof not every such tile need be nailed or otherwise fixed to the battens. Rules for fixing vary from manufacturer to manufacturer

Fig. 7.6 Underside of an interlocking tile.

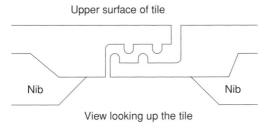

Upper surface of tile

Nib   Nib

View looking up the tile

Fig. 7.7 Schematic of side interlocking joint.

but a reasonable guide would be to nail the first three courses at the eaves, the last three courses at the ridge and three tiles in at every verge or abutment. In addition, every third or fourth course has to be nailed for its complete length. It used to be normal for all tiles supplied to have two nail holes preformed in their head. Many tiles are now made without holes as the manufacturers wish to have their tiles fixed with their own patent clips. See the manufacturers' websites.

Having set the scene it is time to look at each individual roof unit and consider those aspects of it and its use which make it unique.

Slates

Natural or man-made, slates are produced as relatively thin sheets of material in a wide range of sizes. Roofs are usually covered with same size slates, although **diminishing courses** are sometimes used for effect. There is no practical advantage to the technique. In essence it means that the length of the slates reduces as the courses go up the roof.

Slates must all be nailed to the roof timbers. They can be nailed direct to roof boarding, although this should be fairly thick natural timber rather than thin plywood or OSB. Alternatively they can be nailed to battens.

As well as the cut course at the eaves, there has to be a cut course at the ridges to maintain the double lapping. The cut edges are concealed with a ridge covering of clay or metal.

Slates can be **head nailed** or **centre nailed**. Head nailing means exactly that – two nail holes are made in the slate about 20 from the

top edge and used to take slate nails (see Appendix G).

Centre nailing is not exactly as it sounds. When the lap and gauge have been calculated for a specific size of slate at a particular pitch, it is possible to further calculate exactly where the head of the slate below is going to be. The slates are then pierced each side just above this point. Look at Figure 7.3 and in particular at unit B. Just above the B there is a dotted line which indicates the top of the cut course below. Centre nailing would be done just above this line – certainly more than halfway up the slate.

We have already mentioned the fact that wind will tend to lift units off a roof. The loose bottoms of the units are most vulnerable and as they lift they lever out the nails at the top of a unit. However, if the unit is centre nailed, the nails are in a better position to resist that lifting force. On the other hand, head nailing allows the slater repairing a roof to ease slates up and give access to the nails of the damaged slate below, cut the shanks and replace the slate.

Some slaters will not double nail slates and only head nail. The advantage here is all for the maintenance slater. As before, head nailing eases the access problem, and single head nailing further allows the slates to be swung aside, pivoting on the single nail. This gives even easier access for replacing slates or even inserting pipes, cables and so on into the roof structure.

Slates for the cut course are cut from whole slates. A minimum pitch for slates is around 25°. Head lap varies with the lap and gauge and the overall size of the slate, but generally 75 is a minimum.

Plain tiles

Plain tiles come as clay or concrete. Clay is usually a warm terracotta colour, while concrete tiles are usually coloured and also textured by the impression of coloured sands into the surface of the concrete. Laying them is done in exactly the same way as slating but always on battens. The correct term is **tile** hanging. Tiles can be hung simply by placing the nibs over the battens; however, the following rules regarding nailing must be applied.

Whether one or two nails are used depends much on local practice. The tiles generally have two nail holes. Double nailing of verges, eaves and at hips should be insisted upon where the roof is exposed, together with single nailing of all other tiles. Sheltered exposures allow single nailing of only two courses at eaves and ridge and single nailing of two tiles in from the verge, plus single nailing of every fourth course over the rest of the roof.

Moderate exposures allow double nailing of two or three courses at eaves and ridge and double nailing of at least three tiles in from the verge, plus single nailing of every third or fourth course over the rest of the roof. Tiles for the first cut course are cut from whole tiles. As well as the cut course at the eaves, there has to be a cut course at the ridges to maintain the double lapping. The cut edges are concealed with a ridge covering of clay bedded in mortar. A minimum head lap of around 75 is required. A minimum pitch for plain tiles is around 25°. Look at the websites quoted in the following sections and read what the manufacturers have to say about fastening tiles.

Interlocking tiles

Interlocking tiles are only manufactured in concrete and can be coloured and textured like plain concrete tiles; indeed manufacturers co-ordinate the colour and texture of at least some of their plain and interlocking tiles.

Figure 7.7 is a schematic view of the side junction arrangements on an interlocking tile. The ridges and gaps channel water down the grooves as well as reducing the air pressure as the wind blows across the tile face.

Laying is always done on battens, although fixing can be with nails or patent clips. Nailing is always through preformed holes near the head of the tile. Clipping is only done on one side of the tile but can be slightly further down the side of the tile from the head. The difference is so slight that it is hard to see any real advantage.

Like the slates and the plain tiles, interlocking tiles are laid to a pattern which breaks bond across the roof. There is no need for a cut course of tiles at the eaves but the bottom course can droop unless the eaves detail is correctly formed. There is no need for a cut course at the ridge but the ridge is concealed with concrete ridge tiles or metal ridge covers.

A minimum head lap of 65 is common but some low pitch types may require more. Pitch can vary from around 20° to the vertical.

For the best in technical information on tiles try the following websites: http://www.marley.com and http://www.lafarge.co.uk

Both sites cover clay and concrete, plain and interlocking tiles in a wide range of profiles and suitable for a variety of pitch angles.

Timber shingles

Timber shingles are not often seen in the UK but are a very practical form of roof covering and with the right timber can give as good a life as clay tiles. The main timbers used are oak and western red cedar. The shingles are split for logs to give a rectangular shape in which length is constant but width varies. All one length are used on one roof. Thickness varies, being thinner at the head of the shingle and thicker at the bottom or tail. This allows shingles to bed down on top of one another when being laid.

Shingles are always laid direct onto sarking and follow the double lap rules for giving a watertight finish. Nailing is generally copper but stainless steel could be an option now. Compatibility with the timber being used must be taken into account. Cut courses at eaves and ridge are cut from whole shingles as required.

Bituminous shingles

These shingles are made from a mineral surfaces roofing felt, a Type E felt being the most common, although an asbestos fibre felt might have to be used for fire regulatory purposes, i.e. proximity to another building, boundary, etc. They are made in strips of around six shingles, only the tails of the shingles being separated to make it look as though there are individual shingles.

The tails each have an adhesive bitumen dot on the underside protected by a peelable paper cover. As the strip of tiles is nailed into position, the strip is removed and the adhesive dot sticks to the course below.

Bit shingles are nailed to timber board direct, preferably a plywood or OSB sheet material. A strip of heavy roofing felt is laid at eaves as a cut course. This is well nailed to the sarking. Nails used are galvanised steel felt nails (see Appendix G).

These tiles are not particularly long lasting and should only be used in situations or on buildings where a relatively short life is expected, 12–15 years maximum.

Pan tiles

Up and down the east coast of the UK there has been a long tradition of trade with the Low Countries. Many ships left the UK with hides, salt, and so on and returned with other goods and sometimes in ballast with a cargo of roofing tiles – pan tiles. These were the original single lap roofing units with a special arrangement at the side junctions to keep the roof watertight. They also had a nib for hanging the tile on a batten. They were heavy so nailing was not considered necessary and therefore nail holes were not formed. In time, potteries on the east coast sprang up wherever a supply of clay and fuel could be found close together, and one of the first roofing tiles to be manufactured was the pan tile. Pan tiles are still made today but while local manufacturers abounded until the 1950s, only one remains today. For all the technical information and some great photographs try the web site at http://www.sandtoft.co.uk

Modern pan tiles are lighter than the originals and are now nailed or clipped in much the same way as interlocking tiles. One original feature about the nailing is that the nail hole is not in the face of the tile but through the nib. This means that the next tile up sits right down into the curve of the previous tile, there being no nail head to create a pressure point which would induce cracking in the upper tile.

Pan tiles are designed and laid to overlap on all four edges, which means that there is the possibility of a four-layer thickness of tile at every corner. This just does not work. Tiles laid this way do not bed down properly. The solution to this problem must have been arrived at many centuries ago and involved making the tiles with two opposite corners cut off at an angle of 45° – the solution is still used today. The website shows how it works.

The minimum pitch angle for pan tiles is 30° and the head lap should be 75.

Spanish and Roman tiles

These are frequently called **over and under tiles** due to the shapes used and the manner of laying them. The shapes are illustrated in Figure 7.8.

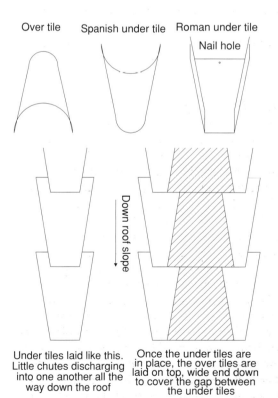

Fig. 7.8 Spanish and Roman over and under tiles.

The under tiles can come holed but only in the Roman version, which being flat lends itself to nailing to a timber roof board. These tiles are heavy so a thick sarking board is necessary. In the Spanish version there is only one shape and the manufacturers will not make half with holes and half without.

Although we are talking about Roman and Spanish tiles, they can be seen in the UK. A Cooperative supermarket in a little town in Fife has clay Roman tiles on its roof and the National Museum of Scotland in Edinburgh is roofed with clay Roman tiles. They present such a pleasant pattern of rounded ridges and deep channels that interlocking tile manufacturers make a mock Roman tile profile in all sorts of colours other than the original terracotta. Lest the reader think that these two roofs mentioned above are covered in concrete mock tiles, both buildings are far older than concrete interlocking tiles.

Millions of these tiles must be laid on roofs in Spain every month. And modern practice is to cast the pitched roof slab in concrete or clay partition panels and then lay a mortar screed. On top of this two coats of bituminous emulsion are brushed and the under tiles are bedded into a mortar layer. The over tiles are then bedded over the under tiles and the mortar joint pressed down and flushed off.

Edges and abutments

Learning about the different units used over domestic timber roofs is fine and we can quickly appreciate the differences and similarities in technique required to lay them over the general roof area. The tricky part is dealing with the edges of the roof surfaces and where these surfaces meet walls, other vertical tiled or slated areas, hips and valleys, and so on. The comprehensive text on the subject has yet to be written and is certainly not going to happen here. But this is not to say that readers could not work out for themselves how to cope with Blogg's Patent Interlocking Tile at a verge, a ridge or an abutment if they had seen a detail or even a photograph of the work done

in slate or plain tiles. The same techniques can be applied or adapted in this trade as in any other.

So the figures which follow are a selection of how to cope with a particular roofing unit in a precise situation. These details include not just the roof covering but also the roof structure. They are complete architectural, traditional details (using mortar for bedding the tiles). Much modern tiling now uses dry systems for finishing eaves, verges and ridges and every manufacturer has their own system. All the websites mentioned above have downloads using Adobe Acrobat to allow the user to print out complete catalogues of all the detailing, or just the pages which interest you. All three sites have good details. From there the roof tiling world belongs to the readers.

Figure 7.9 is a fairly complete detail of an eaves, with masonry wall and trussed rafter roof with sarking. Rainwater disposal is an important feature of all buildings in the UK, far more so than in many other countries where there is a higher rainfall. Here the fascia board provides a mounting for a half round gutter in uPVC. The down pipes are never shown on these details and barely feature on the elevation drawings, only to indicate their position. Gutters are laid to a slight slope, going down to the outlet into the rainwater pipework. Setting out the gutters like this is called **crowning**.

Figure 7.10 shows the detail of a ridge. Again the roof is a trussed rafter construction with sarking board and interlocking tiles on battens and counterbattens. Note that the membrane is taken up one face of the roof and placed over the ridge. Then the opposite face is covered and the membrane again taken over the ridge. The ridge tile is bedded onto the roof tiles with cement mortar which is usually tinted to match the tiles. The space between the tiles is also pointed with the same tinted mortar.

Figure 7.11 shows a verge. The detail is a section across the verge projection, the line of sight being parallel to the slope of the roof surface. The verge is on that same roof but with the wide projections at eaves and matching

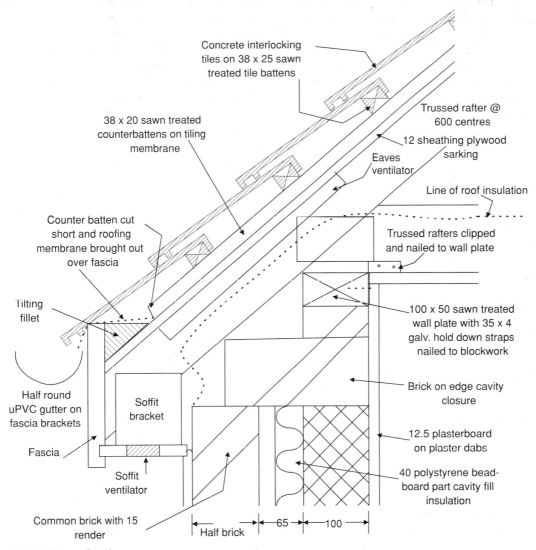

Concrete interlocking
tiles on 38 x 25 sawn
treated tile battens

38 x 20 sawn treated
counterbattens on tiling
membrane

Counter batten cut
short and roofing
membrane brought out
over fascia

Tilting
fillet

Half round
uPVC gutter on
fascia brackets

Fascia

Soffit bracket

Soffit
ventilator

Common brick with 15
render

Half brick

Trussed rafter @
600 centres

12 sheathing plywood
sarking

Eaves
ventilator

Line of roof insulation

Trussed rafters clipped
and nailed to wall plate

100 x 50 sawn treated
wall plate with 35 x 4
galv. hold down straps
nailed to blockwork

Brick on edge cavity
closure

12.5 plasterboard
on plaster dabs

40 polystyrene bead-
board part cavity fill
insulation

65   100

Fig. 7.9 Eaves detail.

verge; the idea of bedding the verge in cement mortar is not really feasible. It was generally only done when the projection was about 75 to 100 and an **undercloak** of asbestos cement sheet or of plain tiles was bedded on top of the masonry and then the roof tiles bedded on top of that. Here there is a wide projection supported on short lengths of timber – **outriggers** – nailed to the side of the first/last rafter. There is a soffit plate and **barge board**, the equivalent of the eaves fascia. The soffit plate is fixed to a batten nailed to the underside of

the outriggers and in a groove in the back of the barge board – sometimes referred to as a **berge** board. A further timber plate is cut to fit the underside of the projection of the roofing tiles – called **scribing** – and then nailed to the barge board with a mastic fillet on top. As it is fixed in position, the roofing membrane should be brought out and stuck between the plate and the tiles.

Figure 7.12 is a photograph of a gable peak just before the masonry skin was built round the house. The roof ladder can be clearly seen;

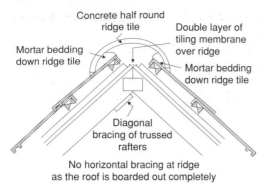

Concrete half round ridge tile

Double layer of tiling membrane over ridge

Mortar bedding down ridge tile

Mortar bedding down ridge tile

Diagonal bracing of trussed rafters

No horizontal bracing at ridge as the roof is boarded out completely

Tiles, battens, counterbattens and roofing membrane as Figure 7.9

Fig. 7.10 Ridge detail.

the barge board and soffit plate have been pre-finished in this timber frame kit. The roof covering edge is finished with a dry verge system, a series of plastic sections which are clipped onto the ends of the tiles – no mastic, no mortar, just push the plastic bits on and the job is done. What could go wrong with that? Well, quite a lot if the timberwork is not square or well finished; the roof tiler will have to follow the timber and then the plastic sections might not fit quite so well.

Figure 7.13 shows a detail of an abutment. The view is taken parallel to the sloping roof surface, just like the verge. The roof is still the same and it is only when the tiles are cut to fit just next to the wall that there is a lot going on under their ends. The heavy lines represent

Fig. 7.12 Verge tiled.

sheet lead. It is there to form a double gutter and is held into that shape by the counter batten on the sarking board, and then nearer the wall another length of tile batten is placed over the rafter. The lead is then dressed up the face of the angle fillet and up the timber plate where it is fixed with copper tacks. That forms the gutter. But rain could get down behind the lead in that upstand and this requires a piece of lead to cover the join.

A groove has been cut in the masonry wall – **a raggle**. Raggles are cut for all sorts of things such as concealing pipes, cables conduits, etc. as well as for tucking in ends of sheet materials such as the **lead flashing**. Before the lead flashing goes in, narrow lengths of lead strip are set into the groove, then the flashing is set in and rolls of lead strip are hammered into the groove, trapping the little strips and the flashing. The strips of lead are called **lead tacks**, not to be confused with the copper tacks. The little rolls of lead sheet are called **wedges** and as they are hammered into place they trap the tacks and the flashing. The raggle should then

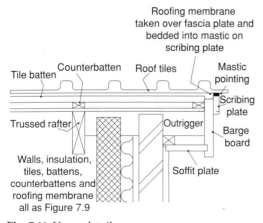

Roofing membrane taken over fascia plate and bedded into mastic on scribing plate

Tile batten

Counterbatten

Roof tiles

Mastic pointing

Scribing plate

Trussed rafter

Outrigger

Barge board

Walls, insulation, tiles, battens, counterbattens and roofing membrane all as Figure 7.9

Soffit plate

Fig. 7.11 Verge detail.

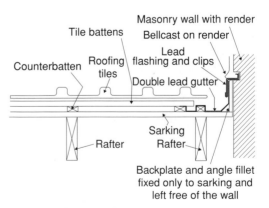

Tile battens

Masonry wall with render

Bellcast on render

Counterbatten

Roofing tiles

Lead flashing and clips

Double lead gutter

Rafter

Sarking Rafter

Backplate and angle fillet fixed only to sarking and left free of the wall

Fig. 7.13 Abutment detail.

be pointed up with mortar but when there is a render to apply to the wall, it will be render which is forced into the gap. The flashing is beaten down over the lead upstand so that they are neatly finished tight together and the protruding ends of the lead tacks are turned up against the flashing and hammered down flat.

The idea of putting in a double gutter is to provide an overflow channel in case the main gutter gets blocked by leaves or other debris – not much seen nowadays but a good idea.

8 Doors

Functions of doors and windows – obvious and not so obvious	182	Pressed panel doors	189
		15 pane doors	190
Types of door	184	Hanging a door	190
Ledged and braced doors	185	Fire resistant doors	193
Bound lining doors	185	Smoke seals	195
Flush panel doors	187	Glazing	196
Panelled doors	188	Ironmongery	196

Functions of doors and windows – obvious and not so obvious

Rather than repeat all the material from this section when we study windows, the comments will be made once here and a cross reference included in the next chapter.

Doors and windows fulfil some fairly obvious functions. There are statutory requirements for both doors and windows in the Building Regulations. These and some of the less obvious functions will be highlighted as the text unfolds.

Doors and windows are always more expensive than the portions of wall which they replace – a point we will return to on several occasions. According to a current edition of a price book, a fairly basic timber casement window with double glazing is 3.75–4 times more expensive than a brick/block cavity wall. A summary of the specifications is shown in Table 8.1. The rates may not be quite up to date or typical of your area but the comparison is fairly valid anywhere in the UK. Similar exercises can be carried out for all parts of a building, costing out alternative forms of construction or the use of alternative components and/or techniques. This is an important part of any building manager's, building surveyor's and quantity surveyor's work, although the latter is the acknowledged expert in this field.

To meet the maximum economic requirement of a client, the total area of doors and windows must therefore be kept to the statutory minimum. Aesthetics can dictate otherwise but always at a cost.

There is no requirement to have more than one entrance door to any house. Internally there does not appear to be any general requirement to have doors at all – holes in the walls would be sufficient. However, the need for privacy in bathrooms and bedrooms does dictate otherwise.

To provide natural lighting, most habitable rooms in a house require to have a door and/or window(s) with an aggregate glazed area either in wall or roof equivalent to **one fifteenth** of the floor area.

Generally, the provision of natural ventilation can be satisfied by use of a **trickle ventilator**. These can be nothing more than a thin slit cut into the top of a window frame although, like most simple things, lots of people have 'invented' the perfect trickle vent complete with insect screening and adjustable opening

Table 8.1 Comparative costing of wall and window.

Masonry wall		Qty	Rate	£0.00	Window		Qty	Rate	£0.00
Facing brick, PC £300/thou	m²	1.00	£75.00	£75.00	S/w casement, knot and prime by manufacturer				
100 mm aac block	m²	1	£35.00	£35.00	1200 mm × 1050 mm	m²	1.26	£240.00	—
Form cavity, 50 mm fibreglass, 3 wall ties/m²	m²	1.00	£18.25	£18.25	So for 1 m²:	m²	1.00	£190.48	£190.48
11 mm Carlite plaster	m²	1.00	£12.50	£12.50	Double glazed unit	m²	1.00	£154.00	£154.00
Mist & 2cts emulsion	m²	1.00	£3.24	£3.24	2 cts oil both sides and sill board	m²	2.20	£6.75	£14.85
					Ironmongery PC Sum	Sum	1.00	£25.00	£25.00
					Close cavity at jambs	m	2.00	£10.00	£20.00
					Pressed steel box lintel	m	1.20	£40.00	£48.00
					PCC slip sill	m	1.00	£28.00	£28.00
					Building in casement with mortar, fixing lugs and mastic pointing	nr	1.00	£40.00	£40.00
					DPC	m²	0.70	£17.00	£11.90
					Timber inner sill and bed mould	m	1.10		£0.00
					plaster on ingoes and soffit <300 mm wide and angle bead	m	3.00	£7.75	£23.25
					Mist & 2cts emulsion on ingoes and soffit	m²	0.30	£3.24	£0.97
				£143.99					£556.45
					Window dearer than wall by a factor of				3.86

which is put in the slit in the timber. These are usually placed in the window frame. The Regulations do not require windows to open for the purpose of ventilation. Otherwise a ventilator with an opening area not less than 4000 mm² is required. Note that kitchens and bathrooms require special treatment.

Most obviously, doors are required to allow access to a building or compartments within a building. External doors must exclude the influence of the external climate – wind, rain, heat and cold – and for that reason many external doors today incorporate a certain level of insulation. External doors must provide security, both when the property is unoccupied and at night when the occupants are asleep, but also a means of escape in case of fire. Security can be enhanced by using a steel face on the door but this is all for nought if the frame, hinges and lock(s) are not up to the same standard.

Internal doors provide privacy and perhaps security within compartments. They are frequently required to provide resistance to fire. Internal doors can be used to control the modification of heating or cooling regimes between compartments, and they can modify the passage of sound between compartments.

All doors can be used to provide or assist with natural ventilation. More usually, the junctions between external doors and frames are provided with seals to reduce the natural infiltration of air and therefore only 'ventilate' when they are opened. Internal doors are seldom sealed at jamb and head but frequently have draught excluders fitted at the threshold.

Doors can be fully or partly glazed to provide a source of natural light or to allow a view to the other side.

The provision of doors and windows poses a large number of problems, which often make conflicting demands on the solution adopted by the designer. For example, the desire to have a large area of glass to obtain a view must be tempered by the very obvious increase in heat loss occasioned by large areas of glass. On the other hand, triple glazing could reduce that loss to acceptable limits but at a large increase in construction cost. Wider windows in particular are more expensive than taller ones of the same area, but the wider window may suit the client's requirements.

Glazing of doors and windows decreases the general level of security. Double glazing with sealed units is supposed to increase the level. Triple glazing should therefore increase it even further. The use of toughened or laminated glass or polycarbonate sheet will increase the level even more. The use of these special glasses may be necessary on the grounds of safety.

Other problems arise with regard to maintaining the integrity of the wall structure at the junction with the frames to doors and windows. Poor detailing at junctions will allow rain and wind to penetrate the fabric of the wall to its detriment, and even into the interior of the compartment itself to the discomfort of the occupants.

Positioning of the openings in a wall can give rise to problems with the structural integrity of the wall. Instability can occur if openings are placed too near corners or in heavily loaded portions of the wall. Additional wall ties must always be built into cavity walls at all door and window openings.

The portions of wall over openings for doors and windows will require support. This is provided by building in lintels made from a variety of materials: precast concrete, prestressed precast concrete, and hot or cold rolled mild steel section, which may also be galvanised and/or coated with plastics. Adequate support for ends of lintels is vital. The provision of this support has already been studied in Chapter 5, Openings in Masonry Walls.

Types of door

Having looked at some of the functions of doors and windows, we will look now at the various types of doors available. Only the more commonly used types will be listed; the number of different kinds is very large, as a quick look at manufacturers' literature will show. Try these websites:

http://www.magnet.co.uk
http://www.johncarr.co.uk
http://www.jewson.co.uk

Doors can be grouped as:

☐ External doors
☐ Internal doors.

They can be further grouped as:

☐ Entrance doors
☐ Pass doors – doors allowing passage from one compartment of a building to another.

They can also be grouped as:

☐ Doors with a specific resistance to fire
☐ Doors with no particular resistance to fire.

We will first study the construction of doors without any requirement for a specific fire resistance.

No matter what the construction of the doors, they are always hung in a frame; on hinges; with a latch or lock to keep the door closed, with stops against which the door closes; sometimes with weather stripping or draught proofing round the edges; architraves to cover the joint between frame and wall; handles or 'furniture' with which to operate the latch and open and close the door; and more often than not now, a threshold plate or a weather bar.

That is a lot to remember. Taking it a bit at a time we will look at vertical and horizontal sections through some common door types, and then frames, adding more complexity until all is explained. Wall interfaces are partly included in Chapter 5 Openings in Masonry Walls, but this will be expanded on here.

The first question which comes to mind is 'What sizes do these doors come in?' Doors can be made any size you wish but bespoke joinery is expensive and there are so many manufacturers making a good range of sizes and types of doors at competitive prices from a variety of materials that few buildings incorporate bespoke doors and especially housing.

Doors are still made in both imperial and metric sizes. The more usual sizes for internal doors are: 1 ft 6 in × 6 ft 6 in, 1 ft 9 in × 6 ft 6 in, 2 ft 0 in × 6 ft 6 in, 2 ft 3 in × 6 ft 6 in, 2 ft 6 in × 6 ft 6 in, 2 ft 9 in × 6 ft 6 in and 626 mm × 2040 mm, 726 mm × 2040 mm,

826 mm × 2040 mm and 926 mm × 2040 mm. For external doors: 2 ft 6 in × 6 ft 6 in , 2 ft 8 in × 6 ft 8 in , 2 ft 9 in × 6 ft 9 in and 807 mm × 2000 mm.

Types of door we will discuss now are:

☐ Ledged and braced doors
☐ Bound lining doors
☐ Flush panel doors
☐ Panelled doors including 15 pane glazed doors
☐ Pressed panel doors.

Ledged and braced doors

The face of this door is made from tongued and grooved boarding nailed vertically to at least three rails, the **ledges**, and with two diagonal braces toe jointed into the rails. Match or floor boarding is frequently used from 12 to 20 mm thick. Boarding should be nailed to the ledges and braces with round or oval brads long enough to penetrate fully and be clenched over.

A ledged and brace door is illustrated in Figure 8.1. These doors are only suitable for sheds, outhouses, garden wall gates and ancillary buildings with minimal security requirements.

To hang these doors, usually T or crook and band hinges are used screwed to a wooden frame or with one end of the hinge built direct to masonry jambs. Note to which edge of the door the hinges and fasteners are fixed. The bracing is put in to prevent the lock edge of the door from dropping. Fasteners can be slip or tower bolts, Norfolk or Suffolk latches, hook and eye or hasp and staple. Some basic ironmongery – hinges, locks and latches – is illustrated in the last section of this chapter.

Bound lining doors

Bound lining doors look to the uninitiated much like ledged and braced doors, but there are a number of crucial differences:

☐ A frame is used all round the outside edge of the door.

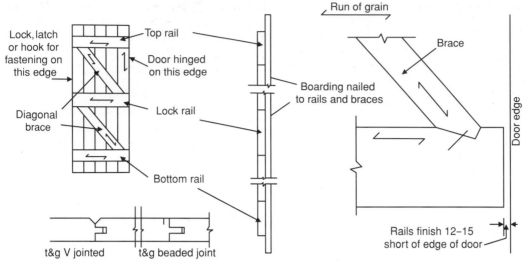

Fig. 8.1 Ledged and braced door.

- The t&g boarding is 10–12 mm thick and is always V-jointed or similarly shaped, and it is nailed to the frame all round the edge.
- The edge frame is generally made of 38–45 mm thick timber and diagonal braces are set between the three horizontal 'rails' which form part of the frame.
- The joints of the frame are 'through morticed and tenoned'.
- The braces are morticed into the top, middle and bottom rails rather than plain toe jointed.

These are much more robust doors and with good ironmongery will fulfil many requirements where basic security is an issue. It was the art of constructing such frames and filling them with light boarding or panel work which gave the craftsmen who made them the title joiners. This was a secret known only to the old guild members and distinguished them from the carpenters who made things from solid timber.

To fully understand the frame of this and panelled doors the student must know about morticed and tenoned joints (see Appendix C).

For external doors it was customary to coat the mating faces of the joints with white lead or red lead paint in the belief that this would preserve the timber in the joint from rotting. The lead paint dried out fairly rapidly and the timber still rotted. Doors with bare joints allow the wood to swell and generally give a watertight joint – and still rot. Timber can now be treated to prevent rot but the timber will still take up moisture and swell to keep joints tight.

There may be some confusion in the student's mind on reading of this door being made with a **frame** and then hung in a **frame**. These *two* frames are required; the first is an integral part of the door, the second is made, supplied and fixed quite separately.

Figure 8.2 illustrates a bound lining door. Note that the lock and bottom rails are thinner than the other frame members to accommodate the thickness of the lining. The lining is nailed with round or oval brads to all the frame members. Nails do not penetrate the frame and are driven at an angle to the face of the door The heads are punched below the surface and filled over before decoration.

The door is hung on one or one and a half pairs[1] of hinges, a single pair being 150 mm and a pair and a half being 100 mm.

---

[1] Hinges are always bought, specified, talked about, fixed etc. in pairs. Three hinges are therefore a pair and a half. The terminology apples to all types of hinges.

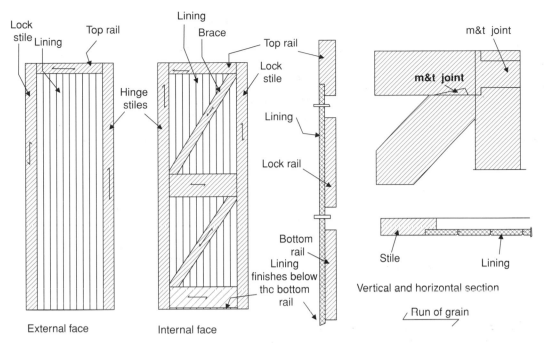

Fig. 8.2 Bound lining door.

Note that the position of the brace determines which is the lock stile and which the hinge stile. For a door of the opposite hand, the brace must be fitted the other way round. Most types of lock or latch can be fitted to this door but the mortice and tenoned joint of the lock rail should be avoided when fitting any lock or latch, but particularly mortice locks.

Flush panel doors

These doors are made from two sheets of principally plywood or hardboard (although any sheet material can be used) separated by a **core**. Cores can be **solid** or **cellular**. Solid cores are made of timber strip or sheets of particle board, various densities being used. Cellular cores are made from an 'egg box' construction of timber strips or cardboard. They can also be made from sheets of particle board which have holes running the height of the door.

The flush panels are glued only to the core material. Wire staples are occasionally used in the core but never in the panel face material. Cellular cores make the lightest doors but offer little security and can only be used as **pass doors**[2]. Solid cores make the heaviest doors, which do offer security and are used for external doors if the glues used are weatherproof.

The long edges of flush panel doors are nearly always covered with a thin strip of hardwood. The short edges can also be covered if required. This cover is known as a **lipping** and it is always glued on without any nails or screws.

Doors with cellular cores of egg crate construction have an extra edge of solid timber – normally softwood – all round between the two panels. This timber is simply butt jointed at the corners. A solid block of softwood is also included at both sides of the door where the lock or latch would usually be fixed. Some manufacturers fix only one lock block in place and mark the edge of that door. Sometimes they mark the wrong edge! Better manufacturers fit two lock blocks, then it doesn't matter – but it can be good for renovation work as the

---

[2] A pass door is another term for an internal door, generally without any security arrangements.

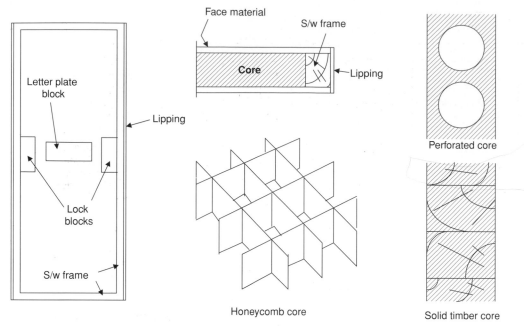

Fig. 8.3 Flush panel door.

door can be hung the other way round without turning it upside down. The edging and the blocks allow ironmongery to be secured.

Figure 8.3 shows the construction of flush panel doors.

Honeycomb cores are made of thin wooden strip or cardboard strip, half checked at all joints. Cores are glued to the S/w 'frame' and to the face materials. Timber honeycomb cores are frequently 'let into' slots in the S/w 'frame'.

The S/w 'frame' is only butt jointed and glued at the corners. S/w 'frames' are not used with solid timber cored doors.

Lock and letter plate blocks are only glued in. Light staples may be used to hold them in position during manufacture.

Lippings are generally of hardwood and are glued in place. Plywood and hardboard face material can be left for painting and plywood can be veneered, pre-finished if required.

Panelled doors

Panelled doors are constructed by forming a frame generally of 38 to 45 mm thick timber

inside which thinner layers of timber, plywood or veneered particle board or even glass or mirror are set into grooves cut in the inside edge of the frame. Most usually the edges of the frame, where the panels are **let in**, are **moulded** to improve the appearance of the door. The most common shape is the **ovolo moulding**, shown in Figure 8.5, although you may find older doors have variations on the ovolo or the **cavetto** moulding.

The panels can be any thickness up to the full frame thickness but are usually 10–15 mm thick in modern construction. The faces of the panels can be carved. The number of panels can vary but two, four or six is usual. Panels can be shaped by reducing the thickness at the edges. The reduction can be a simple rebate or a splayed rebate. The back of the panel can be left plain or can be shaped differently from or in the same way as the face. It is not uncommon to have the upper panel(s) glazed and the lower panel(s) solid.

The joint between frame and panel is sometimes covered by the addition of a shaped or 'moulded' length of matching timber. This is called a **Bolection moulding**. There is no one

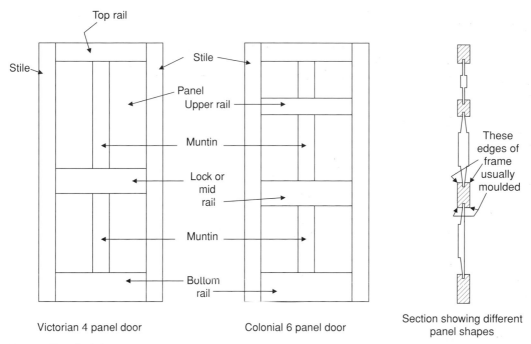

Top rail

Stile

Stile

Panel
Upper rail

Muntin

Lock or
mid
rail

Muntin

Bottom
rail

Victorian 4 panel door

Colonial 6 panel door

These
edges of
frame
usually
moulded

Section showing different
panel shapes

Fig. 8.4 Panelled doors.

profile which bears this name, only the fact that it is rebated on the back to fit over both frame and panel. Bolection mouldings can be fitted on one or both sides of the door, although one side is normal.

Figures 8.4 and 8.5 show panel doors of different types and their construction.

The vertical members called stiles are further differentiated when the door is hung, being called **hinge stile** and **lock stile** respectively. Members are morticed and tenoned

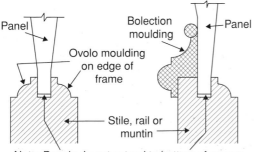

Panel

Bolection
moulding

Panel

Ovolo moulding
on edge of
frame

Stile, rail or
muntin

Note: Panels do not extend to bottom of groove.
This allows panel to expand/contract independently
of the frame

Fig. 8.5 Fitting panels into frame. Bolection moulding.

together in good work, dowelled in cheap work but now very common. In high class work all the mortices are 'blind'. **Muntins** are stub tenoned into the rails.

Pressed panel doors

Pressed panel doors are similar to flush panel doors in that they have a S/w 'frame', to each side of which is glued a sheet of hardboard which has been hot pressed into the shape of a four or six panel door. The 'frame' is generally just large enough to accommodate the screws for the hinges, i.e. no more than 25 long. A further feature is the impressing of a timber grain into the surface of the hardboard. It is done quite well with the grain running in the correct direction in what are supposed to be separate timbers.

Lock blocks are fitted, though not necessarily both sides, with the same consequences as before. Only glue is used in assembly and there are no lippings. Cheap and cheerful, it is usually supplied with a coat of white primer but some are not always easy to decorate.

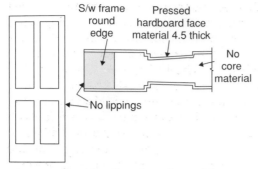

Fig. 8.6 Pressed hardboard mock panel doors.

Figure 8.6 illustrates a pressed panel door.

15 pane doors

15 pane doors comprise a frame of two stiles, a bottom rail and a top rail, the space inside that frame being divided into 15 equal panes, 5 × 3, by the use of **astragals** or **glazing bars**. The construction of the frame is similar to the panelled doors already described, with ovolo mouldings generally used on one side of the frame edges, with a plain rebate and loose **beads** on the other side to receive the glass.

Glazing can be in putty, glazing compound, velvet, wash leather or silicone glazing compound. The use of compounds is based on the idea that a small amount of compound is placed in the rebate in the frame or astragal and the glass is pressed into it. A little more compound is added and the beads fixed in position with glazing pins – small, fine, round wire, lost head nails. These nails can be hammered in but the proximity of the glass has brought in a simple invention, the pin push, which makes broken panes a thing of the past. It is a piston with a magnetic tip running in a cylinder just large enough to take the glazing pin, and the whole thing mounted under a large round push pad. Pop a nail into the cylinder, put the end of the cylinder on the bead and push. The bead is nailed in place in a few pushes. A word in favour of silicone compound here: it sets and sticks to both glass and timber, and it doesn't shrink or dry out so the glass never comes loose and rattles when the door is opened and closed.

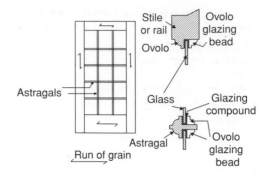

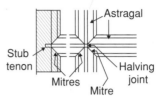

Fig. 8.7 15 pane glazed door.

The astragals are made with double ovolo mouldings and double rebates as shown. The ends of the astragals are reduced to rectangular section tenons and let into mortices cut into the stiles and rails. The mouldings on the astragals match those on the stiles and rails and these are mitred at the morticed and tenoned joints. Junctions of astragals are generally half checked, with the mouldings mitred.

Figure 8.7 illustrates a 15 pane door.

Hanging a door

A door is not fixed or fitted or built in, it is **hung**, generally on hinges, sometimes in patent devices for sliding, folding etc. doors. Ledged and braced doors are generally hung on **T hinges** and if of heavy construction on **crook and band hinges**. The other doors we have described are usually hung on **butt** or **edge** hinges.

The independent frame in which the door is hung can be made up in a number of different ways:

☐ Plain with a **planted stop**. This type of frame is often referred to as a **casing**. In older forms of construction a sawn timber frame was fixed into the opening and the casing fastened to that but that practice has long been dropped largely due to the cost of the additional timber.

☐ **Rebated** to provide an integral stop.

The joints can be **plain housings** or they can be **morticed and tenoned**. A **threshold plate** is common and always supplied when buying in ready-made frames. Figure 8.8 illustrates both types of frame.

Door frames rebated to provide a stop for the door are generally associated with good quality work but are more frequently used for external doors where the frame is fixed according to English practice.

Planted stops are more commonly used for internal work, short stops being used in cheaper forms of construction. Use of a planted stop for external door frames set in a rebated or checked reveal is shown in Figure 5.27. There the stop also becomes a door facing and is generally termed a stop/facing.

Figure 8.9 illustrates the use of a full width casing and stop at an internal door, while Figure 8.10 is a photograph of the opposite side of that door with a more decorative architrave which covers the joint of the frame with the wall finish and the complete door frame.

The door is hung on two or three hinges. Hinges are always referred to as being in 'pairs' so a door is hung on a 'pair' or a 'pair and a half' of hinges. Hinges are always screwed to door and frame. Joiners will always sink edge or butt hinges into the edge

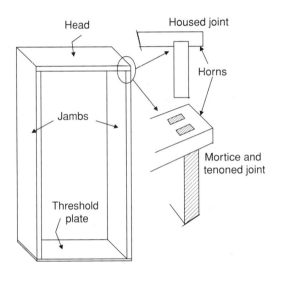

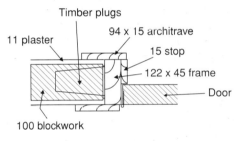

Plan section through hinge stile of door, frame, stops and architraves

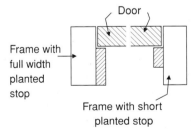

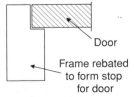

Fig. 8.8 Door frame construction.

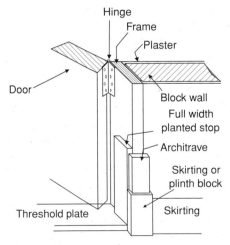

Fig. 8.9 Hanging a door – arrangement of all the timbers.

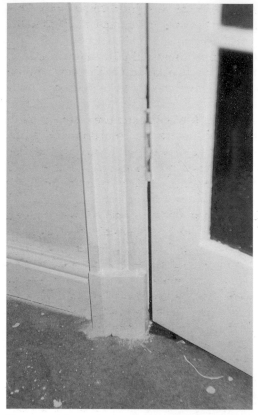

Fig. 8.10 A hung door.

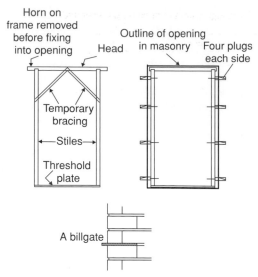

Fig. 8.11 Fixing door frame using plugs or billgates.

of the door. They are sunk into both door and frame in good quality work.

In Figure 8.9 the frame is shown fixed in the opening to a set of plugs or dooks driven into the blockwork. The width of the frame is set by measuring the thickness of the blockwork and adding the nominal thickness of the plaster both sides. The frame is fixed in place, the plaster is applied and finishes applied to the frame, then the door is hung and the stops and architraves added.

Figure 8.11 shows the fixing of a frame in such an opening:

☐ The frame is made up as the left hand drawing; it is squared and braced.
☐ The plugs are driven into the mortar joints and left protruding from the masonry.
☐ The door frame is offered up to the opening, i.e. it is held against the ends of the plugs, the jambs are set plumb with a builder's

level and the plugs marked with a pencil at the back of the frame.
☐ Plugs are cut off on the pencil mark.
☐ Door frame pushed between the ends of the plugs and nailed to them once it is plumb with the wall faces each side.

An alternative – but very inferior – fixing to the driven plugs is the use of billgates. Pieces of wood 10 thick and the same width as the wall and about one brick long are built into the wall and left protruding. Fitting the frame to them is the same as for plugs. Problems arise when the billgate dries out and shrinks, losing its grip in the wall. Modern fixings tend to use framing anchors and packing pieces to keep the frame plumb inside the opening.

One pair of hinges is generally sufficient for light internal doors. For external or heavy doors a pair and a half should be used. Hinges should be 100 mm for all pass doors and may be 100 or 150 for external doors. Consider solid brass or stainless steel hinges for external doors. Plated hinges are not really suitable for external doors. On heavy doors, brass hinges should be 'washered'.

Hanging a pass door on a pair of hinges allows the joiner to set the alignment of the hinge pins so that as the door opens, the lock stile rises, allowing it to clear carpeting or

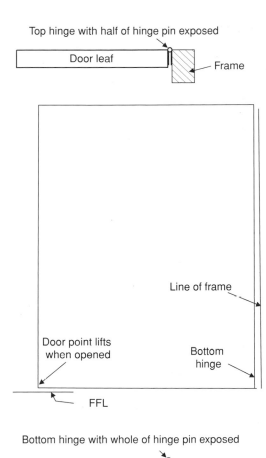

Top hinge with half of hinge pin exposed

Door leaf

Frame

Line of frame

Door point lifts
when opened

Bottom
hinge

FFL

Bottom hinge with whole of hinge pin exposed

Door leaf

Frame

Fig. 8.12 Making the door rise to clear a floor covering.

other floor covering. On moderately heavy doors this will also give a measure of 'self closing'. It is not really feasible to try this with more than a pair of hinges, nor is the self closing feature a substitute for a proper door closer where that is required.

Hinges are described as being 'full' or 'half ball'. Figure 8.12 illustrates the idea.

Fire resistant doors

Doors with a designed resistance to fire are required in quite specific circumstances:

☐ An external quality door would be used in a common entrance stair

☐ An external quality door would be used between a house and a garage attached to the house

☐ An internal quality door would be used in a three-storey house for each room opening off the staircase.

That list is not exhaustive but serves to illustrate where fire resistant doors might be used. Fire resistance is measured as the time for which the door, its frame and stops will resist the passage of flame, hot gas and heat transfer.

Before going any further, one material which is new to fire resistant technology must be explained and that is intumescent material. Intumescent material is a plastic which when heated will foam up to fill a gap or provide a protective layer over another vulnerable material. For example, many exposed steel frames are now painted with an intumescent paint. In the event of fire, the plastic skin foams up and insulates the steel from the heat of the fire and delays any distortion it might suffer and which could lead to failure of the structure. In doors and door frames, strips of intumescent plastic can be inserted into grooves in the components and in the event of fire they swell up, sealing off the gap round the door. Two types of this seal are made. One exerts pressure on both door and frame, keeping the door in place and also preventing it from warping under the heat. One merely seals and insulates between the two surfaces. The longer periods of resistance required benefit from the use of pressure seals.

So what makes a door fireproof? Well, there is nothing which is fireproof. That is an absolute which cannot be achieved and is why we talk of fire resistance, and measure that resistance in terms of the time delay from the start of the fire to the time when smoke, hot gasses, etc. will break through the door to the danger of life – a **breach**.

Doors are rated FD20, FD30, FD45, FD60, FD90 and FD120, which all fall within the requirements of BS 8214, *Fire door assemblies with non-metallic leaves*. FD means fire door and the number is the rating in minutes. Faced with a door one can tell what its resistance is by looking on the edge of the door for a colour coded plastic plug driven into the timber, or some other form of label attached to the door. The plug, for example, has two colours: a basic or background colour with a centre core colour. Core colours can be red, green or blue. Background colours are white, yellow, pink, blue, brown and black.

The British Standard has an explanation of the coding system in Table 1 and from this can be gleaned the facts that:

☐ Intumescent seals should be fitted for certain ratings at the time of hanging the door – red core with all the ratings above given a unique background colour
☐ Intumescent seals should be fitted for certain ratings at the time of manufacture of the door – green core with all the ratings above given a unique background colour
☐ The blue core only has two ratings:
  ▪ no intumescent seal fitted – blue core, white background, giving an FD20 rating;
  ▪ intumescent seal fitted – blue core, white background, giving an FD30 rating.

While it is certainly not the little plastic plug, what gives a door the required resistance? it is its construction using materials which have built in fire resistance and the way the frame and stops are built into the building, and finally the way the door is hung and secured in a closed position by hinges and other iron-mongery.

The door

Materials of construction cover the face materials, the frame materials, core materials and glues:

☐ Face materials can be any timber-based sheet material. The Standard states that any face material is likely to be consumed fairly rapidly in a fire.

☐ Frame materials refer to the structural frame of the actual door and although nothing is given in the Standard, it can be assumed that similar criteria will apply to it as they do to the frame in which the door is hung – straight grained timber free from knots, shakes, splits and wane. More about this when we look at the frame. Nowhere does the standard show a preference for softwood or hardwood, although one must assume, on the basis of the evidence for other frames, that hardwood would be used in doors with longer ratings.
☐ Core materials can be any timber or fibre-based sheet material from flax board to solid timber laminations, and these will generally achieve ratings up to FD60. For ratings above FD60 and sometimes for FD60 itself, the core must incorporate inorganic materials such as cement bound mineral fibres or plasterboard layers. In every case the integrity of the core is dependent upon the glues used within it as well as attaching it to the face material. The core thickness is, of course, increased to give an appropriate rating, with doors of up to 55 thick not uncommon.
☐ Glues such as PVA and similar glues may prove to be effective in doors rated up to 30 minutes but beyond that time a WBP glue should be used. The glue is important, for once a face has been burned away, the glue line between the remaining face and the core is responsible for keeping the core in place.

The method of construction must ensure that the component parts are a good fit, for if gaps are left between the face materials or there are faults in frame or core, there will be **burn-through**.

The frame and door stops

For timber ratings FD20 and FD30 there is generally no need to use anything other than softwood but it must be straight grained and free from knots, shakes, splits and wane. For FD60 and beyond, hardwood would seem to be preferred. In each case the density of the timber is the important factor, softwood of at least 420 kg/m^3 and for hardwood 650 kg/m^3. Straight grained timber is less likely to warp and twist in the heat of a fire.

Defects in timber – knots, splits, shakes and wane – introduce holes in the general construction, and holes are weak points. Fire can break through at these points much earlier than the general rating expected from the assembly – a **breach**.

The method of construction must ensure that component parts are a good fit without gaps and voids. Selection of good quality timber is the important factor in this.

With the introduction of intumescent seals, the necessity to have deeply rebated solid frames or extra thick planted stops screwed in place all seems to have gone. Now stops of 12 or 15 material may be planted and fixed with nails. The stops material should match the frame material. The fit to the wall in which the frame is fixed is more important than the stops.

Architraves now take over as an important feature. They should match the frame in terms of material and should overlap the frame and the adjacent wall by at least 15. This applies to both masonry and timber stud walls covered with plasterboard. If there is a gap of more than a few millimetres between back of frame and masonry wall, this should be filled with mineral wool or glass fibre quilt. No density is given but it is imagined that it should be packed in quite tightly. Architraves can be dispensed with if the masonry opening is a good fit to the frame, but the gap requires sealing with an intumescent mastic both sides.

On the matter of rebates in door frames and planted stops, there is one more thing which can be done to improve the integrity of the assembly and that is to double rebate or **over-rebate** the door. Figure 8.13 shows how this is done.

All of these precautions are carried out round the perimeter of the door *except* on the threshold. Testing has long ago ascertained that the threshold is far less susceptible to the effects of fire than the rest of the door

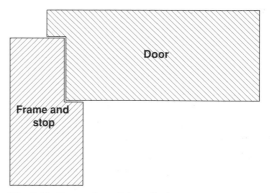

Fig. 8.13 Over-rebated door in frame.

and frame. Gaps between door and threshold should however be kept down to 3. Smoke seals might have to be fitted in the bottom edge of the door.

Smoke seals

Smoke seals are different from the intumescent seals. They are about containing the spread of smoke from one compartment to another. For example, there might be a common stairway rising through a block of flats. Each corridor joining that stairway would have a fire door, but not only would these doors have a fire rating, they would also have to have smoke seals even if there were intumescent seals fitted to make them comply with their rating. Smoke seals are about stopping **cold smoke** or smoke at ambient temperature spreading into escape areas etc. In our mythical block of flats, fire on one floor or in one flat would not break through the fire door for some time, but smoke could pass if the intumescent seals were not activated. This smoke could prevent escape down the stair or harm people attempting to escape.

Smoke seals can be of the wiper style or the brush style. Both are made of synthetic materials. Figure 8.14 illustrates both.

In narrow doors it might not be possible to accommodate both a smoke seal and an intumescent seal in the edge of the door, but there is nothing to prevent one being in the door and the other in the frame.

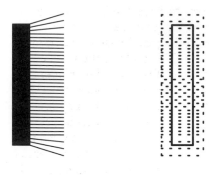

Brush seal – man-made fibres set in a hard plastic backing strip

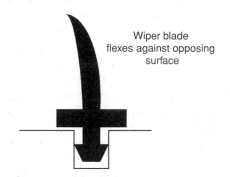

Wiper blade flexes against opposing surface

Other end of seal fits into groove in door or frame

Fig. 8.14 Wiper and brush smoke seals.

Glazing

So far the discussion has centred on doors and door frames, but the Standard goes on to discuss glazed portions of the door and glazed sidelights and so on – the whole thing becoming quite complex. In summary, glazing in a fire door needs a lot of care and attention. When a hole is cut into a door the core is exposed and this core edge requires protection from the effects of fire. Also, core materials do not often give ideal conditions for mechanical fixing – nails and screws – so that bead style solutions to fixing glass in an opening are doomed to fail. There are proprietary systems of high resistance rated materials which can be used to hold glass in such openings, and they

are frequently used in conjunction with a solid timber sub-frame set between the face materials. The glass itself has little resistance to fire and early attempts centred on reinforcing the panes with a wire mesh. Modern glasses are available which do resist the effects of fire. Of course, one of the effects of glass is to allow radiant heat to pass through the door and this can be so intense that it can set inflammable materials alight.

Ironmongery

This could make a book in itself, and a few have been published. If you can ever lay hands on an ironmonger's catalogue take it with thanks; we mean, of course, an architectural or builders' ironmonger.

Among the many items made and sold in a huge variety of styles, materials, colours and complexity, a few have been singled out here for mention, more in the hope that the correct terminology can be learned than that the complete range be described. As the particular requirements for fire resistant doors come up, the reader's attention will be drawn to the point being made. Before getting into the list, however, it should be mentioned that ironmongery is always fixed with matching screws. They may be countersunk, raised or round-headed, but the finish on them always matches the ironmongery.

For doors we might require:

☐ Hinges
☐ Latch or lock or both
☐ Bolts
☐ Hinge bolts
☐ Letter plate
☐ Weather bar
☐ Door furniture.

Hinges

Hinges come in a variety of styles but for the doors described earlier we would use **T-hinges**, **edge** or **butt hinges** and **crook and band** hinges.

T-hinges come in a range of sizes which describe the length of the leg of the T. They can be black japanned finished or hot dipped galvanised. A 300 hinge would be adequate for the ledged and braced door of light construction.

Crook and band hinges are only used for heavy doors. They also come as black japanned or hot dipped galvanised. The crook part can be supplied for building into masonry, or driving into a predrilled hole in masonry or timber, with a bolt end, nut and washer for fixing through timber and with a plate for screw fixing to timber.

Edge or butt hinges are the choice for entrance doors and pass doors. They come in a range of sizes from the impossibly small to the quite large, 20–150, the size being the length of the plate screwed to door or frame. Butt hinges are preferred on fire doors but the screw nails should be a minimum of 38 long. No recommendations regarding gauge of screw are made but 4.5 mm or No. 8 screws would be advisable.

Figure 8.15 illustrates these hinges and Figure 8.16 is a photograph of a few: T-hinges, galvanised and black japanned, and the edge or butt hinges in 50, 100 and 125 sizes. The white hinge on the bottom left is a **rising butt hinge**. This hinge has one leaf which raises up the cam shape you see round the pin as the door is opened. This hinge is in white nylon with a stainless steel pin.

Latches and locks

Latches and locks can be confusing terms, especially when both functions are combined in a single unit. Basically, asking for a latch will give you a unit which pushes a bolt into a keeper in the frame when the door is slammed, and to release it you turn the door knob or depress the lever handle. A dead lock requires that the door be pushed shut and then a key used to push a bolt into a keeper in the frame. To release it a key is required to pull the bolt out of the keeper. A lock combines both these features. There is a bolt operated by a key both to open and to close, and there is a bolt which will slam shut but requires the handle to be turned or the lever depressed to release it.

Locks which use keys can be of two types: lever operated or cylinder operated. The cylinder operated types divide into two more sub-species – the Yale style or the Euro cylinder style. Of all of these the lever operated style is the oldest but unless there are at least five levers, it lacks security. The key style is familiar to everyone – a long(ish) bar or rod with a handle at one end and a knobbly bit at the other. The knobbly bit is cut into a distinct pattern, the idea being that the different levels of knob each lift a lever by a discrete amount. If all the levers are lifted by the correct amount the bolt will be moved as the householder continues to turn the key. Five levers make it difficult for burglars to lift all the levers and move the bolt to gain access to the property.

Cylinder locks operate on a somewhat similar principle. The key is a flat bar with flutes down the side and a saw tooth pattern on one edge. The fluted pattern on the side of the flat bar is designed to prevent other manufacturers' key blanks from being used but it does contribute to the security by making things awkward for any lock picker. Without a key a series of spring loaded pins in the cylinder stick out into corresponding holes in the body of the lock. As the key is inserted the saw tooth pattern pushes up the spring loaded pins a small amount which depends on the size of the tooth or gap under them. If the correct key is inserted, the pins align in such a way that they are free of the mating hole and the cylinder will turn. Four, five and six pin cylinders are available, but five or six are recommended for security.

Yale-style cylinders are forever associated with the **night latch**, which at its most basic was a latch on the way out of the door – just pull the door and the latch slammed shut, but needed a key to get back in – unless someone had pushed the little lever or knob on the latch casing which stopped the bolt moving! These devices are installed in millions of homes but now they are considered a security risk rather than an asset. One benefit they do have: in the event of fire there is no need to hunt around

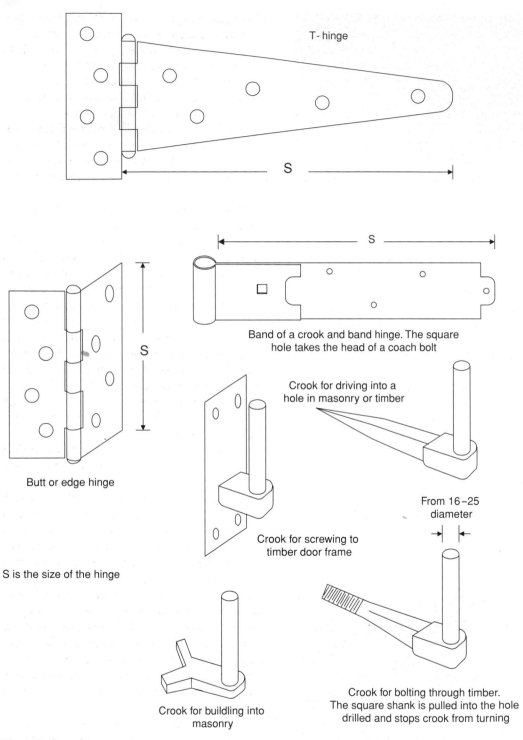

T- hinge

S

Butt or edge hinge

S is the size of the hinge

Band of a crook and band hinge. The square hole takes the head of a coach bolt

Crook for driving into a hole in masonry or timber

From 16–25 diameter

Crook for screwing to timber door frame

Crook for buildling into masonry

Crook for bolting through timber. The square shank is pulled into the hole drilled and stops crook from turning

Fig. 8.15 Some hinges.

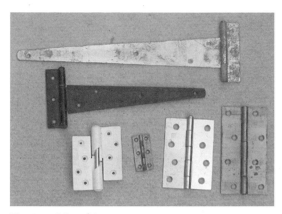

Fig. 8.16 More hinges.

for a key to make your escape through the door. A lot of people have died looking for the door key and Fire Officers everywhere now recommend that any lock fitted to your entrance door should have a **thumb turn** fitted on the inside. Fitting a key-operated dead lock does not allow this but fitting a Euro cylinder lock does.

Figure 8.17 shows a variety of locks. Starting at the top left and moving along the top row of units there is a bunch of cylinder keys for the Euro cylinder and Euro lock to the right. To the right of the Euro lock is a cylinder by Legge, and the two items below that are the keeper and night latch which are used with the cylinder. Both the night latch and the keeper are face mounted, on the door and door frame respectively. The hole for these cylinders is standardised at 32. At the extreme right of the

photograph are a tubular latch and its keeper. This simple latch is used on pass doors. Simply bore a hole in the edge of the door, take out a little rectangular recess for the front plate and screw the latch into position. If you remembered to drill a hole through the door for the spindle, fit a handle. The keeper is pressed steel and has to be fitted to the door frame with a recess for the latch bolt. All metal versions of this cheap latch can be very effective and quite long lasting, those with plastic in their construction much less so.

On the other side of the night latch is a five lever mortice lock which uses a conventional key. The hole for the latch spindle is clearly seen, as are the dead bolt for the lock and the latch bolt. This lock and the Euro lock are morticed into the door with holes cut for the cylinder or conventional key to pass through, and the spindle for the latch. Above the five lever lock is a fore-end plate and above that the keeper. All mortice locks and dead locks have similar fore-end plates. The keepers of higher security locks are of welded box construction and are fixed into a mortice in the door frame, and the cover plate is put over the exposed edge of the lock.

During all this discussion we are assuming that these dead locks and ordinary latches are of the mortice variety – i.e. a slot or hole is cut in the edge of the door and the unit slid into place and secured with screw nails into the edge of the door. Older locks were called **rim locks** and were fitted to the inside face of the door and the adjacent edge. It is still possible to purchase these older style locks but not nearly so large as the one illustrated in Figure 8.18. It is of heavy brass, 270 long × 150 high and 40 thick. It incorporates a lock, a latch and a sliding bolt, which is only operable from inside the house. It has one lever, so in terms of modern security requirements it is very much below par. A complicated ward [3] is

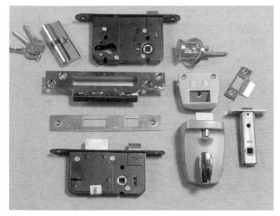

Fig. 8.17 Some locks and latches.

---

[3] A ward is a system of metal strips inside the cover of the lock case, which prevents a solid key from turning in the lock unless the key has a matching pattern cut in its blade. Wards are still used today together with anything from three to six levers to provide security; five lever locks are recommended for external doors with or without wards.

Fig. 8.18 An old rim lock.

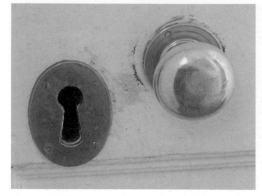

Fig. 8.19 The escutcheon plate for the lock in Figure 8.18.

incorporated, which was a security feature at the time it was made – around 1820 – that was supposed to stop the use of picks. Another interesting feature is the round head screw nails used to fix the lock to the door. They look like brass but it is only a cover of brass squeezed tight onto a heavy round headed steel screw. Once this cover is fitted, the slot is cut in both brass and steel. That may seem to be a lot of trouble, but there are a couple of points which follow. First, steel screws are much stronger and the slot is less likely to be damaged inserting a heavy screw if it is made from steel. Second, brass plating wasn't possible 180 years ago so covering with brass was the only way to get screws to match the ironmongery. The key has long since disappeared but it would have been massive and not at all easy on the trouser pocket.

A further feature of locks is that there is frequently a separate piece of ironmongery to reinforce the edge of the timber round the keyhole. It is called an **escutcheon plate**. This reinforcement can also incorporate a hinged flap to reduce draughts and provide privacy, especially when fitted on the room side of the door. With mortice locks it is usual to fit an escutcheon plate on the inside of the door with a flap, and one without on the outside. Sometimes this is unnecessary if the handles fitted have a large back plate which includes a reinforcement for the keyhole cut-out. Not many of these incorporate a flap. If you look closely at the photograph in Figure 8.18 you will see

that the lock itself has a built in cover for the keyhole opening, thus preserving the occupants' privacy. There is a plain escutcheon plate on the outside (Figure 8-19) which measures 80 from top to bottom.

Bolts

Bolts are frequently used to bolster security of a door, but only if the occupier is inside the house! With double leaf doors, they are used to secure one leaf while a lock is used for the other leaf. They are available in a variety of types to fix to the face of the door or the edge of the door. They have keepers for the bolt or **shoot**. Frequently the keeper is only a hole drilled into the frame or into the floor or slip step. Bolts are also used in smaller sizes on WC and bathroom compartment doors, to ensure privacy for the occupant.

Bolts are available in black japanned steel, galvanised steel, brass, aluminium and even plastics. Sizes are indicative of the overall length and range, from around 25 to 600. On fire doors, bolts which hold the door leaf shut must be of metal and be screwed to the face of the door, never on the edge. A minimum screw length of 38 is recommended. Screw manufacturers have standardised on 40 so 40 × 4.5 should be the recommendation.

Figure 8.20 shows some of the more 'normal' bolts used in domestic work. From the top they are:

☐ A 225 galvanised tower bolt

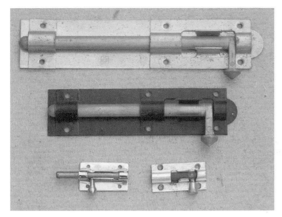

Fig. 8.20 Tower and slip bolts.

□ A 180 black japanned tower bolt
□ Two 50 slip bolts, chromium plated steel on the left and satin anodised aluminium on the right.

Hinge bolts are a security device and despite their name have no moving or sliding parts. They are passive – until someone tries to beak down the door. Mating holes are drilled in frame and door edge as close to the top and bottom hinges as is reasonable but between those hinges. The 'bolt' is hammered into the hole on the door and the keeper plate is screwed to the frame. The hole under it is widened slightly to allow easy movement of the bolt when the door swings open and shut. Any attempt to force an entry has to overcome not just the resistance of the hinges but also the hinge bolts set much deeper into both door and frame. Figure 8.21 shows a sketch of a hinge bolt in position, and the two components. Bolts are normally around 12 o.d. (outside diameter).

Letter plates

Letter plates are a must on the front door, and the only thing to say about these has to do with their insertion in a fire door such as one would find in a common stairway. British Standard 2911 applies to these items, and one with a maximum aperture of 250 × 38 should be fitted. It should have a strong spring or good gravity action and be made of steel or solid

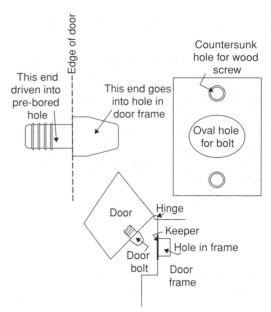

Fig. 8.21 Hinge bolt.

brass. Internal as well as external flaps are a must, but all of these precautions will give a rating of 30 minutes. Note that plastic flaps are not even considered.

Weather bars

Weather bars are the constant subject of the 'invent a better mouse trap brigade'. There are at least six companies in the UK and Eire which make their own styles of weather bar – not just one style – one company making dozens of different styles. Weather bars are only used on external doors or doors in common stairways. A common form is a hinged flap attached to the door, and as the door is closed a simple cam mechanism forces the flap down so that a flexible seal on the flap presses down on the threshold. These fail primarily at the cam mechanism for it has to operate at the hinge end of the door, and enormous forces are at work to shut the flap frequently at an oblique angle to the cam mechanism. A further failure can occur with the flexible strip or cushion, but that is usually easily replaced in the better-designed products. The second general type is the one piece, no moving parts

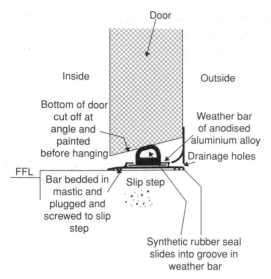

Fig. 8.22 Threshold/weather bar schematic.

Fig. 8.23 Knobs and levers.

type, usually an anodised[4], aluminium extrusion fitted to the threshold. The door closes against an upstand of the threshold and the gap is taken up with a flexible seal of either the wiper type or the cushion type.

Figure 8.22 is a sketch of a cushion type seal in a one-piece threshold/weather bar. Note how, as the door is closed, the cushion top on the seal is squeezed forward and seals against the sloping bottom of the door. It is important to paint or otherwise seal the bottom of the door before it is finally hung in place. Try this website, which will give you a lot of information about the various types available: http://www.stormguard.co.uk

Door furniture is the term used to describe the knobs and levers, pull handles and push handles with which a door is opened and closed and which operate the latches. Eschewing the architectural, the usual domestic set-up is to use either knob or lever furniture. Knobs and levers of course come in a variety of shapes and patterns. They also come in a variety of materials and finishes.

Materials include cast iron, brasses, stainless steel, wrought iron, aluminium, plastics, wood and porcelain. Finishes include enamels or japanning (applied to cast iron and wrought iron), BMA (bronze metal antique – applied to brass ironmongery), SAA (satin anodised aluminium) and plain old-fashioned polishing and lacquering (applied to brass ironmongery).

To operate, all latches have a square spindle which locates in a square hole on the latch mechanism. They can be secured there by a variety of devices, ranging from soft steel pins driven into place, to nuts on the spindle, grub screws, cotter pins, split spring washers and so on, some good, some poor but there is not the space to discuss the merits of all the systems used. Faced with a choice, a spindle with a thread run on it and a nut or handle with thread internally is possibly the most secure – we are not talking security, but the least likely to come apart in the user's hand. Grub screws are 'iffy'; those tightened with a screwdriver tend to have the slot or cross head badly distorted when someone tries to fit it or retighten it. Those which use an Allen[5] key are much better. Figure 8.23 shows a selection of knobs and levers with which to operate the latch mechanisms of locks and latches. From left to right there is a brass lever handle, Bakelite ball handles, wooden ball handles and brass ball handles. The brass handles are hollow.

---

[4] Anodising is a process applied to aluminium and its alloys whereby the metal is oxidised within a controlled environment and the oxide layer formed can be left with a natural colour or dyed to give a gold, silver or brass appearance. The oxide layer can prevent further deterioration of the metal substrate.

[5] An Allen key is an L shaped piece of hexagonal section hardened steel. Either the short or long leg can be inserted into a hexagonal section blind hole in the end of the grub screw. It is a very positive method of connecting tool to screw or bolt.

In general terms, for fire resistant doors, the fitting of ironmongery can degrade its potential seriously if the holes cut are too large and leave spaces within the door core; or if there is too much metal conducting heat into the core and therefore burning out the core round the metal and so on. Fitting ironmongery to these doors has to be done with precision, and any void left must be filled with incombustible core material or an intumescent putty.

Timber casement windows	205	Timber sash and case windows	211
Depth and height of glazing rebates	206	The case	212
Timber for casement windows	206	The sashes and case together	214
Draught stripping materials	206	Vertical sliding sash windows	214
Hanging the casements	207	Glazing	218
Joining the frame and casement members	209	For ordinary glazing work	218

Windows are made from a variety of materials, in a variety of types and in a variety of configurations:

- Materials – wood, plastics and metals
- Types – casement, sash and case, pivot and sliding
- Configurations – plain, mullioned, oriel, bay, etc.

Our detailed study will cover only timber windows configured as casement windows and sash and case windows. Other materials for these two types will be discussed briefly. Only plain windows will be discussed as these are the basis upon which all other types are built.

The British Standard 6375 Parts 1 and 2 is still current in the Regulations in relation to windows but has been partially superseded by three BS EN documents: 12207:2000 on Air permeability; 12210:2000 on Resistance to wind loading – Classification; 12208:2000 on Water tightness – Classification. Generally these specifications of windows are on a performance basis and are not of the type which give precise minimum dimensions for component parts or minimum standards for materials and workmanship in the manufacture of the window. These Standards are a specification for the operation and strength of the windows, which is also performance-based.

A wide range of tests is given which windows must pass to be deemed to be up to Standard. This leaves the specifier free to choose a product from a manufacturer who will design and make windows which either meet or exceed the performance requirements; thus the choice is made on performance and cost.

Windows have one thing in common – glazing. This can be done in a variety of ways and will be dealt with in the last section of this chapter when we have discussed the actual windows.

Alternative materials used for the manufacture of windows are:

- Plastics, particularly uPVC, which may be white or coloured and even impressed with a timber 'grain'. The sections are hollow and may be of a single skin or more likely with internal ribs. Sizes of sections are similar to timber sections. Sections are mitred at the corners and 'welded' using RF (radio frequency, i.e. microwave) radiation. Metal reinforcement may be introduced at the corners or even the full length of the sections with the metal 'crimped' at the corners as well as having the plastic 'welded'.

- Metals such as steel and aluminium can be used. Steel windows use hot rolled sections and these are now hot dipped galvanised after fabrication, i.e. cutting to length with mitred, electric arc welded joints. Aluminium windows are now made using extruded hollow sections. Some sections may be of a single skin construction or may have internal ribs. Corner joints can be made by inserting an L-shaped corner piece and crimping the sections to it or by drilling and screwing with plated steel self-tapping screws.
- Metal windows can be self finished, coated with plastics, e.g. powder coating, in a variety of colours, including metallic colours. Additionally, aluminium windows can be 'anodised' – oxidising the surface of the aluminium and dying and polishing the oxide coating.
- Plastic coated timber. Here the timber is machined to shape and a thin skin of PVC is drawn over the section prior to fabrication.
- The author did see casement windows of precast reinforced concrete displayed at the National Building Exhibition some 20 years ago. They were not a success!

All these windows can be draught stripped and can be glazed in the same ways as timber windows, although glazing in beads or gaskets are the most common methods.

Timber casement windows

Traditionally, casement windows of whatever material are made with a **'frame'** and a number of fixed or opening **casements**, these casements being nothing more than a simple frame fitted into the main overall frame, or the main frame may be glazed directly, which is more often the case for those portions of modern windows without an opening sash.

Fixed sashes are screwed into the frame. The opening sashes are usually hinged (hung) on one vertical side, although it is common practice to include one or two top hung sashes for use as ventilators, sometimes opening in, sometimes outwards. The majority of case-

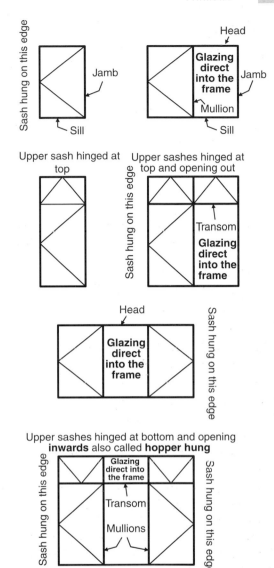

Fig. 9.1 Typical layouts of fixed and opening sashes for casement windows.

ment windows used in the UK have the side hung sashes opening *out* whereas on the Continent inward opening sashes are more normal.

Figure 9.1 shows some typical layouts of fixed and opening sashes for casement windows. Note the conventional symbolism regarding the opening sashes and the side on which they are hinged, and the proper names of the members of the casement window frame.

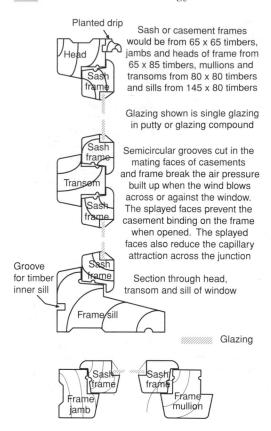

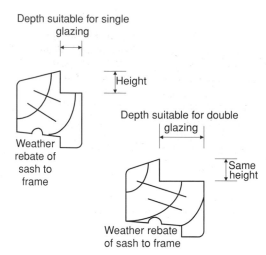

Fig. 9.2 Typical casement window sections.

Fig. 9.3 Ordinary and deep rebates in casement sashes.

Depth and height of glazing rebates

The sketch in Figure 9.3 shows what is meant by height and depth of glazing rebates.

Timber for casement windows

Figure 9.2 shows typical sections for a basic casement window frame and casements.

There are a few points to make about these figures:

☐ The timber sizes given are fairly typical of modern joinery practice and are not necessarily used by different manufacturers. Older windows may have heavier timber sections.
☐ The depth of rebates used to take the glass can be of different sizes to allow single, double or triple glazing.
☐ The method of glazing also affects the depth of the rebates, e.g. in putty or with beads or in gaskets.
☐ The height of the rebates remains at 12–15 mm

Timber used may be softwood or hardwood. The majority of casement windows made in the UK are of softwood, European redwood being the most common. It is now common to treat the timber used by the vac-vac process or similar – see Appendix C – *after* the timber has been machined to shape for the various casement members.

The rebates used at the opening sashes/frame junctions are the minimum required to exclude weather and should not be reduced even if **draught stripping** is used. **Draught stripping** involves the introduction of a flexible or compressive strip of material which 'seals' the joint. Figure 9.4 shows examples of this, some of which we have discussed in earlier chapters, some of which are new.

Draught stripping materials

Figure 9.4 shows that draught proofing materials can be classified as compression or wiping

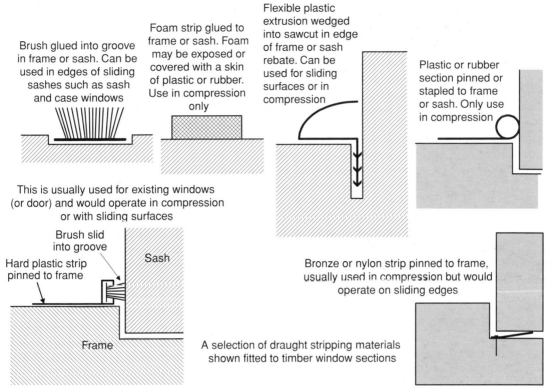

Brush glued into groove in frame or sash. Can be used in edges of sliding sashes such as sash and case windows

Foam strip glued to frame or sash. Foam may be exposed or covered with a skin of plastic or rubber. Use in compression only

Flexible plastic extrusion wedged into sawcut in edge of frame or sash rebate. Can be used for sliding surfaces or in compression

Plastic or rubber section pinned or stapled to frame or sash. Only use in compression

This is usually used for existing windows (or door) and would operate in compression or with sliding surfaces

Brush slid into groove

Sash

Hard plastic strip pinned to frame

Frame

Bronze or nylon strip pinned to frame, usually used in compression but would operate on sliding edges

A selection of draught stripping materials shown fitted to timber window sections

Fig. 9.4 Brush, wipe and compression seals.

depending on where they are placed in relation to sash and frame. Foam strips are best limited to compression seals whereas brush and plastic or metal strips can be used in either way. Fitted against the sash opening, plastic or metal strips allow wind to penetrate and can jam the sash in a closed position or make it difficult to open. Fitted with the sash opening, they make the sash difficult or impossible to close and can be noisy in high winds if loosely fitted, acting like a reed in a musical instrument.

Hanging the casements

Fixed sashes are retained in the frame by placing screws through pre-drilled holes in the glazing rebate. Mastic bedding is frequently introduced. Opening sashes are hung on hinges and have fasteners to retain them in the closed and open positions. Fixed sashes

add considerably to the overall cost of the window and a cheaper alternative is to glaze the frame direct. Glazing direct interrupts the line of the glazing across multiple sash windows.

A particular requirement of the Building Regulations is the facility to clean windows from inside the building. Hinges can be designed to throw the sash clear of the frame to facilitate this. An early form of hinge made to throw a sash clear of the frame is shown in Figure 9.5. The clearance was minimal and not very effective.

Modern hinges work on a pantographic principle and can allow a tilt and turn facility in a vertical plane. The hinges are made of galvanised or sherardised steel, stainless steel or cast in an aluminium alloy. Screws match the ironmongery being fitted. Figure 9.5 shows how the sash is projected clear of the frame in the open position.

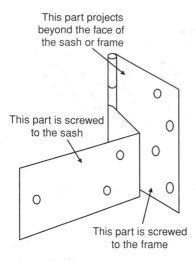

This part projects beyond the face of the sash or frame

This part is screwed to the sash

This part is screwed to the frame

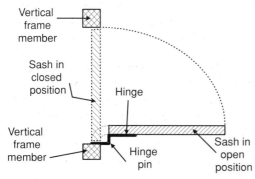

Vertical frame member

Sash in closed position

Hinge

Vertical frame member

Hinge pin

Sash in open position

Fig. 9.5 Easy clean hinges.

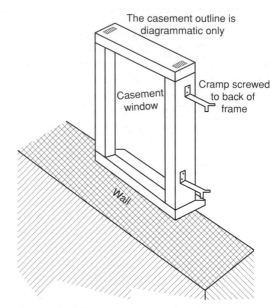

The casement outline is diagrammatic only

Casement window

Cramp screwed to back of frame

Wall

Fig. 9.6 Building in a casement window.

Fixing casement windows into openings in walls varies according to the position of the window in relation to the various layers of the wall structure, as well as the traditional building practice in the wider areas of the UK. Whether windows are built forward in the opening or to the back of the outer leaf but built in as the walls are built up, window 'cramps' are fastened to the back of the frame. *The back of the frame is the edge nearest the masonry.* Cramps are of galvanised steel, with holes for at least two screws, and fishtailed for building into a mortar bed. They should be twice fixed with screws, *not* nailed.

Fixing casement windows into openings can be achieved in other ways. There is a tradition that casements should be wedged over the

frame stiles into openings with checked reveals and head. The sill is bedded onto the masonry sill – traditionally with a haired[1] lime mortar – but in recent times in a cement/lime mortar. It is impossible to nail the wedges into position to prevent them coming loose due to vibration, so a little glue should be placed on the mating surfaces prior to hammering them home. The glue not only secures them in position but also lubricates the moving surfaces when putting them in place. With plain reveals and head, the frame is nailed or screwed to plugs in the reveals, but more modern practice uses frame fasteners and packing pieces. The number of fasteners put through each jamb into the reveal depends on the height of the casement window. Two would be a minimum but it could be up to four.

These methods are illustrated in Figures 9.6, 9.7 and 9.8. What is not shown on any of the figures is the possibility of filling the gap between wall and frame with self foaming plastic. We have pointed out before that this reduces draughts and cold bridging as well

---

[1] Cow hair was most commonly used. It was readily available, being a by-product of the tanning industry.

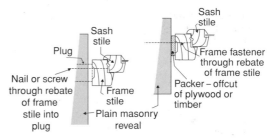

Fig. 9.8 Plugging in and framing anchors.

Joining the frame and casement members

The sections of all the members of a casement have been shown in the figures above, but what has not been covered is how they are joined together. Many casements in the past have had all the joints carried out as mortice and tenon joints with the rounded and splayed edges mitred. With modern machinery cutting mating faces and the joints themselves, variations of the finger joint are now commonplace and gluing is the favoured 'fastener'. Figure 9.9 shows a typical joint of two casement sections. The joint of jambs to sill of the frame can be plain housings, glued and nailed from the back, or they can be finger jointed but the sill is not cut to its full depth.

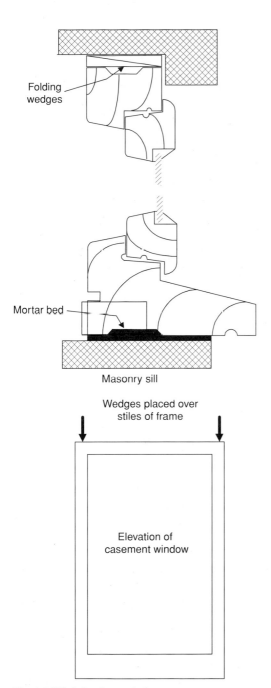

Fig. 9.7 Wedging in a window.

as adhering to both masonry and frame, so increasing the efficacy of holding the frame in position. To obtain maximum adhesion a light spray of water on both surfaces is advantageous.

Fig. 9.9 Finger joints in a casement.

Fig. 9.10 Typical casement window.

Fig. 9.11 Casement sill and frame stile joint.

Normal length fingers are cut on the stiles, and blind or through sockets are cut in the sill. The glue does the rest. The cutter used cuts both the fingers and the scribe to match the adjacent timber.

Figures 9.10 to 9.14 are of a typical casement window. Figure 9.10 is a picture of two casement windows, one leaning against the other. The outermost one shows only the opening casement with the frame stile on the left and a mullion on the right. Note that glazing beads are pinned into the rebates in the left and bottom rebates. Figure 9.11 is a picture of the junction of frame stile to the frame sill. Note that the sill is in two pieces: the stile is finger jointed to the sill with through fingers, and there is a groove on the back of the sill to take the rebate on the timber inner sill.

Figure 9.12 is a picture of the joint of the frame head to the frame stile. Note that this

is a finger joint and in particular note how the fingers on the right-hand side incorporate a scribe to fit the ovolo moulding which is run on the frame members. There is also a weather moulding planted on the frame head and it appears to be only glued in place. This is not usual. It is usual to have a rebate on the moulding to allow fitting it into a groove in the head.

Figure 9.13 is a picture of the junction of the mullion to the frame sill. It has been done with a through tenon on the mullion into a mortice in the sill. Note on the left of the mullion there is the opening for the casement, and the rebate has been fitted with a P-type compression weather seal. The light-coloured block on the

Fig. 9.12 Casement head and frame stile joint.

Fig. 9.13 Mullion and frame sill joint.

right-hand side of the mullion is a soft plastic setting block for the double glazed units which will be used. Finally, Figure 9.14 is a picture of the corner of the casement showing a very deep rebate, and it is on the outside. The whole window was set up for glazing with a sealed unit in beads from the outside. The householder will get a window which has the security value of a sealed double glazed unit completely negated by having it glazed from the outside.

Fig. 9.14 Casement joint.

Timber sash and case windows

Sometimes known as **vertical sliding sash** windows, this latter term should only be applied when the frame holding the sashes is not forming a **case** to contain the **counterweights** for the sashes. This will become clearer as the sketches are studied.

The British Standards previously listed can be used to set standards for sash and case windows. The list of alternative materials to wood given at the beginning of this chapter also applies to this style of window, although the windows are usually vertical sliding sash type, there being no **case** and no counterweights on cords.

Although they can be manufactured in hardwood, the traditional material for sash and case or vertical sliding sash windows is European redwood, which in modern practice is treated with preservative using a vac-vac process or similar *after* the timbers have been machined.

The most basic window comprises:

☐ A frame with head, sill and stiles, the stiles forming open-backed boxes with frontal grooves in which the sashes slide
☐ Two sashes, upper and lower, which slide past each other in grooves in the frame and are counterbalanced by weights on **sash cords** or **chains** over **pulleys** set in the stiles of the frame.

Weights are commonly cast iron although lead can be used. They are made up in a wide range of standard weights, and small lead 'makeweights' from 250 g to 1 kg are also used to adjust the counterweights for their particular situation.

A sash is counterbalanced by a pair of weights connected to the sash over a pair of pulleys by two lengths of **sash cord**. The sash cord core should be of hemp, or now polypropylene, with a braided cotton outer cover. It is made in a number of strengths and qualities. For extremely heavy sashes a chain can be substituted for the cord. The chain is similar in style to a bicycle chain. Obviously

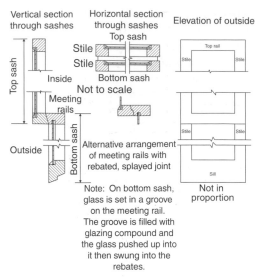

Fig. 9.15 Schematic of sashes in sash and case window.

the pulleys for chain are different from those used for sash cord and should have a metal sheave on a stout metal axle.

The sashes shown in Figure 9.15 are made in pairs to fit into a single case. Points to note:

- The **meeting rails** can be splayed or splayed and rebated, the latter being a superior finish.
- The sketches show the sashes in their relative positions when closed. Note that the top sash is the outermost sash. They are retained in these positions by the components of the case and the counterweights
- While the system of rebates and beads in the case provides weather proofing for the completed window, it is modern practice to include seals, similar to the ones we have already seen in Figure 9.4.
- It is common practice to have an ovolo moulding on the internal exposed arris of sash members. Decorative mouldings have been omitted from these drawings for simplicity. They are shown in later sketches.
- To keep the upper sash closed, i.e. at the top of the case, the counterweights should be a half to one kilogram heavier in total than the weight of the sash and its glazing. To keep the lower sash down, i.e. in the

closed position, the counterweights should be a half to one kilo lighter than the sash and its glazing. The sashes are weighed by the joiner, who then adds the calculated weight of the glass.

The case

The case is made up with a sill to which are attached a pair of vertical members for each side or jamb of the window. A head is fixed across the top of the jambs. The jambs and head have the general shape shown in Figure 9.16. Note that this shape is made up from *five* separate timbers.

The sill of the case can be formed from a single piece of timber, or two or three pieces tongued and grooved together to give an overall shape much as shown in Figure 9.17. The double rebated shape where the lower sash sits is required for weather resistance.

The head and jambs of the case are made from five pieces of timber, which are joined in the specific ways shown in Figure 9.18. The sketch shows a typical jamb and head. Note how the inner facings of the jambs have to be made wider than the outer facings when the masonry opening has splayed reveals, a detail common on old thick stone walls.

Figure 9.19 shows three ways in which the sill can be made – in one, two or three pieces. The three piece is the least desirable option. Note the ovolo moulding on sash rails.

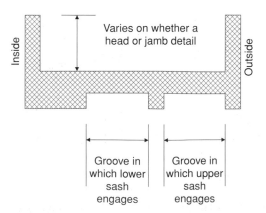

Fig. 9.16 Outline of case head and stiles.

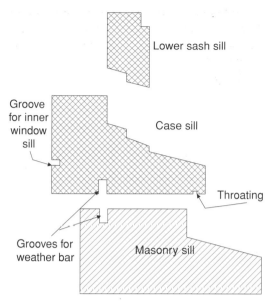

Fig. 9.17 Outline of case sill and sash sill in relation to masonry.

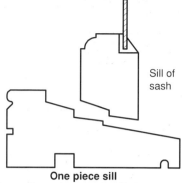

One piece sill
Generally used where the sill does not extend beyond the outer facing of the case

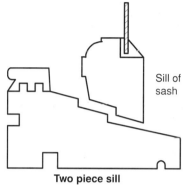

Two piece sill
Used to give extra depth to the rebate for the sash sill and so increase the weatherproofing effect. The additional depth of the double splayed rebate is beyond the capacity of some four-sider machines – see Appendix C

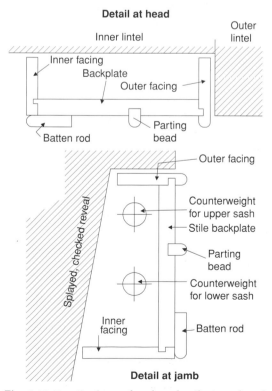

Fig. 9.18 Detail of case head and stile in splayed, checked reveal.

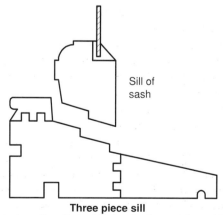

Three piece sill
Used where the sill needs extra depth as well as width beyond the outer facing of the case. The vertical joint in the sill is not a good feature as it could allow water to penetrate. The joint should be glued with a WBP, thermosetting glue

Fig. 9.19 Alternative sill build ups.

The sashes and case together

We must know how the pieces of the sashes and the case are joined together:

☐ Stiles, sill and rails of sashes are morticed and tenoned together with a single shouldered tenon, or they are finger jointed, modern practice using a waterproof glue and wooden dowel or metal star dowel.
☐ Case facings are tongued, grooved, glued and nailed to the backplates with oval brads.
☐ Case sills are trenched for the backplates, and these are glued and nailed (and sometimes wedged) into the sill.
☐ Parting beads and batten rods are nailed and glued to the head backplate.
☐ Parting beads are a friction fit into the groove in the stile backplates.
☐ Batten rods are usually nailed to one stile back plate and screwed to the other, thus allowing removal of one batten rod. With this removed the lower sash can be swung into the room. Removal of the parting bead allows the upper sash to be swung into the room.
☐ To take the weight of the sashes, **duplex hinges** are screwed to the fixed batten rod, and round head screws fitted to the face of the sash stile, which is then lowered into the slots in the duplex hinges. With the sash cords trapped in the cord grips, the lower sash can be swung inwards.

Figure 9.20 shows a schematic of this operation.

Figure 9.21 shows details of the upper part of a sash and case window, showing how the counterweights, pulleys and sashes all fit in the case. Typical sizes of components for modern sash and case windows are:

☐ Backplates – 20 mm thick, width depends on circumstance but 185 mm would match minimum sill
☐ Facings – 15 mm thick, width depends on circumstance
☐ Batten rods – 15 mm thick, 36 to 45 mm wide
☐ Parting beads – 15 mm thick, 19 mm deep

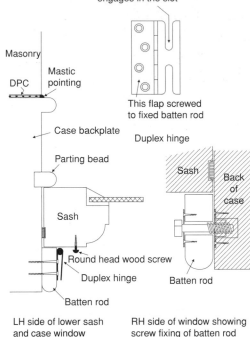

Fig. 9.20 Duplex hinges and operating schematic.

☐ Case sill, two piece
☐ Sash head – 65 × 65 mm overall
☐ Sash stiles – 65 × 65 mm overall
☐ Meeting rails – 80 × 45 mm overall
☐ Sash sill – 95 mm thick, 185 wide, minimum.

Vertical sliding sash windows

The basic difference between these and the sash and case window is the use of a 'spiral balance' on each side of both sashes to support the sashes in either the open or closed position. The spiral balances are made from alloy tube with one end having a hole for screw fixing to the case stile. From the other end of the tube there protrudes a flat metal strip which has been twisted into a spiral. The end of the tube is finished with a rectangular slot such that when the spiral is withdrawn or pushed in, the spiral will revolve. The hidden end of the spiral is attached to a strong spring fixed

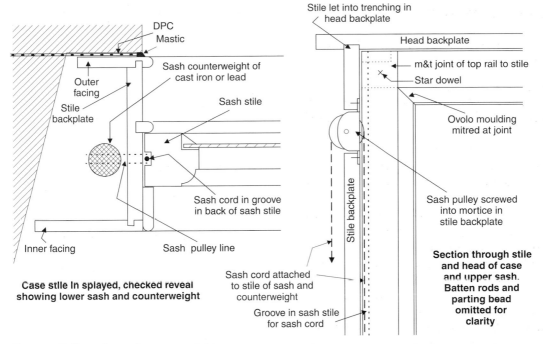

Fig. 9.21 Pulley stiles and counterweights.

inside the tube. As the spiral revolves, the spring is wound ever tighter and, if the free end of the spiral were fixed to the sash, would support the weight of the sash.

Fixing of the balance to the sash is done with a metal bracket which allows the balance to be released and refastened to the sash easily and effectively. The balances are manufactured with a range of spring strengths which accommodate all but the largest sashes. The critical factor in many instances is the use of double or triple glazing.

Figure 9.22 illustrates a typical stile detail of a vertical sliding sash window and a sash balance. The balance can be housed in a groove in the sash – most commonly used when fitting balances to old corded windows. In the detail shown of a new or replacement window, the balance is frequently housed in a groove in the backplate, which is of much thicker timber than a conventional sash and case window.

Note how the parting bead is a plastic extrusion with multiple blade wiper seals between the sashes, and that the sashes have brush seals between them and the backplate.

The only wood-on-wood contact is where the sashes touch the outer facing and batten rod. This reduces friction considerably and makes the sash less likely to stick in damp conditions. Paint must not be allowed to clog any of these seals.

The balance has a hole for an ordinary wood screw fixing to the stile, and the spiral has a

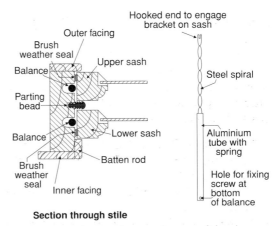

Fig. 9.22 Spring balances.

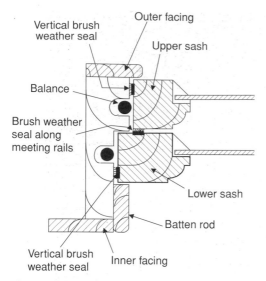

Fig. 9.23 Hinge down sashes with spring balances.

Fig. 9.24 General view of sash and case window.

split hook which fits into a bracket fastened to the corner of the sash. A pair of balances is used for each sash. Balances are purchased which match the weight of the sash but which are adjustable by twisting the spiral before hooking it to the bracket.

Facings are the same sizes as for sash and case windows; however, the backplates and heads are generally out of 50 or 65 thick timbers.

Spiral balances can be fitted to windows made from plastics and metals which do not lend themselves to the fitting of counterweights, pulleys and cords.

An alternative arrangement for this type of window is shown in Figure 9.23. Here the idea is to allow the traditional vertical movement of the sashes but to add the possibility of tilting the sashes inwards (from the top of each sash) for ease of cleaning. The system has to incorporate a pair of stays for each sash to prevent them tilting completely over, and balances are available which incorporate them. These can be accommodated in rebates cut into the backplates. The batten rods are hinged on both frame stiles and can fold back clear of the stiles. They are held in the operational position by captive thumb screws. When the batten rods are hinged back, first the lower sash

and then the upper sash can be tilted inwards for cleaning.

Figures 9.24 to 9.30 are from various old and fairly new sash and case windows of traditional pattern. Figure 9.24 is a general view of the upper part of an old window with two sashes. Only the top sash opens and has sash cords and sash weights. Figure 9.25 is

Fig. 9.25 Sash pulleys and cord grip.

Fig. 9.26 Batten rod screws.

Fig. 9.27 Pocket in stile open.

taken from inside a more modern window and shows two pulleys in the case backplate. The left-hand cord runs through a cord grip which allows the user to pull the cord and raise the sash weight, allowing the sash to be released from the cords. Figure 9.26 shows the batten rod removed, and the captive screws which hold it in position can be clearly seen in the rod and the plate they screw into in the backplate. The parting bead is still in position. Figure 9.27 shows the parting bead removed and the pocket – the access plate to the weights – which is cut into the backplate. Note how the pocket cover is cut in the groove for the parting bead. When the parting bead is fitted, this together with the lower sash keeps the cover in place. Maintenance work can be carried out such as replacing cords, adding make-weights, shortening off stretched cords, etc. Figure 9.28 is a picture of the joint of the sash stile and sash sill. The joint is a mortice and tenon joint with a dowel, which can be seen proud of the face under the paint. Figure 9.29 is the same joint viewed from the side, and the end of the tenon can be seen with the wedges either side to hold it firmly in place. Figure 9.30 is the joint of sash stile to meeting

Fig. 9.28 Sash sill to stile junction (1).

Fig. 9.29 Sash sill to stile junction (2).

rail. The meeting rails on this old window had a hook arrangement machined on the mating faces. The joint is a hand-cut finger joint and one finger and slot is dovetailed.

Glazing

The fitting of glass into openings in doors and windows has already been mentioned on several occasions but without much explanation beyond what is necessary to understand the requirements of our door and window discussions.

Fig. 9.30 Meeting rail to stile junction.

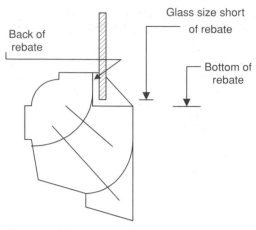

Fig. 9.31 Rebate nomenclature in glazing.

Glass is generally fitted into rebates in the timber joinery work we have already described. The glass is always cut a little less than the opening measured to the bottom of the rebates. This allows fitting without forcing the glass into position and so allows the glass to expand and contract without itself or the joinery being put under stress. It also allows the joinery to move (moisture content, central heating, etc.) without putting the glass under stress. Figure 9.31 illustrates this point.

For ordinary glazing work

Glass was at one time measured by weight – ounces per square foot – but is now measured by thickness in millimetres. There were a large number of different qualities of glass but these have now been cut down to just **float glass** and **polished float glass**. Float glass is used for general glazing such as we have previously referred to. Polished glass is used for mirrors, table tops, etc. Glass comes either **clear** or **obscured**, the degree of obscurity depending on the pattern which has been impressed into the molten sheet during manufacture. There are not quite as many patterns available now.

The basic method of glazing is done in joinery which is to be painted, and requires a simple rebate primed to seal the timber and prevent it from absorbing the oils in the glazing

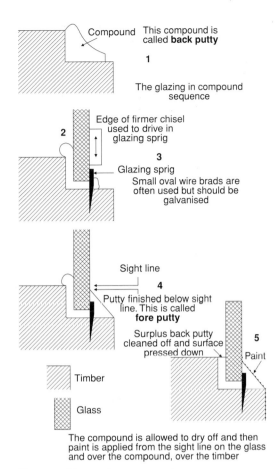

Fig. 9.32 Glazing in compound.

materials. These can be linseed oil putty or glazing compound – an improved version which does not dry out and crack quite so quickly. There are a number of distinct steps described below and illustrated in Figure 9.32:

(1) Place a generous bead of compound in the rebate.
(2) Press the pane of glass into that bed of compound ensuring that it is evenly bedded in and that the compound is squeezed up between the back of the rebate and the glass.
(3) Drive glazing sprigs into bottom of rebate to secure glass in position.
(4) Apply more compound in front of the glass and smooth off to an even slope, pressing it in to give an even, firm density.

(5) Remove surplus compound from the inner face of the glass.

Once this is completed the glazier's work is done but the painter has to paint over the compound and up the front of the glass in order to seal the compound and the joint to the glass against weather.

Glazing in beads is commonly done where there is a need for greater security, the glass is very heavy, in an exposed situation or the timber has to be varnished or polished. Sometimes the beads and the glass are bedded in compound, sometimes in silicone glazing compound, sometimes in wash leather[2] strips or velvet ribbon, and sometimes in a synthetic rubber or plastic gasket.

In compound, the steps are:

(1) Place compound in the rebate as back putty.
(2) Press glass into the back putty and press home.
(3) Spread a little compound against glass as a fore putty.
(4) Press bead against the fore putty, squeezing it up the face of the glass but not trapping any under the bead. Fix beads with glazing pins or wood screws.
(5) Clean off surplus compound.

The effect is to surround the glass with compound and have the beads secure it all in position.

In velvet or wash leather, the steps are:

(1) Place strip of velvet in rebate with enough upstand to cover the back of the rebate. The pile of the velvet should face the glass.
(2) Place glass against the velvet and secure with two temporary sprigs on the sides.
(3) Turn velvet up the front of the glass on an edge without the sprigs and press bead into place and secure with pins or screws.

---

[2] Wash leather was originally natural leather from the chamois, an Alpine antelope, but was also manufactured from the leather made for other animal hides which were split to give the correct texture. There is a synthetic chamois leather available now.

(4) Turn velvet up on an adjacent edge and secure with the bead, having removed the temporary sprig.

(5) Treat adjacent sprig-free edge.

(6) Remove last sprig and treat last edge.

Note that this is an entirely dry exercise – no compound, glue, silicone or anything like that. The glass is literally gripped firmly between the pile of the velvet or the soft wash leather. The velvet comes as a ribbon but the wash leather has to be cut with a sharp knife into strips.

Velvet or wash leather is no use for external glazing and is used mainly for joinery work to cupboards and cabinets.

In silicone glazing compound, the steps are:

(1) Place spots of compound in the corner of the rebate. For a 15-pane door, four spots per pane would be sufficient.

(2) Press glass into the rebate.

(3) Press beads into the rebate, pushing the compound off the back of the rebate and up against the glass. Fix beads with pins or screws.

(4) Clean off surplus silicone.

This must be one of the favourite methods of fixing the small panes in a 15-pane door. It is quick and easy, especially if pins are used to fix the beads and are driven home with a pin push.

In gaskets with beads, the steps are:

(1) Cut gasket to length for each rebate and mitre the corners *or* cut length of gasket for complete glazing length and form mitred fold cut-outs.

(2) Place gasket around the edge of the glass. If an external window or door, bed the gasket onto the edge with a suitable compound; usually a silicone-based compound is used.

(3) Place bedding compound layer on back and bottom of rebate.

(4) Bed the glass and gasket in place.

(5) Fit beads against gasket, compressing the gasket, and fix in place with pins or screws.

The type and design of the gasket will determine whether or not to apply a sealer at each stage. There are gaskets which can be fitted entirely dry.

In gaskets without beads the gasket is in two pieces. The first is pressed home into a groove in the rebate and has a front which hinges over to allow the glass to be pressed into place. Finally a locking strip is pressed into the hinged portion which prevents the gasket hinging open and so retains the glass. The locking strip is a very tight fit to compress the gasket against both the timber and the glass.

This is very unusual in timber framed openings but is common enough in plastics and metal windows where, instead of being set in a groove in a rebate, the gasket fits over a thin fin much like the glazing methods used in the automobile industry.

From a security point of view, fixing with beads and gaskets with locking strips can only be done if the rebate is on the *inside* of the door or window. There have been instances of burglars gaining entry by simply unscrewing the fixings for the beads and taking out the pane of glass.

Figure 9.33 shows some of the above methods.

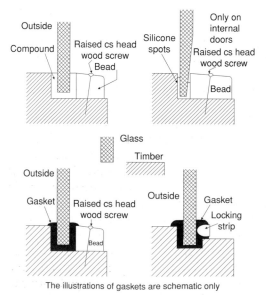

The illustrations of gaskets are schematic only

Fig. 9.33 Glazing alternatives.

10 Stairs

Landings	222	Measurements	224
Steps	222	Joining steps to stringer	225
Balustrades	223	Winders	227

This chapter will be confined to simple domestic stairs within a single dwelling. While this limits us to basic timber stairs it is nonetheless quite complex because of the Building Regulations and in terms of the joinery and precision measurement required.

Like doors and windows, stairs in housing were an early contender for the factory-built approach, and while bespoke stairs are still built in numbers, that is nothing compared to the thousands built in factories up and down the UK. But factory-built or not, built down to a price or not, stairs are still manufactured in much the same way as they were 100 and more years ago.

Some definitions to begin with. A **stair** is the correct term for that which we use to gain access to a different level within our homes. A stair can comprise a **flight** or **flights** of **steps**, and if flights are involved there has to be a **landing**(s). A flight is a series of steps supported each end by deep but relatively thin timbers called **stringers**. A single step comprises a horizontal board – the one you stand on – called a **tread**, which in turn is attached to a vertical piece of timber called a **riser**. The tread and riser are connected by a rebated and grooved joint, and the ends of the step so formed are **housed** into an L-shaped groove cut in each stringer, where they are glued and wedged into position. **Angle blockings** are glued all round the underside of the step between tread and riser, tread and stringer, and riser and stringer to reinforce the joints. Screws may be used in these blocks but nails should not be used in any part of a stair; when the timber dries and shrinks away from the nail it will squeak as people use the stair. The tread overlaps the riser in each step and the projecting edge of the tread is rounded off; there may also be a **cavetto** moulding applied at the joint. The projection of the tread is called a **nosing**. Figure 10.1 illustrates the points made so far.

The strings support the steps and in turn require support. This is provided by the joisting, trimmers and trimmed joists of the opening in the upper floor and, at the bottom of the flight, by the joists etc. of the lower floor.

In narrow stairs that construction is usually adequate without further support, but where this is required one or even two supports may be added. This comprises a strong piece of sawn timber running parallel to the underside of the stair flight, with shaped timber brackets fixed on alternate sides and pressing on the underside of the tread. The ends of the timber are supported by the floor structures mentioned above

The compartment which contains the stair is the **staircase**. Do not adopt the habit of referring to the stair as the staircase – you cannot climb a staircase, not without a stair.

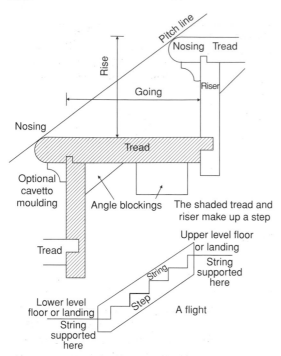

Fig. 10.1 Illustration of some definitions associated with stairs.

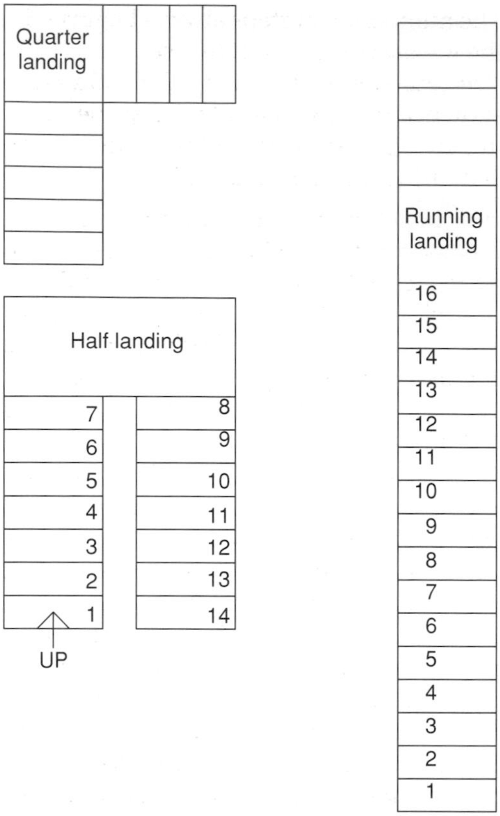

Fig. 10.2 Flights and landings.

Landings

Where a stair is in two or more flights for whatever reason, there has to be a **landing(s)**. These are intermediate platforms where the user can rest or the stair will change direction. Landings obviously must be the same width as the stair but must also project that same amount in the direction travelled on the flight. Landings are constructed in the same way as floors, with joists, a floor boarding and a ceiling finish. The joists and a trimmer joist on the open edge are sometimes built into the staircase walls, or runners are fixed to the walls and the joists housed into the runners. A flight pressing against the open edge of a landing exerts considerable pressure and it is prudent to arrange the joists supporting the landing to run in the same direction as the travel on the stair. These joists then prevent the open edge of the landing from being pushed back by the load on the stair. This is particularly important on a **half landing**.

Landings may be **quarter landings** or **half landings** (see Figure 10.2). The Regulations only allow 16 steps maximum in any flight. Should a stair be in excess of 16 steps, it must be broken into two or more flights with **running landings** in between. The length of a running landing must be at least the same as the width of the flight.

Steps

All steps in a stair must be of exactly the same shape and dimensions.

There are conventions on how stair information is presented on drawings. The three most immediately important are:

☐ Steps *only* are always numbered. Landings and main floor levels are *not* numbered

- ☐ The numbering of steps always begins with the lowest step in the staircase
- ☐ The direction of travel up the staircase is shown at the lowest level on any one drawing by using an arrow pointing *up* the stair and with legend *up* attached.

Figure 10.2 illustrates the above points.

What should be obvious to the more observant reader is that any flight starts and proceeds with whole steps, until it reaches the next level up, be it landing or upper floor. At this point an additional riser is inserted. For example, a flight of *7 steps* will have *8 risers* and *7 treads*.

Stairs may be built into the staircase in such a way that the strings may touch the walls on both sides, on one side, or be completely unattached. Strings against the wall are called **wall strings**; strings not touching the wall are **free-standing strings**. Wall strings can be attached to the wall over their length and so can be manufactured from thinner timber. Free standing strings can be shaped so that the end of each step projects over the string. Such strings are termed **open cut strings** and they can be done in many different ornamental styles. Figure 10.3 is a photograph of an open cut string stair being fixed into position. Note the projection of the tread over the string and the curved plate mitred to the riser. See also Figure 10.4.

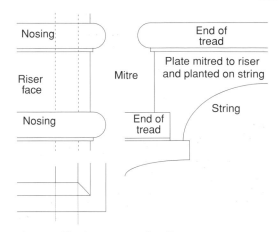

Fig. 10.4 Open cut string details.

Balustrades

With a free-standing string a rail is required to ensure the safety of the users. This is called a **balustrade** and comprises a **handrail** supported on **balusters**, all exposed and frequently elaborately turned on a lathe. Alternatively, a frame may be erected on the string and panels placed in the frame, or it is boarded over with sheet material, matchboarding, etc.

With balustrades of either style there is a need to support the ends of the handrail, and this is done by fitting **newel posts** top and bottom of the flight. These are generally bolted at the bottom to the joist work of the floors or landings, with coach bolts, the heads being set in counterbores and pelleted over.

Newels can also be used to support the free-standing string top and bottom of the flight. Mortices are cut into the newels, and tenons on the string are arranged to fit these rather than housings in the trimmer face of a landing etc.

With two wall strings only a handrail needs to be provided, and here the handrail is fixed to metal brackets screwed (and plugged) to the wall. Sometimes a handrail profile is used which can be screwed to the wall without brackets. Handrails must be fitted in such a way that catching a hand, fingers or clothing is not possible. Apart from the injuries which can be caused, this could also impede anyone trying to use the stair as an escape route in the

Fig. 10.3 Open cut string stair being installed.

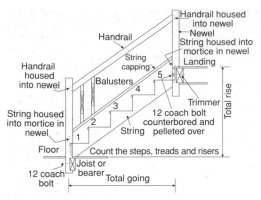

Fig. 10.5 Balustrade, newels, total rise and total going.

event of fire. Rounding off ends, turning ends into the wall, leaving sufficient gap between wall and handrail and so on can all be used to ease a potentially dangerous situation.

Figures 10.5 and 10.6 illustrate some of the above points.

Measurements

Measurements on a stair are critical. If one draws an imaginary line along the centre of the flight, touching the nosing of each step in turn, we have created the **pitch line**. The angle this line makes with the horizontal is the **pitch angle**. In domestic internal stair design, the pitch angle is kept to between 38° and 43°.

Two more measurements are the **going** and the **rise**, both of individual steps and of the complete flight, the latter being frequently

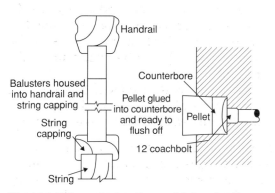

Fig. 10.6 Balustrade detail, newel fixing detail.

referred to as **total rise** and **total going** (see Figure 10.5). Going is measured from the face of one riser to the face of the adjacent riser. Rise is measured from the surface of one tread to the surface of the adjacent tread. The ratios of rise and going are used to control the pitch angle under the Building Regulations.

When measuring up for, making and installing a stair it makes quite a difference to the ease of using it if the treads are not truly horizontal. It is better if they slope in the direction of travel down the flight. The amount of slope is miniscule – an *eighth part* the author was told – the reference being to one eighth of an inch or 3 millimetres across the tread width. The Regulations state that treads should be level, but the assumption has to be that this means laterally – from string to string.

The relative dimensions of going and rise for individual steps in the Regulations for England and Wales are:

☐ Rise shall be no less than 155 nor greater than 220
☐ Going shall be no less than 245 nor greater than 260
☐ Selection of any sizes shall not make the pitch angle greater than 42°.

The Scottish Regulations are quite similar but there is a classification of stairs into **private** and **any other stair**. Stairs should have a minimum width of 1000 between handrails and be free of obstructions; however, the stairs we are discussing here fall into the private category and again there is a choice. A private stair may be as little as 900 between the handrail and clear of all obstructions, may be only 600 when it serves only one compartment (but not a kitchen or living room) and/or sanitary accommodation, and 800 elsewhere. Maximum rise for a step in a private stair is 220 and minimum going is 225 with a maximum pitch of 42°. Combining these extremes of rise and going results in a pitch line in excess of the maximum allowed – approximately 44° 20′.

Once one gets through all the civil service speak in the Regulations, there doesn't seem to be much difference between them.

Joining steps to stringer

Earlier, methods of joining the steps to the stringer were mentioned, but they warrant some further illustration and explanation. Figures 10.7 and 10.8 are photographs of a stair sitting on its side in a house waiting to be installed.

The string is around 32 thick, the tread is around 25 thick, and both are of solid timber. The riser is of hardwood plywood and is only 9 thick. The plywood is nailed and glued to the back edge of the tread and glued into a groove on its underside. Blockings are fitted but are glued and power nailed. Despite appearances on the side that shows, this is not a conventional stair by any means. Tread and string are fairly good but the riser of plywood and the nailing, even with a power nailer, are cause for concern with regard to a creaky staircase. That said, the nails could be of an improved pattern and therefore less prone to slippage in the timber.

On Figure 10.7 the joint of steps to stringer is illustrated. Remember that a step comprises a tread and the riser immediately below it, glued and screwed together with triangular blockings. One can see the groove which has been cut in the inside face of the string and into which each step is placed – with glue. Once all the steps are in place they are joined together by nailing the riser of the step to the back of the tread below – with more glue. Then

Fig. 10.8 Ends of stair stringers.

long wedges are driven into the remainder of the groove, which forces the tread and riser of each step in turn to close up to the upper edge of the groove, and also puts the glue line under pressure. This is all done with the stair in the position shown in the photograph.

The next stage is to fit the other string. Glue is spread in the grooves and the string is brought down over the ends of the steps. Sash cramps are placed across the stair and tightened, which forces the step ends into their respective grooves. The stair can be turned over now and the wedges driven into the second string. Once the glue has set, the sash cramps are removed and the stair is ready for delivery.

A couple more points about that stair:

□ When the steps were made, angle blocks were fitted between tread and riser but they were nailed as well as glued. One can be seen in Figure 10.7 at the centre/top of the picture
□ The wedges used are quite wide and would prevent the placing of angle blocks at the junction of tread to string and riser to string.

In Figure 10.8 we have a more general view: part of the upper surface of the stair, a view of the bottom end and a view of the two string ends. The string at the bottom of the picture

Fig. 10.7 Step to stringer joint.

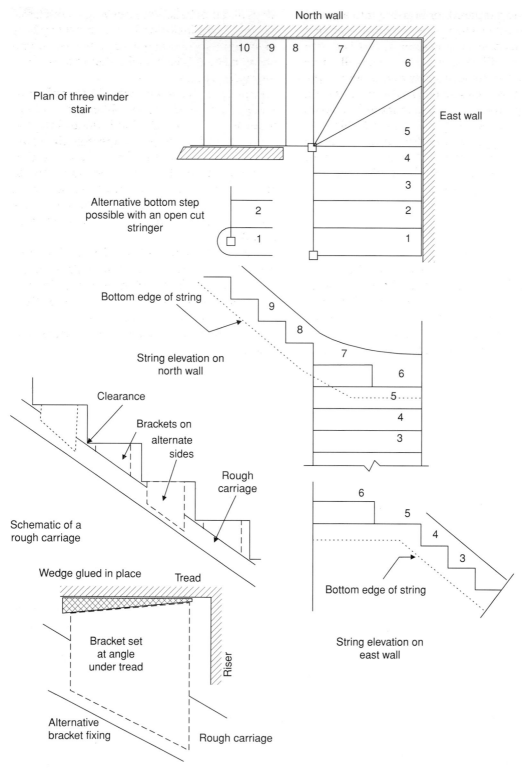

North wall

10 9 8 7

6

East wall

5

4

3

Plan of three winder stair

2

1

Alternative bottom step possible with an open cut stringer

2

1

Bottom edge of string

9

8

String elevation on north wall

7

6

5

4

3

Clearance

Brackets on alternate sides

Rough carriage

Schematic of a rough carriage

6

5

4

3

Wedge glued in place   Tread

Bracket set at angle under tread

Riser

Bottom edge of string

String elevation on east wall

Alternative bracket fixing

Rough carriage

Fig. 10.9 Winders and rough carriage.

has its end prepared for housing into a newel post; the other string has plain ends angled off to simply sit on the floor as this is a wall string and will be fixed back to the wall for its support. The tread nearest the camera is the bottom-most tread of the flight and one would normally expect to see a riser there. However, as one string is going into a newel, this riser has been supplied loose for fitting when the flight goes into place. One wonders about the blockings and the gluing of the riser into the tread – the groove can be seen.

Winders

A single stair flight can change direction without being broken into two flights with a landing. The device used is the **winder** or **kite** step. The treads are of triangular shape and the string round the outside of the turn is shaped to accommodate the wider ends of the kite steps and the shallower **pitch angle** which this causes. At the inside of the turn the two pieces of straight string are normally housed into a newel post. This is what the stair in these photographs will do.

Winders are covered in the Building Regulations. Rise presents no problems, but where to measure the going? That is done along the centre of the tread, and thereafter the same rules about ratios apply. One provision is that the winder steps shall not be less than 50 wide at the narrow end. Since tread measurement doesn't include the nosing projection, we can only assume that it is not included in the 50.

The usual number of winders is three but the author has had a stair built with an open cut string and (complying with the Regulations) with four winders. It was wider than the minimum we now have of 900.

Figure 10.9 shows some more construction details.

11 Mutual Walls

Transmission of sound	228	Wall types	229
Calculation of surface density	228	Fire resistance	231

Mutual walls are those walls which are shared between adjacent houses in a terrace – party walls, and occasionally known as mutual gables since the division is built complete to the very near underside of the roof covering. The law everywhere has a lot to say about party walls but here, thankfully, we are only concerned with those aspects of safety and comfort covered by the Building Regulations.

Two aspects are covered in the Regulations:

☐ Transmission of sound
☐ Fire resistance.

We will attempt to distil some of the main requirements of the Regulations in order to give the beginner to construction an insight into what will be finally required of him or her.

Transmission of sound

The transmission of sound has to be limited otherwise neighbours would be complaining about each other. This does still happen but only when one party is extremely noisy, something against which it is impossible to legislate. So, what do the Regulations require? They require that the amount of noise which passes from one house to another is modified down to a particular level. The amount of sound is measured in decibels. The levels required are not important in this text as we shall concentrate simply on the stock solutions

available. Modifying the level of sound transmitted is called **attenuation**.

Noise can be carried through the party wall or round the wall. The first is known as **direct transmission** and the second as **flanking transmission**. Just to complicate things further, sound can also be generated by impact upon the structure, e.g. footsteps, vacuum cleaning, children playing, etc. and this can be directly transmitted or a flanking transmission. Then there is noise generated by people talking, radios, hi-fi systems and so on, which can also be transmitted as direct or flanking noise.

We will look at the more obvious permutations a little later as we look at possible solutions to transmission problems.

Calculation of surface density

To attenuate direct transmission, the first effort is usually placed on putting mass into the wall, and this should be relatively easy with a masonry construction. However, the Regulations are never quite straightforward and they require minimum densities for the bricks or blocks and the mortar used in the wall's construction. In the Regulations this is $375 \, \text{kg/m}^2$ for solid brick and $415 \, \text{kg/m}^2$ for solid concrete block walls, and there is a procedure laid down which has to be followed in calculating that mass – which may or may not include any plaster finish or isolated panels of finish on each side. 'Isolated' means the panels will not touch the masonry core and should only be fixed at floor and ceiling.

What is meant by a mass per square metre? This is termed a **surface density** and is the product of the **volumetric (bulk) density** of the material used and its thickness. For example, we wish to use a 200 thick concrete block to build a party wall. The block has a volumetric (bulk) density of 1750 kg/m^3. We propose to finish the wall each side with 13 thick gypsum plaster. The Regulations allow the calculation of plaster surface density as:

(1) Cement render   29 kg/m^2
(2) Gypsum plaster   17 kg/m^2
(3) Lightweight plaster   10 kg/m^2
(4) Plasterboard   10 kg/m^2

The Regulations provide formulae for calculating the surface density from the above data. For composite walls – i.e. those with a finish applied to a core – the formula is:

$$M = T(0.93D + 125) + NP$$

where M = the mass of 1 m^2 of leaf in kg/m^2
T = the thickness of unplastered masonry in metres
D = the bulk density for the masonry units in kg/m^3
N = the number of finished faces
P = the mass of 1 m^2 of wall finish in kg/m^2

So, if we put some of the above values into the formula we would have:

$$M = 0.200 \times ((0.93 \times 1750) + 125) + (2 \times 17)$$
$$\text{So, } M = 384.5$$

This proposal does not comply with the regulatory surface density of 415 kg/m^2. It is not far short, so what can be done? The use of a cement render would bring the surface density to just over the limit of 415 kg/m^2, as would the use of a slightly denser concrete block. For the latter we could consider the use of a block of 1800 kg/m^3:

$$M = 0.200 \times ((0.93 \times 1800) + 125) + (2 \times 17)$$
$$M \text{ is now } = 393.8$$

This is not a lot different and still fails to comply. In fact, the best solution is to use a cement-based render with the original blocks, or use a very much higher bulk density of block.

Wall types

Wall types are given in the Regulations (the Scottish Regulations' 'Deemed to Satisfy' constructions) and if followed will ensure compliance.

Wall Type 1 is a solid masonry wall of brick or concrete block with a minimum volumetric density of 1500 kg/m^3. Wall finish either side is not required but would of course be usual. Obviously, from our calculation above, a wall with the minimum density of block, 1500 kg/m^3, would require a leaf of more than 200.

Wall Type 2 is a cavity masonry wall. There are a number of possibilities:

☐ Half brick thick leaves with a 50 cavity between and 13 plaster each face
☐ 100 concrete block leaves with a 50 cavity and 13 plaster each face
☐ Half brick thick leaves with a 50 cavity between and 12.5 plasterboard lining – any fixing method, provided there is a step and/or a stagger in the alignment of the adjacent houses
☐ 100 concrete block leaves with a 50 cavity and 12.5 plasterboard lining – any fixing method, provided there is a step and/or a stagger in the alignment of the adjacent houses.

Figure 11.1 illustrates the construction of Types 1 and 2 and the idea of steps and staggers in schematic form.

Wall Type 3 is a masonry wall between isolated panels. The masonry core provides a certain amount of mass but the isolating panel brings in another method of attenuation, the flexible and absorbent layer.

Four alternatives are listed, each with a minimum surface density and clad with one of two possible isolated panels:

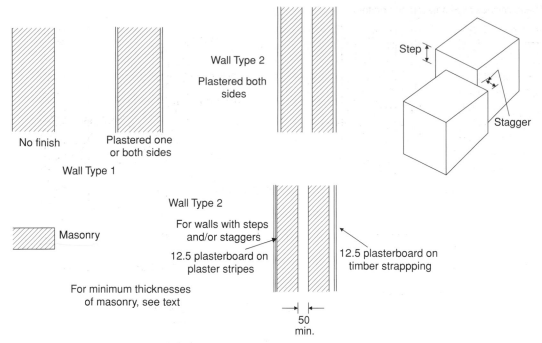

Fig. 11.1 Wall Types 1 and 2.

(1) Brickwork   300 kg/m²
(2) Concrete blockwork   300 kg/m²
(3) Lightweight aggregate blockwork
                                           200 kg/m²
(4) Aac blockwork   160 kg/m²

The panels:

(1) Two sheets of plasterboard joined with a cellular core   18 kg/m²
(2) Two layers of plasterboard placed in position broken jointed and with or without a frame. If there is a frame, two sheets of 12.5 plasterboard are sufficient, but without a frame the total thickness of plasterboard must be 30, i.e. 2 × 15 thick.

Wall Type 4 is one example only of the types of wall which can be used with timber frame housing. The type given is basically the timber frame walls of each house separated by a clear space of 200 between their facings and with a curtain of mineral wool suspended between both timber frames. The mineral wool is absorbent and so attenuates the transmission of sound very well. It should have a density

of 12–36 kg/m². The facings would normally be of plasterboard in two layers, fixed broken jointed and with a minimum thickness of 30. The inner face of the timber frames may be covered with plywood if there is any structural requirement.

A masonry core can be built between the two timber frame panelled areas but it can only be physically connected to one frame and its role is limited to one of structural support. It plays no part in the attenuation of sound. It can, however, play a part in the fire resistance.

Figure 11.2 illustrates the construction of Types 3 and 4 mutual walls.

In addition to the requirements discussed so far, the construction of these walls does require attention to other details.

Solid connections across cavities or voids must be avoided. In masonry cavity walls this might pose a problem with the provision of wall ties. However, tests have shown that while solid galvanised steel wall ties are not acceptable, butterfly wire ties are flexible enough not to upset the precautions against sound transmission. It must presumed that

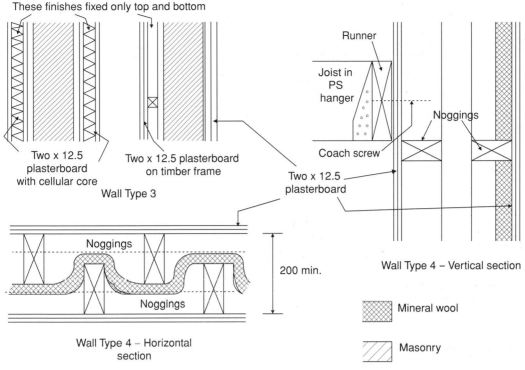

These finishes fixed only top and bottom

Two x 12.5 plasterboard with cellular core

Two x 12.5 plasterboard on timber frame

Wall Type 3

Two x 12.5 plasterboard

Noggings

Noggings

Noggings

200 min.

Wall Type 4 – Horizontal section

Runner

Joist in PS hanger

Coach screw

Noggings

Wall Type 4 – Vertical section

Mineral wool

Masonry

Fig. 11.2 Wall Types 3 and 4.

flexible plastic wall ties might perform in a similar manner although they may not be allowed from a fire resistance point of view.

Holes are not allowed in mutual walls such as we are considering here. This of course includes the incidental formation of holes due to incomplete bedding of masonry. Beds and perpends must be fully filled. Holes and cracks may also form when building materials or components shrink and crack. Where there is a danger that this might happen, the use of a flexible filling compound might be considered appropriate, and where the detail is combined with a need for fire resistance, intumescent compounds would be even better.

Equally, cavities which have not been designed in must be eliminated. For example, bricks with large frogs must be laid frog up, and perforated or cellular bricks, where mortar cannot penetrate into the cell or hole, should not be used even if laid cell facing up.

Much of what has already been written applies to sound passing through the structure directly to the next house, but flanking sound can be structure borne as well as air borne. The Regulations give detailed guidance to the requirements necessary to avoid this.

Fire resistance

Having dealt fairly simply with building a mutual wall which provides sufficient attenuation of sound travelling across it we must also know how to resist the passage of fire.

Looking again at the various wall types, we can see that Types 1, 2 and 3, with their masonry core or leaves, are inherently fire resistant, and provided attention has been paid to the sound-related problem of holes and cracks which would allow fire to penetrate, then there should be no problem providing adequate fire resistance. Type 4 walls might seem inadequate when compared with the others but the idea of separating the frames of the houses with a mineral wool quilt, or building that into the panel structure of one house, is perfectly adequate, especially when taking the effect of char on the substantial structural timbers. The mineral wool quilt is generally made up with a galvanised wire mesh reinforcement which holds the fibres together, preventing the quilt form disintegrating too soon. In the Regulations the Type 4 wall is shown as two distinct sets of studs, each in a different plane. This is done to emphasise the need to maintain the 200 gap between the face materials. However, provided that gap is maintained there is nothing to prevent alternate studs in both frames from interlinking. Provided the noggings don't touch the studs opposite and the mineral wool blanket can be woven between the studs, it will be all right. Figure 11.2 shows the arrangement.

From a fire point of view, the support of the upper floor(s) is best done by the other walls, but where other design considerations take over, then there are definite rules about what do to support the floor joists and other timbers. Runners supported on brackets or Rawlbolted to the wall are good, provided neither brackets nor bolts penetrate the skin of masonry. Runners can also be used on timber frame mutual walls, and although the detail cannot be backed up independently, the author would suggest that the runner should be placed on top of the face material and not break the run of it down the timber frame wall. Heavy gauge joist brackets built into the masonry wall are another solution. All of these have been shown earlier in Chapter 4, Timber Upper Floors. The one thing that cannot be done is to build the joists into the masonry, and it is for that reason that it would seem logical to keep joists from getting close to a timber frame mutual wall. Figure 11.2 shows how the runner might work on a timber frame mutual wall.

For an informative web site try http://www.british-gypsum.com, where you will find up-to-date information on the use of their products in solutions for all types of sound proofing and fire resistant construction. It would appear that Lafarge manufacture a honeycomb core panel under the name Panelplus, although their website does not mention it. However, for more general information try http://www.lafargeplasterboard.com

12

Plumbing and Heating

Pipework	233	Overflows	246
Pipe fittings – couplings and connections	234	Water supply from the main	246
Range of fittings	239	Equipment	247
Valves and cocks	241	Cold water storage cisterns	248
Services generally	243	Hot water storage cylinders	248
Hot and cold water services	243	Feed and expansion tanks	251
Soil and ventilation stacks	246	Central heating	252

Remembering that we are confining our study to domestic walk-up construction, our text in this chapter will be confined to a fairly basic look at the current materials and techniques applied to the provision of hot and cold water supply, water-borne heating systems and systems for waste disposal within one- and two-storey housing.

All water installations are subject to the provisions of the Model Water Byelaws. To discuss the provisions of these in detail would be impossible in this text; however, the work described here will comply with these provisions. The byelaws are particularly concerned to prevent waste, misuse, undue consumption or contamination of the water supply. Further byelaws are published by the various supply companies or authorities across the UK, and these supplement or expand on the Model Water Byelaws. It is important to comply with both sets of byelaws. For further detail on the byelaws refer to the website of the local supplier or authority.

Pipework

To begin, we will look at the materials and components used. The most common material is pipe and the fittings required to join pipe to pipe and pipe to appliances and equipment. Pipe is manufactured in a wide range of materials, of which those listed in Table 12.1 have relevance in modern plumbing.

On cost, the use of cast iron is confined to those parts where the strength of the material is important, e.g. exposed lengths of pipe which might be the subject of vandalism. PVC pipe can be damaged quite easily whereas cast iron is much more resistant to attack. At the time of writing, the cost of cast iron pipe was roughly twice that of uPVC. Cast iron was also used for waste disposal, the size of pipe being 2 in (50), with smaller bore copper pipe (32 and 40) being used to connect sanitary appliances to the 2 in cast iron. The advent of plastics waste pipework at a much reduced cost has superseded the use of cast iron and copper in this way. Just for comparison, 50 copper pipe work is roughly three times as expensive as 50 ABS pipework.

On top of the obvious cost savings in favour of using plastics for waste disposal pipework, there are two other factors involved. First, the fittings for cast iron and copper pipe work are very expensive compared to plastic fittings, and second the plastic pipework and fittings can be put together much more quickly and

Table 12.1 Materials used for pipe.

Material	Use	Sizes used (mm)
uPVC (unplasticised polyvinyl chloride)	Disposal pipework from WCs and ventilating stacks	100 nominal, 80 nominal
	Ducts through walls for service entry	32, 40 and 50
	Disposal pipework from sinks, WHBs, showers and baths	
Cast iron	Disposal pipework from WCs and ventilating stacks	100 nominal
cPVC (chlorinated PVC)	Hot and cold water supply Central heating pipe work	15, 22, 28
ABS (acrylonitrile butadiene styrene)	Disposal pipework from sinks, WHBs, showers and baths	20, 32, 40 and 50
Polypropylene	Disposal pipework from sinks, WHBs, showers and baths	20, 32, 40 and 50
PE-X (cross-linked polyethylene)	Hot and cold water supply Central heating pipework, particularly underfloor heating*	15, 22, 28
Polybutylene	Hot and cold water supply Central heating pipework*	15, 22, 28
Copper	Hot and cold water supply Central heating pipework Gas supply pipe	15, 22, 28
Stainless steel	Hot and cold water supply Central heating pipework Gas supply pipe	15, 22, 28
Black iron	Central heating pipework	15, 22
Galvanised iron	Gas supply pipe	15, 22 and 23

*When used for central heating pipework, ordinary plastics pipe will allow oxygen to diffuse in through the pipe wall. A special grade of pipe is manufactured with an oxygen barrier in the wall of the pipe.

without the use of heat. This is of prime importance to any developer who must keep a contract moving swiftly to a conclusion. Any practice which takes up a lot of time has to be eliminated and the use of cast iron and copper for waste disposal does fall into that category.

Stainless steel pipework has fallen out of favour for domestic pipework, largely due to the cost of the special fittings required. Being a much harder material, ordinary pipe fittings used for copper pipe cannot be used on stainless steel. Also, the price advantage that stainless steel enjoyed has now disappeared.

Pipe fittings – couplings and connections

When joining pipe to pipe, plumbers use **couplings**. When joining pipework to appliances or equipment, they use **connections**.

Couplings fall into a number of categories, depending on the way the coupling works and, occasionally, the type of pipe to be joined:

- Compression fittings for copper, black iron and plastics pipework
- Capillary fittings for copper pipework
  - End feed soldered joints
  - Integral solder ring joints
- Push fit fittings
  - For copper and plastics pipework
  - For plastics waste pipework
- Solvent welded fittings
- Socket fusion and butt fusion joints
- Electrofusion.

Compression fittings

Compression fittings are made of either brass or bronze, the latter being more expensive but necessary to overcome 'dezincification' caused in some areas with 'aggressive' water. Brass is an alloy of copper and zinc, and where the water in a supply is aggressive it will remove the zinc from the alloy and leave behind a porous copper structure which will initially weep water then fail completely. Also, fittings on external supply piping which is buried in the ground and has aggressive ground water, will be attacked. The resistant fittings are termed **DZR**.

Compression fittings are, principally, of two types. The first is where the end of the pipe being joined is manipulated in some way, the most usual method being flaring the pipe and squeezing this flared end between the parts of the fitting. A variation on this is to insert a tool into the end of the pipe which, when turned, raises a ridge on the pipe which gives the joint mechanical strength. The second is a non-manipulative fitting. Here the pipe is merely cut square across and the fitting assembled on the ends, the mechanical strength and water seal being made by squeezing an internal ring in the fitting on each pipe end. As a general rule, the manipulative fitting can be used on copper pipe and the non-manipulative fitting on plastic and copper pipe. Care has to be taken to select a fitting which the pipe manufacturer recommends.

Jointing compound or sealing tape must be wound round the metallic parts of the seal when joining copper pipe, and tape only is used with plastics pipe. The tape used is a thin 12 wide tape of polytetrafluorethylene (PTFE). The tape is soft and stretches as it is wound into place. When pressure is applied as the joint is tightened, it flows into any spaces and seals the joint completely. Note that ordinary PTFE tape is unsuitable if the pipework is to carry natural gas. A special tape is manufactured for that purpose.

Figure 12.1 shows an exploded photograph of a non-manipulative compression fitting being assembled on copper piping on the left, and plastic piping on the right. Note the body of the fitting fitted over the end of the copper pipe and on each pipe the compression ring and the compression nut ready to slide up and squeeze the ring down onto the pipe once the sealing compound or tape has been

Fig. 12.1 Compression fitting on copper and polybutylene piping.

put in place. The end of the plastic pipe is left exposed to show the stainless steel sleeve insert, another of which is shown separately just below. Below this again is a cutting of copper pipe taken from an old joint, showing the compression ring squeezed onto the pipe with the PTFE tape around it.

Non-manipulative compression fittings have been used on black iron pipe for central heating pipework. Although the pipe is made by folding a strip into a tube and welding it closed, the ridge on the outside of the pipe is ground off, leaving a circular section which can be sealed by a fitting and either joint sealer or PTFE tape.

Both of the above joints are suitable for use where the pipework is under pressure, such as hot and cold water services or gas supply piping.

The non-manipulative compression fittings used on plastics waste pipe (they can also be used on copper pipe of appropriate diameter) follow the same general design principles but without the need to withstand more than a modest pressure. The wedge-shaped section design of the sealing ring is only about obtaining a seal and a modest grip on the pipe. A liner is not required as the plastics of the pipe is generally quite rigid.

Figure 12.2 shows, from left to right, a straight coupling with the pipe in the body of the coupling, then, on the pipe, the sealing ring, a hard fibre washer and the compression nut. The fibre washer provides a

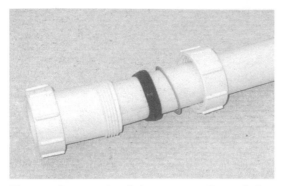

Fig. 12.2 Compression fitting on waste disposal pipe.

smooth surface between the compression nut and the sealing ring. The sealing ring, being soft synthetic rubber, would stick to the nut and would turn with it on tightening and become distorted, destroying the seal.

Capillary fittings

Capillary joints can be made on copper and stainless steel pipe; however, the remarks here apply only to copper pipe. Capillary joints are soft-soldered joints made by pushing the ends of the pipe into plain sockets on the fitting and applying heat and/or solder. To solder successfully, the ends of the pipe and the pipe sockets must be clean[1] and must be coated with **flux**. Flux is a chemical which, when the heating of the joint takes place, forms a layer over the metal which excludes the air trying to oxidise it and also helps the flow of molten solder over the metal. Some fluxes are also chemical cleaners, removing the oxidised layer which forms over the surface of the pipe and fittings between manufacture and final installation. It is still best to clean the pipe with steel wool. Heat is usually applied using a gas torch fuelled from a disposable or refillable cartridge; there is, however, a tool which clamps round the joint one socket at a time and is heated by electricity. The only problem is the need to change the size of the clamp every time a different socket size

is encountered. End feed fittings have plain sockets which fit closely round the pipe, and when the clean pipe is inserted and heated, the solder is introduced at the junction of pipe and fitting, the molten metal being drawn in between the two by capillary attraction.

Integral solder ring fittings are made with an annular groove in each close-fitting socket which, during manufacture, is filled with solder. All that is necessary is to clean and flux the pipe and fitting, push together, and heat until the solder appears as a bright ring at the junction of pipe and fitting.

It is important to note that solders can be made of alloys with a high lead content which is unsuitable for use in hot and cold water services. The lead content of the solder can be leached out and so poison the users. Where capillary fittings with an integral solder ring are used in H&C services, they should be **potable water fittings** which have tin-based solder. Tin-based solder must also be used with end feed capillary fittings in hot and cold water supply piping. All manufacturers make fittings with both types of solder. There is nothing wrong in using lead-based solder in the installation of central heating, etc.

Although it might seem obvious, it has to mentioned that all the pipe ends must be soldered into any one fitting at the same time. While it is possible to solder in one pipe end and then solder in another at a later stage, this would melt the solder in the first joint and it would not be possible to guarantee a waterproof or gasproof joint.

Figure 12.3 shows a capillary fitting, an elbow, with integral solder ring joining two copper pipes together.

All these joints are suitable for use where the pipework is under pressure such as hot and cold water services or gas supply piping.

Push fit fittings

Push fit joints are of two types. The first has a simple sealing ring which is squeezed between the fitting and the pipe being inserted. This type is only suitable for disposal

---

[1] Cleaning is usually done by rubbing the end with steel wool.

Fig. 12.3 Capillary fitting joining two lengths of pipe.

Fig. 12.5 Push fit fittings for hot and cold supply pipework.

pipework and is the most common type found on large pipework for soil waste and on ventilation pipework, and is also found on the bores 32, 40 and 50 waste pipe.

The second type of joint has a sealing ring squeezed between the pipe and the fitting, but also a grab ring which resists any tendency for the joint to pull apart. There is no such tendency with disposal pipework but there is with any pressurised system such as hot and cold water supply and central heating. Push fit joints are not used for gas supply pipework.

Figure 12.4 shows push fit fittings suitable for waste disposal pipe. The seal shows as a black ring inside the socket. In the fitting shown it is circular in cross-section and

is known as an O-ring seal. Seals on larger diameter pipes have been made of D-section material and as a flat section with a tapered wiping seal attached resembling a T-section.

Figure 12.5 shows push fit fittings suitable for hot and cold supply pipe work. On the left is a copper pipe, on the right a plastic pipe and in the middle the body of the fitting. On the copper pipe, starting from the left, there is a nut – not a compression nut, but simply a nut to keep all the other components alongside it inside the body of the coupling. After the nut there is the 'O' ring seal, a spacer of plastic, the grab ring of stainless steel (the 'barbs' on the grab ring are clearly visible) and finally another plastic spacer. The grab ring is at the point on the pipe where it would be when all the components were in the body of the pipe and the nut screwed back in place on the body. The components on the plastic pipe are exactly the same and in the same order.

Solvent welded fittings

Solvent welded fittings are mainly used now for disposal pipework. The fittings are made with sockets which fit closely round the pipe for which they are intended. There is no cleaning to be done unless the fitting or pipe is very dirty. Cut the pipe square, coat the socket with solvent weld cement and push the pipe into the socket with a quarter twist. Leave the joint undisturbed for a few minutes until an initial set of the weld has taken place, and continue with the rest of the joints and pipework. The principle is quite simple. The

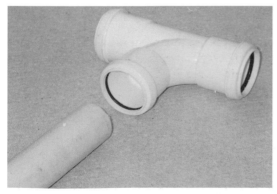

Fig. 12.4 Push fit fitting for waste disposal pipe.

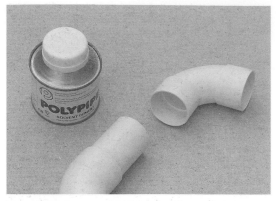

Fig. 12.6 Solvent weld fitting for disposal piping.

'cement' comprises plastic in a plastic solvent. The dissolved material acts as a filler, the solvent literally softening the surface of both pipe and socket, and the three elements – pipe, socket and filler – then fuse together when confined in the space between socket and pipe, the quarter turn mixing all the dissolved components together. Figure 12.6 shows a solvent weld fitting for disposal piping.

Fusion joints

Socket fusion, butt fusion and electrofusion jointing are well beyond the scope of this text, as they are relatively expensive processes to employ for simple domestic pipework. They are, however, widely used by the gas supply industry, especially in the mains supply in the street and to the house, where plastic gas pipework is now common and where mixed materials installations can lead to problems with corrosion etc.

Joining pipes to appliances and accessories

Having given a brief explanation of joining pipes we must now look at the joining of pipe to appliances and accessories. In the domestic market there are three possibilities which arise when joining a pipe to an appliance:

☐ The appliance has an externally threaded end protruding – **a male thread**
☐ The appliance has a protruding socket with an internal thread – a **female thread**
☐ The appliance has no threads or sockets protruding but may or may not have a hole for a fitting.

In the first situation above, the most common method of joining the pipe depends on the **bore** of the protruding end. This may or may not accept one of the standard diameters of copper or plastic supply piping, 15, 22 and 28.

Should the end accept pipe, then a variation of the non-manipulative compression joint is used. This is common when connecting pipe to taps. Should the end not accept pipe, then a special connection must be used. This has an end which accepts the pipe and can utilise any method to accept the pipe – compression fitting, grab ring, push fit fitting, capillary fitting, and so on, but the other end is an internally threaded socket – a female thread. Such a fitting is described as a **female iron to copper connection** – 'female iron' – because it is a female thread and is a form of thread first developed to make threaded joint on iron pipe, and 'to copper' because it was to copper pipe that these compression type joints were first developed – even if it is a grab ring, push fit joint.

The joint is, of course, mechanically strong to resist pressure and the seal of the threaded portion is made by using jointing compound with oakum or cotton waste or with PTFE tape to seal the spaces between the threads. If the male thread is very long, the connection will bottom on the spigot, and to ensure that the jointing material applied to the thread is kept tight inside the joint a **back nut** is screwed on to the male end before the fitting and then it is screwed back tight against the fitting, trapping the joint material between itself and the fitting. If that isn't complicated enough, the connector can be straight or bent, i.e. the pipe and fitting are in line or the pipe and fitting are at right angles to each other, respectively.

Fig. 12.7 Female iron to copper fitting connection.

Figure 12.7 shows a female iron to copper fitting connecting to the threaded spigot on a tap. There is a back nut there ready to screw down on the fitting, trapping jointing compound or tape and ensuring a watertight seal.

Figure 12.8 shows a male iron to copper fitting with a back nut ready to go through a thin wall piece of equipment such as a water cistern. Washers would be put in place to make a watertight seal.

Range of fittings

The range of fittings available is quite staggering, and the text would be lengthened consid-

Fig. 12.8 Male iron to copper connection with a back nut.

Fig. 12.9 Selection of bends, elbows, straight couplings and Ts all with compression joints.

erably by including a detailed description of each one. They all come as straight couplings or connections, bent, elbows, Ts, swept Ts, crosses, and all of these could have flanges for fixing to some kind of background. Even manufacturers have their unique extensive range, and yet it is sometimes necessary to use different makes of fitting to make a neat job of an installation. The best we can do here in the limited space is provide some photographs of generic types of fitting and refer the reader to manufacturers' catalogues and websites.

Figure 12.9 shows a selection of bends, elbows, straight couplings and Ts, all with compression joints. From the top left and moving clockwise: a straight coupling; an elbow; a male iron to copper connection; a female iron to copper connection; a T with branch reduced; a plain T.

Figure 12.10 shows a selection of straight and bent capillary couplings and Ts. At the bottom right is a straight tap connector and immediately above it a blank end.

Figure 12.11 shows a selection of bent and straight, male and female iron connections all with compression joints.

Figure 12.12 shows a selection of push fit couplings for hot and cold water pipe work. In the centre is a male iron to copper straight

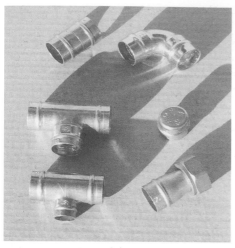

Fig. 12.10 Selection of straight and bent capillary couplings and Ts.

Fig. 12.12 Selection of push fit couplings for hot and cold water pipework.

connector and beside it is a back nut. The fitting on the top right with the four branch coupling is called a manifold and might be used in central heating installations where larger bore pipework supplies hot water to a central point and then the four branches distribute this to smaller bore pipes connected to the emitters. Moving clockwise there are: a cross with branches reduced; a tap connector; a T with end reduced; a T with both ends reduced; a straight coupling and an elbow.

Figure 12.13 shows a selection of compression couplings for disposal pipework. From the left and clockwise there is an elbow, a swept T, a straight coupling and an obtuse bent coupling.

Figure 12.14 shows a selection of solvent weld couplings for disposal pipework. From the top left and working clockwise there is a swept bend, a female iron to waste pipe connection, pipe clips (not exclusive to solvent welded installations), a straight coupling, an elbow, an obtuse bent coupling and a 45° T piece.

Manufacturers generally print catalogues and mini-handbooks of their fittings. The latter are about A5 or A6 size and are ideal for the student to obtain. Try your local plumbers merchant for some literature or visit these websites. Some will allow you to order at least some general literature:
www.yorkshirecoppertube.com
www.imiyf.com

Fig. 12.11 Selection of bent and straight connections, all with compression joints.

Fig. 12.13 Selection of compression couplings for disposal pipework.

Fig. 12.14 Selection of solvent weld couplings for disposal pipework.

www.hepworthplumbing.co.uk
www.polypipe.com/bp
www.peglerhattersley.com

Valves and cocks

The carrying of water to equipment and appliances has to be capable of being shut off so that pipework and appliances etc. can be repaired or maintained. This is achieved by inserting valves or cocks at various points in the pipework. The websites above will include information on the manufacturers' own ideas about valves and cocks. There are a number of different types of valve and cock known by a variety of names, and some by the same name as a totally different product. Table 12.2 attempts to sort out the confusion.

All of these valves, in their most basic form, are supplied with **tails for copper**. Essentially this means they join to the pipe at either end with a non-manipulative compression joint. Manufacturers such as IMI Yorkshire Fittings Ltd make them with capillary joints as well as compression joints. Some of the manufacturers of plastic H&C pipe, such as Hepworth and Polypipe, manufacture valves with push fit ends with a grab ring. Drain down valves can be obtained where they are part of another fitting, such as a bent coupling

Table 12.2 Valves and cocks, nomenclature.

Common name	Correct name
Screw down valve, stop cock	Screw down valve
Gate valve	Gate valve
Ball valve	Ball valve
Ball valve, ball cock, float valve	Float valve
Plug cock, stop cock, stop and frost cock	Plug cock
Drain down valve	Drain down valve

or elbow or even a T. Valves can also be obtained with flanges for fixing to some type of background.

Depending on the manufacturer, brass and DZR versions of the valves are made, and there can be one or both ends with male or female threads. So one could have one end for copper and one male or female end, or both male ends or both female ends. To the best of the author's knowledge no-one has made a valve with one male and one female end. Having valves with single male or female ends means that a connection can be made to an appliance and then the pipe connected directly to the valve.

Figure 12.15 shows schematics of the distinct valve and cock types. Follow these sketches as you examine the photographs which follow.

Figure 12.16 shows a selection of valves, starting at the top right and moving clockwise: a plastic plug cock with push fit ends for copper or plastic pipe; a drain down valve on a (long) straight coupling; a drain down valve male iron end; a ball valve operated by using a screwdriver in the slot – the slot shows it to be open; a gas cock; a gate valve; a screw down stop valve.

Figure 12.17 shows a plug or stop cock with compression tails for copper. The position of the rectangular top shows it to be closed.

Figure 12.18 shows a pair of float valves, the upper one is a diaphragm float valve, the lower one a plunger float valve, known as a Portsmouth pattern valve. The floats have been taken off.

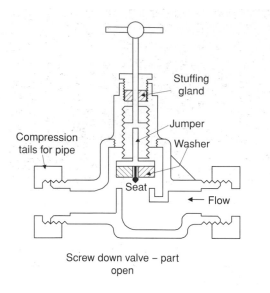

Stuffing gland

Jumper

Washer

Compression tails for pipe

Seat

← Flow

Screw down valve – part open

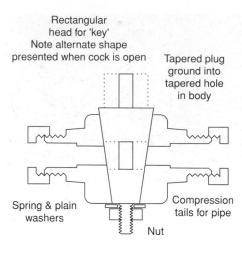

Rectangular head for 'key'
Note alternate shape presented when cock is open

Tapered plug ground into tapered hole in body

Spring & plain washers

Compression tails for pipe

Nut

Plug cock – shut

Alignment of hole through ball and screwdriver slot to show valve in open or closed position

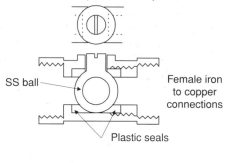

SS ball

Female iron to copper connections

Plastic seals

Ball valve – closed

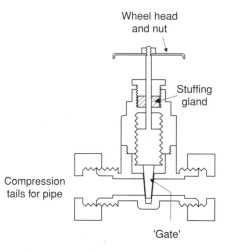

Wheel head and nut

Stuffing gland

Compression tails for pipe

'Gate'

Gate valve – part open

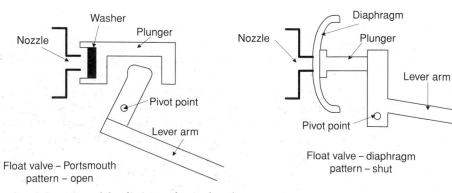

Washer

Plunger

Nozzle

Pivot point

Lever arm

Float valve – Portsmouth pattern – open

Diaphragm

Nozzle

Plunger

Lever arm

Pivot point

Float valve – diaphragm pattern – shut

Fig. 12.15 Schematics of the distinct valve and cock types.

Fig. 12.16 Selection of valves.

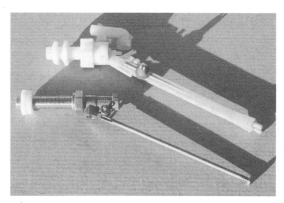

Fig. 12.18 A pair of float valves.

Figure 12.19 shows an exploded version of the Portsmouth pattern valve. Across the bottom is the lever arm with the float and at the other end the hole for the pivot pin and the ear-like appendages which engage in a groove in the plunger. Above, from left to right, a pair of back nuts and a washer to secure the valve in the tank or cistern; the male iron spigot with the nut which attaches it to the valve body; the valve body with the nozzle and the hole for the pivot pin; below the body, the pivot pin – a brass split pin; the plunger which has a soft washer in the end and a groove along the side for the ears on the lever arm; a cap to keep the plunger in the body. What cannot be

seen is the nozzle in the body against which the washer on the plunger presses to seal off the water flow.

Services generally

Before we go on to look at appliances and equipment, we should look at the classification of services which are piped around a typical house and the kind of pipe most commonly used. Table 12.3 summarises that information.

Hot and cold water services

Cold water is brought into a building from the water main, which is situated outside the boundary of the property. The supply

Fig. 12.17 A plug cock with compression tails for copper.

Fig. 12.19 Exploded version of the Portsmouth pattern valve.

Table 12.3 Service and pipe used.

Service	Type of pipe
Hot and cold water supply	Copper, polybutylene for refurbishment work
Central heating	Copper
Waste piping under 50	ABS, polypropylene, uPVC
Soil pipe over 80	uPVC
Overflows	ABS, polypropylene, uPVC
Pipe sleeves*	Any plastics pipe

*A pipe sleeve is a cutting of pipe built into a masonry wall through which a service pipe is passed. It protects the service pipe and allows it to expand and contract or move lengthwise in the sleeve without damage to itself or the wall. It always bridges cavities in walls and so prevents pipe contents from spilling into the cavity in the event of a leak. A pipe sleeve is mandatory for gas service pipes passing through a cavity or any other walls. There should never be any pipe joints secreted in a pipe sleeve.

authority arranges for the tapping into the main, fitting an underground valve with an access cover – a **toby** – made of cast iron sitting on a brick or precast concrete seat and leaving a length of water pipe just inside the property to which the plumber will connect the supply pipework. Depending on the region in which building takes place, the water authority may also provide a water meter so that the cost of water used is charged per litre rather than being an unmeasured supply charged for with the Council Tax. The water authority will make a charge for the work involved in a connection to their main supply pipe.

The plumber will use a blue polyethylene pipe to bring the water supply from the tail left by the supply authority into the house. This is done via a duct under the foundation and up through the solum or concrete ground floor of the house. This pipe ends with a valve, which can be used to shut off the water supply to the house. If there is a storage tank in the roof space, a pipe is run to the tank from this first valve. This pipe is known as a **rising main**. Generally there is a branch off this pipe to the cold tap in the kitchen which supplies fresh water for drinking and cooking. Ideally the branch should have a valve as close to the rising main as possible so that the branch pipework and the kitchen cold tap can be isolated. If no tank is supplied for the

storage of water then pipework must be taken to all cold water draw-off points in the house. These would include kitchen sink, bath(s), shower(s), WC(s), WHB(s), water heater(s), hot water storage cylinder(s), central heating header tank, etc. If a tank is supplied, then pipework is run from the tank to all the cold water draw-off points in the house, listed above. That pipework is known as the **cold feed**.

The reason for having a storage tank in the roof space is quite simple. In the event of a temporary cut in the mains water supply, the tank will still supply water for flushing WC(s), hand washing, dish washing and for the hot water system. Without a storage tank any devices to control the water flow into appliances or equipment must be capable of dealing with water at mains pressure.

Hot water supply can be arranged in two ways. A **hot water cylinder** can be installed and pipework put in to take hot water to all appliances which require it. Various means can be adopted to heat the water in the cylinder. Hot water is drawn off the cylinder via a pipe from the top of the cylinder and cold water is fed in near the bottom from the **cold feed**. A valve should be placed on the cold feed to the cylinder. Shutting the valve stops hot water being delivered at any tap as there is no pressure of cold water to force it out of the cylinder. It is normal to run cold feed pipe

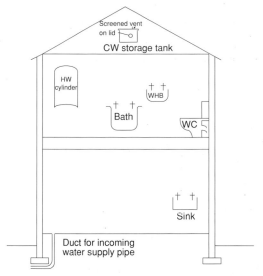

Fig. 12.20 Identification of the appliances and equipment to be connected up.

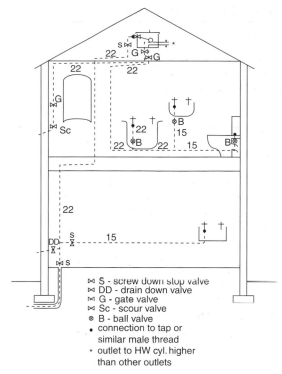

⋈ S - screw down stop valve
⋈ DD - drain down valve
⋈ G - gate valve
⋈ Sc - scour valve
⊗ B - ball valve
• connection to tap or similar male thread
* outlet to HW cyl. higher than other outlets

Fig. 12.21 Schematic of the cold water supply pipework.

through a tee piece at the bottom of the cylinder, the branch being a male iron spigot (tees like this are often referred to as boiler tees as that is where they were most commonly fitted). This branch is connected to the cylinder, and the other end of the tee connects to a **scour valve** (usually a plug cock) with a pipe going to the outside of the building. Opening the scour valve empties the hot water system completely above the level of the scour valve. If the valve on the cold feed is left open, the cold water cistern is emptied.

The arrangement in Figure 12.20 will be used for schematics of various classes of pipework and identifies the appliances and equipment to be connected up.

Figure 12.21 is a schematic of the cold water supply pipework. Note that screw down valves are used where there is mains pressure against the valve when it is closed. Gate valves are used on cold feeds since they have a straight through opening and pressure is not lost in any significant quantity.

Figure 12.22 is a schematic of the hot water supply pipework. Note that a gate valve is placed on the hot feed to the sanitary appliances but there is *never* any valve put on the open vent. If a valve was placed there, closed

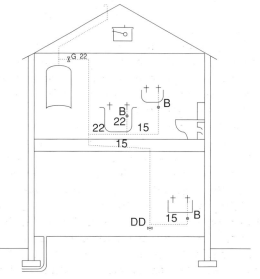

Fig. 12.22 Schematic of the hot water supply pipework.

and the cylinder heated, the system would explode.

The schematics above show only the appliances with the hot and cold water supply piping. The basic drawing from Figure 12.20 will be repeated showing individual pipework networks as they come up for discussion, such as soil waste and ventilation pipework, waste pipework, central heating pipework and overflow pipework.

Soil and ventilation stacks

Disposal of waste from the sanitary appliances shown in the previous schematics must also be considered. It is usually carried out in plastics pipe as previously explained. It is allowable, but not often done, to make the main pipe from the WC an 83 pipe. More usual is the use of 100 nominal bore pipe. The term 'nominal' is applied here for although all pipes are measured outside diameter, 'od', the individual manufacturers use different wall thicknesses and slightly different bores so that it is not always possible to mix and match fitting and pipe from different sources – some will, some won't.

Figure 12.23 shows a schematic for the appliances already piped up for water supply. Note

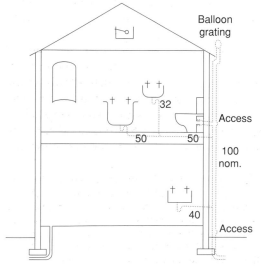

Fig. 12.23 Soil, waste and ventilation pipework.

the access on the main vertical pipe at the bottom at ground level and opposite the junction with the WC branch. Note also that the pipe continues vertically up past the eaves level with just two bends. It is allowable to have *only* two bends above the last WC junction on this part of the stack. The top of the pipe is open and protected from debris and birds entering by being fitted with a grid or grating of some type. The open end is important for ventilating the drainage network and the sewer network. In old installations the drainage was kept sealed off from the sewers by a trap called a **disconnecting trap**. The trap incorporated a fresh air inlet for the house drains. Sewers were ventilated at their highest point and the Victorian drainage engineers frequently disguised these large ventilation stacks as flag poles, factory chimneys, etc.

Overflows

Overflows must be fitted to any storage cisterns which are fed through a float valve. Should the valve stick in the open position, not close properly because of dirt on the seal, or otherwise fail, the cistern would quickly overflow and the water could cause much damage to the fabric and decoration of the building. Overflows allow any excess water to be discharged outside the house, preventing spillage in the house. The discharge point should be arranged so that it is in as conspicuous a place as possible. Over a kitchen window or a front door are good places. Figure 12.24 shows typical overflows required. Pipework is usually plastic, the larger bores being in the same material as the waste piping, the smaller pipe for the WC overflow being a one-off.

Water supply from the main

External services will not be dealt with here but it should be noted that a supply from the water main in the street is arranged by the supply company or water authority in the

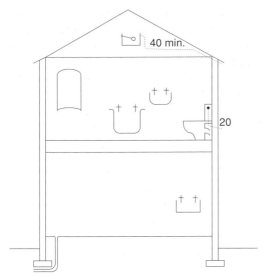

Fig. 12.24 Overflow pipework.

Fig. 12.25 Cast iron toby cover in the street.

area, and they will arrange to tap the main and lay a blue plastic water pipe to the house site boundary, where a toby is built into the ground, fitted with a cast iron cover. At the bottom of the toby cover, the plumber will join a blue plastic pipe to the main with a 'stop cock', in fact a **plug cock**. The rectangular head means that a special key is required to open and close the plug cock. Old installations used a special plug cock which had a hole in the side of the cock body and a three-way port in the plug such that when the straight through hole in the plug was in the off position – against the mains pressure – the third port in the plug was opposite the hole in the body. This allowed the water in the rising main to drain back down and out into the earth surrounding the stop cock. What it meant was that there was no water left in the pipe to freeze and burst the pipe. This is not such a problem with plastic pipework.

Figure 12.25 shows a cast iron toby cover in the street.

Equipment

Equipment in a domestic situation normally includes only a **hot water cylinder**, a **water**

storage cistern and a **feed and expansion tank** where **open vented central heating** has been installed. The HW cylinder is usually placed on the highest floor in the house, and the cistern and tank in the roof space.

Assuming that there is a water storage cistern in the house, this is placed on a boarded area supported on timber bearers which span across several ceiling ties in the roof structure so that the load is spread as far as feasible. The rising main is connected to a **ball cock**[2] mounted on the side of the tank near the top of the rim. Cold feeds are taken out of the tank near the bottom, one for the general supply of cold water and one for the supply of cold water to the hot water cylinder. General cold water feeds are taken off the tank about 50 from the bottom, but the feed to the hot water cylinder is taken off about 60–75 above the bottom. This means that the pressure forcing hot water out of the cylinder will stop before the supply of cold water to taps fails, which should obviate scalding when occupants are

---

[2] It is easy to confuse the two valves – ball valves and ball cocks – but the ball valve only appears in the run of pipework and is an actual ball, whereas the ball cock is a plunger or diaphragm valve actuated by a float, commonly in the shape of a ball. The correct name for this latter valve is a float valve, and there are at least five patterns to choose from. Many people in the trade use the terms indiscriminately.

using mixer taps on baths to supply a shower, or those rubber push-on attachments on bath or basin taps. Additional individual cold feeds are taken off the tank to supply cold water to showers which rely on the hot water cylinder for their hot water supply, and again these should be taken off the tank below any separate feed to a hot water cylinder.

Cold water storage cisterns

Cold water storage tanks or cisterns have been made from asbestos cement and galvanised steel sheet, both of which required painting with a black bituminous compound after the holes were cut and before fitting in place. They are now invariably made from plastics – poly-olefin or olefin copolymer.

They are available in rectilinear or cylindrical form. Plastic cisterns are very flexible when empty, but so long as they have full support on their base the weight of the water makes their sides quite rigid. The sides could deform where the ball cock enters due to the lever action of the ball and arm, but manufacturers supply a galvanised steel reinforcing plate with each tank to overcome this. Other connections pose no such problems. The benefit of installing a plastic cistern is that if it fails it can be brought out of the roof easily through a relatively small ceiling hatch – and the new tank goes back into the roof just as easily.

Cisterns are made in a range of capacities from 18 to 500 litres, the water line on the two smaller tanks being 100 below the rim and on all others 115 below the rim. To comply with the appropriate British Standard, the water line must be marked on the inside of the cistern. It is between the rim and the water line that the hole for the ball cock and the hole for the overflow must be drilled, the overflow being nearer the water line than the ball cock. All cisterns must be installed with a lid to prevent the entry of dirt and vermin. The lid must incorporate a ventilating cap with a screen to keep out insects, and the overflow must turn down inside the tank below the water line.

Generally a capacity of around 250–300 litres is adequate for the average three-bedroom house; however, in areas where the supply is not always reliable or there could be pressure drops etc. a larger capacity might be advantageous. If this cannot be accommodated in a confined roof space as a single cistern, two or more cisterns can be linked together. This is done with normal water pipe and fittings joining one cistern to the next in series. The rising main is connected to the ball cock in the first cistern and cold feeds are taken from the last tank in the series. Connected like this there is a constant flow through all the cisterns whenever water is drawn off and no cistern goes stagnant. All tanks should have an overflow pipe connected to a common pipe discharging at eaves level.

Figure 12.26 shows a cold water storage cistern and linked cisterns.

Hot water storage cylinders

Hot water storage cylinders have been made from galvanised steel and from polypropylene but by far the most common material is copper. HW cylinders are made in capacities from around 90 to 500 litres, and their shape

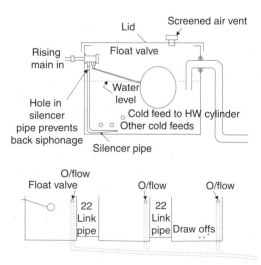

Fig. 12.26 Cold water storage cistern and linked cisterns.

is an upright cylinder with a domed top and bottom, the bottom dome being internal and the top being external.

Each cylinder has a number of solid brass connecting points for pipework brazed to the copper. These may be female iron or male iron, the former being the more common. Copper cylinders are manufactured in three grades, 1, 2 and 3. The difference is the thickness of metal used and therefore the pressure exerted by the head of water and when the cylinder is heated. Grade 1 cylinders have the thickest metal and their maximum working head of water is 25 m. Grade 2 have a maximum working head of water of 15 m and Grade 3's maximum working head of water is 10 m.

HW cylinders are available in capacities from 65 to 210 litres, a capacity of around 150 litres being suitable for a typical three-bedroom house. HW cylinders can be purchased with or without a layer of insulation, but if without they must be insulated when installed with a proper insulating jacket. HW cylinders are manufactured in a number of configurations depending on the source of heating used for the water and bearing in mind that people will come in contact with the heated water.

Pretty well all cylinders manufactured now have provision for a top entry immersion heater or circulator powered by electricity. An immersion heater is a simple heating coil inserted into the cylinder and so surrounded by all the water in the cylinder. A circulator has a heating coil as well but it is enclosed in a close fitting tube, open top and bottom. This heats the water very fast and brings the hot water to the top of the cylinder without it mixing with the rest of the cold water. Thus a circulator will give a faster initial quantity of hot water than an immersion heater. It does not appear to be any faster or slower to heat the remaining water in the cylinder. Whether an immersion is installed or not is a matter of choice for the occupier or builder, but it is seldom the only way to heat the water.

The most simple way is to connect two pipes to the side of the cylinder, one near the bottom and one near the top. The other ends are connected to a boiler in the same orientation. The water in the boiler rises in the upper pipe into the cylinder pushing cooler water down the lower pipe into the boiler, where it is heated until the whole of the HW cylinder has hot water – and then the water could start to boil.

Heated water expands, and this expansion takes place up the open vent pipe fitted to the top of the cylinder and into the feed and expansion tank (usually) placed in the roof space. Unless the heat to the boiler can be controlled, this is what will happen and it is not uncommon to see expansion pipes spurting hot water and steam because the heat source is a solid fuel fire or stove.

A cylinder with this method of heating the water is known as a **direct cylinder**. Note that there are no controls when a solid fuel fire is used for heating the water and the water only transfers from boiler to cylinder and back by gravity: the hot water rising, the cooled water sinking. In Figure 12.27, the boiler is a rectilinear 'box' of thick copper plate with four 22 female threaded connections, two in each end. One takes the cold feed from the cistern in the roof, and the other, higher, connection takes heated water up to the top of the HW cylinder. At the other end the upper connection is blanked off and the lower connection receives cooled water from the cylinder. The cold feed to the boiler is also the cold feed to the HW cylinder. The open vent pipe is the open vent and expansion pipe for the whole system.

Figure 12.27 shows a direct HW storage cylinder connected to a back boiler such as one would find in an open fireplace or closed stove. The pipe which carries hot water to the HW cylinder is termed the **flow pipe** or simply **the flow**; the pipe bringing cooled water back to the boiler is termed the **return pipe** or simply **the return**.

A further development was the **indirect cylinder**, and this is made in two versions: **single feed** and **double feed**. Figure 12.28 is a photograph of an indirect HW storage cylinder with factory applied insulation. Note the immersion heater fitted on the top, and that the two connections for the boiler circuit water

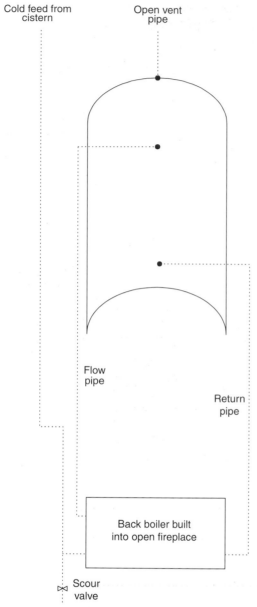

Cold feed from cistern

Open vent pipe

Flow pipe

Return pipe

Back boiler built into open fireplace

Scour valve

Fig. 12.27 A direct HW storage cylinder connected to a back boiler.

Fig. 12.28 An indirect HW storage cylinder with factory applied insulation.

have protective plastic sleeves fitted over the male iron ends.

Taking the **double entry indirect cylinder** first is not the most obvious choice but it is the easier one to understand. In this type the cylinder is supplied with water from the cold cistern, and the HW cylinder contains a coil of copper pipe – a calorifier – the ends of which are connected to a boiler, which is fed with water separately from the HW cylinder. Thus the water in the cylinder does not mix with the water in the boiler. This means that the boiler water can be treated with inhibitors which prevent corrosion and reduce the electrolysis due to mixed metals in the system, as well as containing an anti-freeze, thus allowing the boiler, connected pipework and calorifier to be left full of water in the winter time.

Figure 12.29 shows a schematic for a double feed, indirect, HW storage cylinder.

Now the **single entry indirect cylinder**: in this type there is a calorifier but it is open to the water in the cylinder and so, when the cylinder is fed water for the first time, it fills the

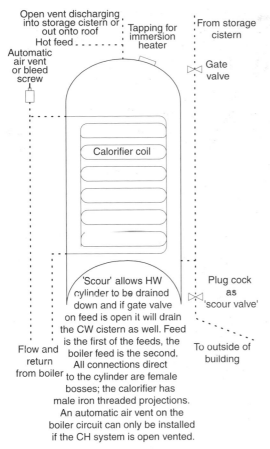

Fig. 12.29 A double feed, indirect, HW storage cylinder.

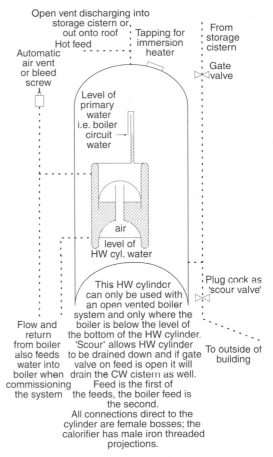

Fig. 12.30 A single feed, indirect, HW storage cylinder.

calorifier and the boiler and boiler pipework as well. When the boiler is fired up, the water in the boiler circuit will heat up and move round the pipework and the calorifier, but because of the way the entries in the calorifier are made it will not mix with the rest of the water in the cylinder. Expansion of the boiler circuit water is taken up by the cushion of air moving out of the calorifier into the cylinder and finally up the open vent pipe; contraction draws water in from the cylinder.

Figure 12.30 shows a schematic for a single feed, indirect, HW storage cylinder. The most common HW cylinder used in domestic work is the double entry indirect cylinder with a central heating boiler fired by oil or gas and controlled by thermostats and time clock(s) to give an adequate supply of hot water at

around 65°C. The aspect of control will be discussed later in this chapter when we look at domestic central heating systems.

Feed and expansion tanks

All that has been said about CW storage cisterns applies to **feed and expansion tanks**. These are only installed when a boiler is fed separately from the HW cylinder, either as a double feed indirect HW cylinder or as part of a central heating system which also heats domestic hot water. They are generally of much smaller capacity, 18 litres being sufficient for the normal domestic heating system. Only where larger boilers and more radiators for larger houses are installed, would it be necessary to provide more expansion space. For that is what these tanks are all about.

They receive water via a ball cock from the mains supply – a branch off the rising main. Initially this water fills the boiler circuit and thereafter acts only to top up the boiler circuit losses from evaporation, radiator bleeding, etc. The water level marked on the cistern is ignored, the ball cock lever being set so that water will enter only in sufficient quantity to cover the cold feed to the boiler. This cold feed pipe comes from near the bottom of the tank, and while supplying the boiler with cold water also acts as an expansion pipe for the water in the boiler circuit. This expansion forces water back up the cold feed pipe into the tank, which has to be large enough to cope with the expansion and not allow water to escape out of the overflow. This would not be a disaster if it happened but if that boiler circuit is filled with inhibitors and antifreeze, then when the boiler cools, fresh water will enter through the ball cock and the boiler circuit water is diluted. A typical, modern, three-bedroom house will have an expansion of around 6–8 litres when the boiler is fully operational. Allowing around 5 litres permanently in the cistern, there is a small safety margin before the contents overflow.

Central heating

Equipment for central heating can vary enormously but, ignoring the fuel used, for modern domestic heating the low pressure hot water system commonly used can be either a **fully vented** or **pressurised** system. Both systems will heat emitters of various kinds as well as towel rails, and will also heat domestic hot water on an indirect basis. So the terms fully vented or pressurised apply only to the boiler circuit[3].

It is not proposed to attempt a discussion on boilers and the merits of the various options

[3] It is possible to have the domestic hot water supply system pressurised but this is rare in speculative housing in the UK, although very common in continental Europe and North America.

that manufacturers make available. However, the principal features are:

☐ A choice between pilot light ignition and electric spark ignition (the first is giving way to the second)
☐ A user adjustable thermostat to control the temperature of the water delivered by the boiler
☐ A foolproof method of keeping any pressurised system topped up with water and at the right pressure.

Naturally the manufacturers try to make the overall internal design feature intrinsically fail safe. This is helped by the control box wired up with the system which ensures that the pump is always running before the boiler fires, and it will only run if the motorised valves on the flow pipes are in the open position. The boiler controls are all now using microchips, and the program in them ensures that the fan on the flue always runs to purge the fire box of all gases except fresh air; the over temperature switches are in the 'on' position; the pump is running sensed by a pressure switch; there is correct water pressure sensed by another pressure switch; fuel is available (another sensor); and finally the fuel ignites and a sensor detects a flame and the system heats the water. If no flame is sensed the system shuts down. In addition there should be a pressure relief valve which when active shuts the system down. If ignition is by pilot light, a flame out sensor is required on that which shuts down the fuel supply to both main burner and pilot light.

Fuel can be natural gas from the mains supply in the street; LPG (liquefied petroleum gas) from a storage tank permanently sited adjacent to the building or from refillable cylinders; oil from a permanent storage tank adjacent to the building; or a solid fuel stove burning small particle bituminous fuel and thus self-stoking. The last are vary rare but are the only solid fuel appliances which can be controlled anywhere nearly as closely as the other fuels. All liquid or gaseous fuels can be shut off with fairly unsophisticated valves.

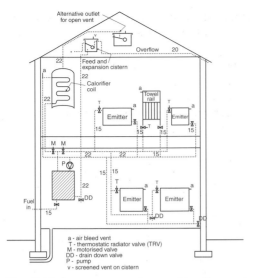

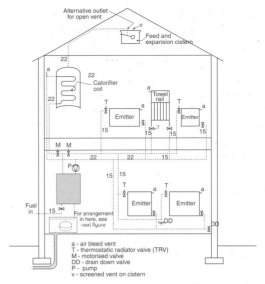

Fig. 12.31 An open vented central heating system – gas or liquid fuelled.

Fig. 12.32 A pressurised central heating system – gas or liquid fuelled.

These cannot be made to work with small particles of coal or coal derived fuel pellets.

Figure 12.31 shows an open vented central heating system – gas or liquid fuelled. The pipe which carries hot water to the emitters or to the HW cylinder is termed the **flow pipe** or simply **the flow**; the pipe bringing cooled water back to the boiler is termed the **return pipe** or simply **the return**.

The most common type of pressurised system is one where the central heating system is pressurised but the domestic hot water supply is left open vented. This does away with the need for an expansion cistern in the roof space but still retains a cold water cistern.

Figure 12.32 is a schematic of a pressurised central heating system – gas or liquid fuelled. Figure 12.33 shows the pressure vessel and valve arrangements of a pressurised central heating system. Figure 12.34 shows the schematic of the pressure vessel and valve arrangements for a pressurised central heating system.

Piping for central heating systems

Looking at the texts which concentrate on building services, the reader is regaled with all sorts of pipe layouts – drop feed, rising feed, ladder layout, etc. – all related to feeding hot water through pipes to emitters by gravity. Here we will deal only with pumped systems. These have taken over the domestic market, giving the plumber almost complete freedom on where to site the boiler, run the pipework and site the emitters. Good plumbers, however, observe some fairly basic rules:

☐ There must be an air vent at the top of every vertical run of pipework unless it

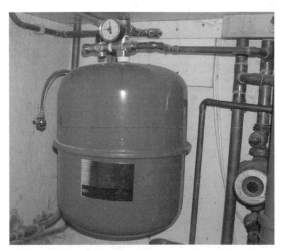

Fig. 12.33 The pressure vessel and valve arrangements of a pressurised central heating system.

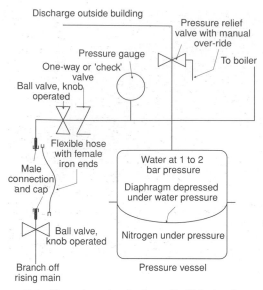

Note: The mains water pipe is used to fill the heating system but cannot be permanently connected to the system in case contamination of the water supply takes place. Flexible hose is used for filling and topping up but must be disconnected and the caps replaced on the pipe ends. The one-way valve prevents boiler water flowing back into mains if pressure is less in the main

Fig. 12.34 Schematic of the pressure vessel and valve arrangements for a pressurised central heating system.

ends with an emitter (with a built-in air bleed valve)

□ There must be drain down valves on the lowest point(s) of a system

□ There must be sufficient isolating valves to allow repair and/or maintenance of parts of the system without having to drain down the complete system

□ When draining down there must be no low or dead legs which will not drain.

This still leaves quite a choice of pipe and pipe layouts, wide enough to cover nearly every situation that the plumber is likely to come up against.

For new build work the favoured installation is **small bore pipework**. The runs of pipe to and from the emitters are carried out in 15 pipe, usually copper. Where a pipe is feeding, say, a floor area, pipe from the boiler to some convenient and central location might be 22, branching off to some of the emitters

in 15 and then reverting totally to 15. Locally to the boiler there might only be 22 pipe or at the most some 28 from pump and boiler, branching into 22.

For installation in existing buildings, copper is still favoured but polybutylene barrier pipe finds favour when getting into awkward spaces, feeding pipes through several holes in timbers or partitions, etc. It is, however, a more expensive option overall.

A further option for new build but certainly for work in old buildings is the use of microbore tubing. The boiler and pump are installed as normal, with 22 pipe taken to central points on each floor (two points still in a single-storey house) where both pipes are connected to **manifolds**. From the manifolds, flow and return pipes are run in 6, 8, or 10 mm fully annealed copper pipe to each individual emitter. Arrangements at the emitter can be the same as that for small bore pipework, or a single valve with double entries can be fitted. Polybutylene pipe can be used for the microbore portion and is well suited to installation in older properties.

Underfloor heating is growing in popularity once again. In old systems installed in the late 1950s to the early 1970s, electric cables were installed in floor screeds on a fully embedded or fully withdrawable system. In the fully embedded system a single cable was laid in a zig-zag pattern over an insulated concrete sub-floor and the screed was poured over it. The cable was monitored continuously during the pour so that early detection of any damage and effective repairs could be carried out. For the fully withdrawable system, a system of ducts was formed in the screed using an inflatable hose. Cables were drawn in after the screed had set. Control was through a wall thermostat and a time clock, the power usually being off-peak. The hot water version uses 15 mm polybutylene pipes solidly embedded in a floor screed over an insulated sub-floor, each room being controlled by a thermostatically operated valve. The pipe(s) from each room are brought to a manifold which in turn is connected to the boiler by a larger pipe, 22 or 28 mm.

This form of heating has a number of advantages:

☐ The room attains an even level of heating over its whole area.
☐ There are no hot or cool spots.
☐ There is heat at foot level, unlike emitters which circulate the air to the upper part of the room and the cool air sinks to the floor level.
☐ The screed stores heat, giving a long slow build up but a long slow cool down. This is particularly advantageous in lightweight buildings where the thermal lag is generally quite fast.
☐ It is claimed to be more efficient and therefore more economic to run than an emitter-based system.

For suppliers of underfloor heating system materials and controls visit the websites of Polypipe and Wavin – see later section.

Emitters

Emitters is the preferred term for the individual room heating devices normally associated with central heating.

What we normally refer to as radiators only radiate part of the heat which is given off. Much of the heat is given off by convection. One only has to look at the back of the radiator to see that many of them have shaped metal baffles welded to the panels, which heat up and in turn heat the air surrounding them, which rises into the room drawing cool air from floor level. Double panel emitters are also now available with factory-fitted side and top covers which boost the convection of warm air. Emitters are also available which have the water panel enclosed in a casing along with a fan. When heat is called for by a built-in or remote thermostat, the fan runs and the air is heated as it is blown across the panel. The panels are heavily finned. There may or may not be a valve on the pipe into the panel so that hot water is not circulating when no heat is called for.

Fig. 12.35 A typical emitter.

They are common as kitchen heaters where they are designed to sit behind the kicking board under fitted kitchen units. At the temperatures at which radiators operate there is not a lot of radiant heat available. One has to be within 600 of the panel to feel anything in most installations, and where there are children or old people the water temperature from the boiler has to be kept low enough to avoid burning vulnerable people should they touch the emitter surface. Figure 12.35 shows a typical emitter and Figure 12.36 shows a typical thermostatic radiator valve (TRV).

Appliances

The term 'appliances' covers things such as sinks, wash hand basins, baths, showers and

Fig. 12.36 A thermostatic radiator value (TRV).

WCs. We will not consider appliances used in public buildings or large institutions.

Appliances come in a wide range of materials: metal, particularly brass, bronze and stainless steel; ceramics such as porcelain, vitreous china and earthenware; glass; fibreglass reinforced resins; acrylic sheet; and so on. No matter what the material the methods of connection do not change greatly except for the type of washers inserted to avoid damage to the appliance material, or necessary to make a watertight connection on especially thin or thick materials. We will consider kitchen sinks in mineral-filled resin and stainless steel, WCs and WHBs in vitreous china, and baths in acrylic sheet or fibreglass reinforced resin. Together with the appliances we must include things like taps, traps and wastes, flushing mechanisms and so on.

There seems little point in including sketches or photographs of these appliances as we all use them daily. What we don't all do daily is really look at them or appreciate how they function – that is, until they break down and we have to call the plumber. I don't expect readers to go off and attempt their own repairs after reading the text, but I do expect that they should look at the sanitary appliances they have in their own houses and those of their friends and really attempt to understand how they function. Not all of them will be modern appliances or up-to-date in the way they are connected to supply or disposal pipework, but a lot will be learned.

The website to visit is www.thebathroom info./relax.html

Kitchen sinks

Kitchen sinks in stainless steel and mineral-filled resins are made in a variety of configurations: twin bowls, single bowl, bowl and a half and all with or without single or double draining boards, holed for a pair of taps or holed for a monobloc tap, etc. The permutations are endless when one adds sinks which can fit into corners of worktops, those which can be fitted with a waste disposer, etc. Look at a website such as www.plumbingsupply.com/elkay and www.titanic-sinks.co.uk

WCs

Water closets have been made in a variety of materials – earthenware, porcelain, stainless steel and vitreous china. The ceramics used are generally referred to under the general title of **ware**. For domestic purposes the most common material is vitreous china, available in white and a range of colours, the shade of which depends on the manufacturer – one pink doesn't necessarily match another pink. Of more relevance here are the two basic styles of WC available and the three combinations of **flushing cistern** and WC available.

WCs in the UK are most commonly of the wash-down type without any assistance from syphonic action to produce a clear flush. A more expensive option is to have a siphon built into the flushing action and trap, a feature which uses the flushing water passing through and out of the trap to create a negative pressure immediately between the flushing water and the contents of the trap. This negative pressure literally sucks the contents of the trap over and into the disposal pipe work. Siphonic WCs are generally quieter in operation than the plain wash-down type.

Examination of a WC will show that it has a trap built into it during manufacture and that the outlet in the older models could either be parallel to the floor or at right angles to the floor on which the WC was mounted – i.e. the traps were either of P or S configuration. Modern practice has almost completely obviated the necessity to manufacture in two configurations. Most are now P traps with the connector to the disposal pipework being anything from straight through a succession of angles up to 90°. Figure 12.37 shows ware to pipe connectors: on the left a long tailed bent connection and on the right a short obtuse angled connection. Both have hard plastic pipe with soft (black) synthetic rubber seals pushed

Fig. 12.37 Ware to pipe connectors.

into or onto the pipe. The seals have several fins which seal to the ware and soil pipe.

Flushing cisterns have been made from a variety of materials – lead-lined wood, cast iron, plastic, earthenware, porcelain, vitreous china, enamelled steel, and so on. Plastic and vitreous china predominate in the modern market. All are now supplied with a lid, and a siphon is built in, operated by a lever, to send the flushing water through the WC. Their configuration with the WC is what sets each cistern apart. They can be:

Fig. 12.38 Low level flushing cistern and WC.

□ High level. The cistern is mounted on the wall above the WC at least 2.00 metres above the floor level. It is connected to the back of the WC flushing rim by a pipe of about 40 bore. In old installations this pipe would have been of lead or enamelled steel but is now plastic. This type of cistern is not now used in domestic installations.

□ Low level. The cistern is mounted on the wall between 300 and 400 above the WC and is connected to the WC flushing rim by a plastic pipe of around 30 bore. Cisterns are either of plastic or, as in the illustration, vitreous china and of a colour to match the WC. Figure 12.38 shows a low level flushing cistern and WC.

□ Close coupled. The WC has an extended shelf at the back of the flushing rim with

an aperture over which the flushing cistern is mounted with a soft rubber ring seal between the WC and the cistern. The two are clamped together. The cistern is of vitreous china in a colour to match the WC. Figure 12.39 shows a close coupled flushing cistern and WC.

WHBs

Wash hand basins in a bewildering variety of styles are now made in vitreous china for domestic use and incorporate an overflow system in their construction. This is an opening in the basin, high up at the back with a duct in the body of the basin leading down to an annular space in the hole provided for the connector to the disposal pipework. The connector to the

Fig. 12.39 Close coupled flushing cistern and WC.

disposal pipework comprises a **waste**, usually of chromium plated brass with a flange and grid showing in the WHB and a length of externally threaded pipe passing through the ware. Underneath the wash basin, jointing compound and washers with a large back nut are used to seal the waste to the WHB. Slots in the waste align with the annular space for the overflow.

Baths

Baths were traditionally of enamelled cast iron but are now made in enamelled pressed steel or plastics – acrylic or fibreglass reinforced resin. They have an overflow but rather than being built into the body of the bath, an opening is left and an overflow grid secured with washers and a back nut, plus a flexible pipe connecting to the bath trap, are installed. The trap on the bath is specially designed to fit into the relatively shallow space available under modern baths, and it has a connection for the overflow pipe.

Showers

Showers can be of a number of types. Shower mixer taps can be fitted to a bath. Basically these taps deliver hot or cold water through a single nozzle to the bath or when a plunger is operated delivery is through a flexible hose to a shower head mounted either in a cradle over the taps, like a telephone, or the shower head can be placed in a holder on the wall above the bath. The water temperature is not controlled other than when the taps are turned off and on. It is important when these types of shower are installed that the cold feed to the HW cylinder from a storage system is above the cold feed to the cold water system in general. This should prevent scalding.

The more expensive but more satisfactory option is a thermostatic shower valve which has separate hot and cold supply pipes, both from a storage system which delivers water through a thermostatically controlled valve which automatically compensates for slight pressure differences, as well giving a more constant stream of water at a more consistent temperature. There is usually a fail-safe shutdown in the event of the cold water failure.

Finally, there is the electrically powered 'instant' shower. This is a shower system with only a cold water supply taken off the mains supply pipe. The water is heated by a small electric 'boiler' and the output is thermostatically controlled and has fail-safe shut off of power and water in the event of system failure or breakdown. Earlier showers suffered from poor flow but as power available has increased beyond the initial 7kW so has the flow, a 9kW unit giving almost as good a shower as a conventional shower. Figure 12.40 shows an instant shower.

Taps

Taps on baths and WHBs are normally of chromium plated brass and can be configured in different ways. Monobloc taps are in

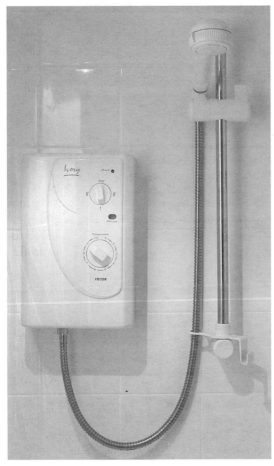

Fig. 12.40 An instant shower.

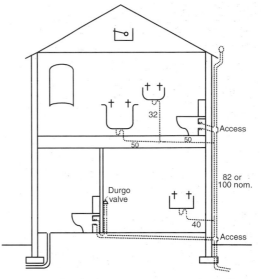

Fig. 12.41 The single stack system with an additional WC on an extended side branch.

fashion on WHBs, and bath taps can be separate hot and cold or combined in one block with or without a shower fitting. Whichever style of tap is fitted they all require to be connected to the supply pipework, and this is done with a tap connector – see earlier photographs of connections for pipework to appliances and equipment (Figures 12.9 and 12.11).

Waste disposal piping and systems

Disposal piping is best looked at from the point of view of the appliance with the need for the largest bore pipe – the WC; 83 pipe can be used for one WC plus other appliances, or, for more than one WC plus appliances, 100 nominal bore piping should be used. This large bore pipework (usually referred to as **soil pipe**) is configured in one of a number of very specific ways. The method illustrated in Figure 12.41 is the single stack system with an additional WC on an extended side branch. The principle of the stack is to allow air to enter the top of the pipe as the WC is flushed, thus preventing the build-up of a siphon in the pipe which would take any sealing water out of any appliance traps. Sewers are no longer vented and now rely on the open ends of pipework to keep a flow of fresh air coming into the sewer, which then rises and exits from the top of the **soil stacks**. So, when we have a pipe receiving the discharge from a WC and rising up above the eaves level of the building it serves, the complete pipework is referred to as a soil and ventilation pipe – SVP. The SVP is usually done in uPVC.

Connected to the soil stack we have the disposal pipe from the other appliances – WHB, sinks, bath, shower, and so on. This disposal pipework is referred to as **waste pipe**. Baths require 50 mm pipe, showers and sinks 40 mm

Fig. 12.42 SVP pipe fittings – a pair of branches, one at 137.5° and one at 92.5° with blank bosses.

Fig. 12.44 SVP pipe fittings – a bend and a pipe with blank bosses.

pipe and WHB 32 mm pipe. This can be uPVC, ABS or polypropylene.

It has already been mentioned that cast iron and copper are no longer used for domestic installations purely on economic grounds. Figures 12.42 to 12.44 illustrate fittings used on soil and ventilation stacks. Figure 12.42 shows a pair of branches, one at 137.5° and one at 92.5° with blank bosses; Figure 12.43 shows an access pipe; and finally Figure 12.44 shows a bend and a pipe with blank bosses.

The fittings in the illustrations are all solvent weld fittings. The fitting, a bend, in Figure 12.45 has a synthetic rubber seal and is a push fit onto the pipe.

Note that all these fittings have a socket at all ends for pipe. The pipe supplied by the manufacturers has no sockets and is merely cut to

length and welded into place, or the ends are chamfered off to facilitate pushing into the socket past the ring seal. Having no sockets means that there is less waste of pipe; the manufacturers make straight couplings so short lengths of pipe can be made up to a longer length to suit circumstances, especially with the solvent welded type.

Traps

The traps for waste pipe are all suitable for sinks or WHBs and are shown in three configurations, P, S and a straight through. One manufacturer shows in excess of 50 trap types in

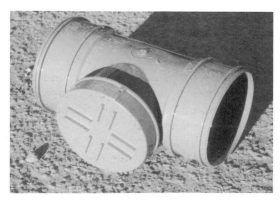

Fig. 12.43 SVP pipe fittings – an access pipe.

Fig. 12.45 SVP pipe fittings – push fit bend with synthetic rubber seal.

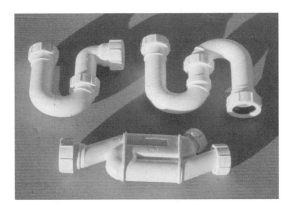

Fig. 12.46 Basic trap shapes.

their catalogue, and the majority of these can be obtained to fit 32, 40 and 50 pipe work. Figure 12.46 illustrates three basic traps: from the top left and anticlockwise, a P trap, an S trap and a straight through trap.

WCs should be sited as close to the soil stack as possible, reducing the length of the branch pipework to a minimum. Should the branch exceed approximately 1200, then some means of ventilating the branch has to be fitted. An expensive solution would be to erect another stack, but apart from the expense it might not be possible to accommodate another stack. The solution is to carry the branch pipework past the WC and turn it up the wall next to the WC. The top of this pipework has to be higher than the top of any other appliance connected to the branch – frequently a WHB in the same compartment. Then cap the pipe off with **an air admittance valve**, sometimes referred to by its original trade name, **Durgo valve**. It is illustrated in Figure 12.47, which shows a general view as fitted to the top of the SVP, and the view into the bottom, which doesn't show a lot but merely indicates that there is more to the thing than just a fancy dome! In fact there is a fairly delicate valve mechanism which opens and closes in the flow of air in and out of the SVP.

Websites to visit for information on soil and waste pipe work and fittings are:
www.wavin.co.uk/main/homepage
www.Marleyplumbinganddrainage.com

Fig. 12.47 Durgo valve.

www.McAlpineplumbing.com
www.Polypipe.com

Before leaving this section we should look at the maximum allowable lengths of the various sizes of waste disposal pipework, and where they may be connected to a soil, waste and ventilation pipe.

For waste disposal, the sizes of pipes and depth of seal of the trap are given against each type of sanitary appliance in Table 12.4. Note that there is a maximum length of

Table 12.4 Disposal pipe sizes, trap seals and maximum branch lengths.

Sanitary appliance	Disposal pipe size (mm)	Trap – minimum seal (mm)	Max. length of pipe (m)	Pipe fall (mm/m)
Bath	40	50	3	18 to 90
	50		4	
Sink	40	75	3	18 to 90
	50		4	
Wash hand basin	32	75	1.7	18 to 90
	40		3	
WC	83	50	6 for a single	18 to 90
	100		WC	
Bidet	40	75	4	18 to 90

pipe measured from the trap on the appliance to the soil, waste and ventilation pipe (SWVP).

Figure 12.48 illustrates other important points regarding the junction with the SWVP.

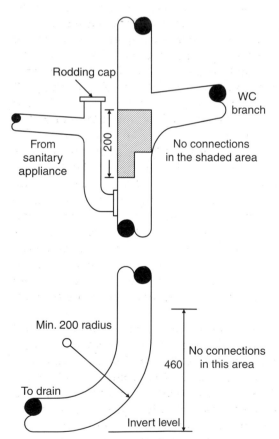

Fig. 12.48 Schematic of disposal pipe connections to an SVP in domestic premises.

Insulation

Insulation of the HW cylinder has already been mentioned, but what of the rest of the plumbing installation? There are two reasons for insulating the installation:

□ To prevent the installation freezing up in cold weather
□ To reduce the loss of heat from storage vessels and pipework carrying hot water.

To prevent freezing up all pipework must be **lagged**, i.e. covered with a layer of insulating material. In the not so distant past 50 mm wide strips of felted fibre on hessian were wrapped round the pipe and held in place with a binding of copper wire. A better approach is the use of preformed plastic insulation. This is formed as a tube slit along its length, and it opens up to clip over the pipe and then adhesive tape can be applied to keep it closed round the pipe. Some manufacturers make an insulation with a zip-type closure – more expensive but simple to apply and seal. The insulation is made in a variety of bores to suit hot and cold water supply pipe and in an economic thickness to comply with the Regulations. As well as preventing pipework from freezing, the insulation will also reduce the heat loss from hot water pipework.

Cold water storage cisterns anywhere, but especially in the roof space, are vulnerable to freezing up and so must be insulated with a proper layer of material which covers top and sides. Where the cistern is low and near the ceiling finish of the upper floor, it is easy to carry that side insulation down to the ceiling. No roof insulation should be placed under the cistern, so any heat rising through the ceiling escapes into the space around the cistern and helps to reduce the chance of it freezing. Where the cistern is high off the ceiling, special measures must be put in place to seal from the side insulation to the ceiling. It is vital to get any heat from the ceiling up round the cistern. A simple box structure of thick polystyrene sheet (say 50 thick) should be built on a suitable timber frame. No gaps must be left in this layer. Proprietary insulating sets are made by all cistern manufacturers but only allow insulation of the cistern.

Not insulation but a suitable substitute in central heating boiler circuits is the introduction of an anti-freeze into the radiators and pipework. A 20% solution can give protection from freezing down to $-21°C$. Note this is only suitable for systems which have a double entry, indirect hot water cylinder.

Corrosion

There are two sources of corrosion which must be dealt with. The first is where the water itself is aggressive and would leach zinc out of brass alloys used for fittings and valves. The use of DZR fittings is the answer in these circumstances.

The second cause of corrosion is the use of dissimilar metals in the systems, whereby electrolysis occurs. The first step is the exclusion of any metals which are detrimental to the bulk of the installation, e.g. aluminium alloys, galvanised pipe and equipment and plain iron or steel. In commercial installations or where these sorts of clashes cannot be avoided, it is common to include a sacrificial anode – usually magnesium or an alloy of aluminium. This is corroded rather than the rest of the installation and literally dissolves away. At that point it has to be replaced.

So for domestic work, the answer is exclusion of metals which would have a bad reaction with the copper pipe and brass or DZR fittings. Plastic pipe takes no part in this process. There is one part of the system which will have steel in it – the central heating system. In fact it might have stainless steel and mild steel in the radiators and some of the pipework. It might even have cast iron in the boiler. Fortunately there is an easy preventative measure – the use of inhibitors. These simply make the water incapable of supporting electrolytic action provided the correct solution has been introduced into the heating circuit. Note this is only suitable for systems which have a double entry, indirect hot water cylinder.

Air locking and water hammer

When water is run into a system for the first time it sometimes happens that air is trapped in legs of pipework which fall against the flow of the water. A competent plumber will run the pipe in such a way that this will be avoided, i.e. the pipework is laid with proper falls to allow it to fill naturally, pushing all air in pipe and equipment ahead of it as it rises in the system. Incompetent plumbers become expert at clearing air locks. Mostly it is a question of identifying where the air lock has occurred and undoing or at least loosening off a coupling to allow the air to escape and the water to fill up to that point. Tightening the coupling and continuing to feed water should solve that problem. Occasionally a sudden build-up of pressure behind the water in front of the air lock will push the air over, and once the water starts to flow past that point in pipework it should then continue.

Air build-ups are more common in central heating pipework, especially when first commissioned. Usually running the pump will push air past the lock points, but of course it has to go somewhere and that is usually the highest radiators in the system. It is vital to get air out of the radiators, for if there is air trapped in them hot water will not circulate and there will be little or no heat. Radiators

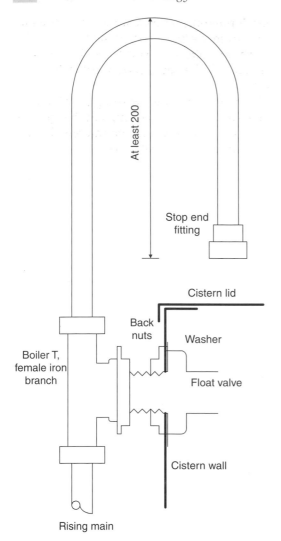

At least 200

Stop end fitting

Cistern lid

Back nuts

Washer

Boiler T, female iron branch

Float valve

Cistern wall

Rising main

Fig. 12.49 Shock absorber loop on rising main.

Fig. 12.50 Plumber's first fixings.

all have bleed screws, and all that is required is to open the screw a little and listen for the air hissing out. As soon as water appears, turn the bleed screw down. Water has dissolved air in it, and when heated this air is given off as tiny bubbles. These bubbles are driven round pumped systems easily enough but come to rest and build up pockets of air at the high points. Look at air vents on the HW cylinder sketches (Figures 12.27–12.30). Air bleed screws on radiators and those air vents at HW cylinders all need regular attention to keep the system free of air.

Water hammer is generally caused by the shock wave generated when a float valve shuts against a fairly high water pressure. It is exacerbated by loose pipework, so the first thing to be done is careful site work where all supply pipework is securely fastened with proper pipe clips. It is important to note that plastic pipe should never be fixed with metal pipe clips. Plastic expands and contracts more than metal, and the sharp edges of the metal clips would quickly wear through the wall of the pipe.

One device which certainly works is to include a shock absorber on the pipework immediately adjacent to a float valve. The one that gives most trouble because it feeds water at main pressure is the CW cistern. Figure 12.49 shows how the shock absorber is fitted. Water from the rising main will push up past the boiler T compressing the air in the bent pipe. It is the compressed air which acts as the shock absorber.

Float valves on WC cisterns seldom give any trouble if fed from the CW cistern. Mains fed float valves on WC cisterns are another matter but it should be possible to fit a smaller version of the device already described.

First fixings

Finally, the photograph in Figure 12.50 shows part of the plumber's first fixing. From left to right the pipes to be seen are:

□ A 100 soil waste and vent pipe connected to the drain going under the foundation and ready to receive a branch for the WC.

□ A 15 copper pipe, insulated where it will be under the floor, as the cold feed to the WC cistern.

□ A group of three pipes, the middle one being a 40 mm plastic pipe for the wash hand basin waste. The cables attached to it are earth bonding wires. On the left is the hot feed from the hot water cylinder and on the right the cold feed to the WHB – notice it is teed off the same pipe feeding the WC cistern.

□ On the right another waste pipe, 40 diameter, for a bath connection, and then just in sight a copper pipe, which can only be hot or cold feed to the bath taps.

13

Electrical Work

Power generation	266		Earth bonding	273
Wiring installation types	267		Final fix	275
Sub-mains and consumer control units	268		Testing and certification	275
Sub-circuits	270		More on protective devices	275
Work stages	272		Wiring diagrams	276
Electrician's roughing	272		Accessories	277

Electrical work embraces a very wide range of techniques and materials designed to meet the needs of a wide range of users, so this text will be short, dealing only with basic domestic installations.

The book of rules for all electrical installations in the UK is the set published by the IEE – Institution of Electrical Engineers. It is currently in its 16th edition with amendments.

Power generation

To explain simple electrical work in context we need to set the scene by describing briefly how we get electricity to our homes, and so we start at the power station. The power station houses generators, each of which is an arrangement of coils and magnets, some fixed, some revolving on a shaft. As coils and magnets pass, electricity is generated and taken away by heavy cables to be distributed to the consumers. The current generated can be shown graphically as a **sine curve**.

When a wire passes through a magnetic field, a current is induced in the wire. Make the wire into a compact coil, the magnet produce an efficient field, and the current generated can be put to practical commercial use.

Figure 13.1 shows in diagrammatic form a scheme for power generation and sine curve of the power generated.

The 'events' marked A to F can be followed: Considering only coil R, at A the coil is halfway between the magnetic poles and no current is being generated. At B, coil R is nearer the north pole and so current is generated and we see the sine wave starting to form – and rise. At C, coil R is opposite the north pole and current generation peaks. At D, it is halfway between the poles and current generation is again zero. At E, the coil is opposite the south pole and generation peaks again, but on the opposite side of the X axis and at F it is halfway between the poles and no generation takes place. The cycle is about to begin once more.

If the coil traces that cycle through the magnetic field 50 times per second, then the sine wave will form 50 times per second, i.e. the current is being generated at 50 hertz. The coils in Figure 13.1 are shown 'star' wound, i.e. one end of each coil is joined and this is connected to the neutral wire in the distribution chain. The other ends of the coils each provide current as they pass through the magnetic field of the magnet: thus we can have three lots of current being generated, each lot being taken off and distributed down the

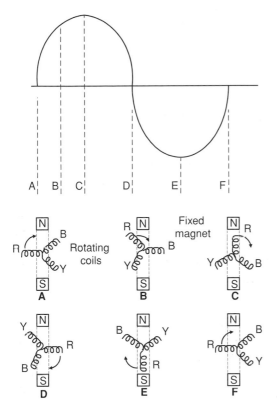

Fig. 13.1 Diagram of sine curve and power generation.

power lines to the consumer, who can have one, two or all three lots of current delivered to the premises. In practice it is only one or three lots which are distributed to the consumer.

The 'lots' are known as **phases**, which are colour coded red, blue and yellow. In theory, because the three phases generated are in perfect balance, the neutral cable should carry no current at all. A further cable is included and this is the earth cable, usually incorporated as the armour steel wire wound round the outside of the mains cable in underground supplies, or as a separate wire in overhead transmission lines.

Generation of the electricity in this way gives rise to a current which flows in one direction, as seen when the sine curve rises and falls on one side of the X axis, and then reverses as the curve rises and falls on the other side. Thus we have alternating current, a.c., in each phase.

Transmission down the supply authorities' cables is at a much higher voltage than that received by the customer, starting off at 33,000 volts and transformed down in stages until it comes into the street where there are customers' premises.

The voltage delivered at the customers' premises on each phase in the UK is 230 volts at 50 hertz. By combining two phases in the premises, the consumer can have 440 volts and with three phases 660 volts. These larger voltages are much more economical for running machinery and industrial processes but have been occasionally used in domestic premises for electrical heating. This is now far too expensive to operate at the general tariff but may be the only option for some householders who can only get off-peak power for storage heating.

We will only consider the use of a single phase supply.

Wiring installation types

Classification of electrical work can be on the basis of the methods and materials used to provide various types of electrical circuit:

☐ Sheathed and insulated cable
☐ Insulated cable in conduit, there being three grades of conduit: slip jointed, threaded and heavy gauge threaded. Also, conduit is available as black enamelled or as hot dipped galvanised
☐ Mineral insulated cable
☐ Armoured cable

And so on. . .

The most simple installation is the first, sheathed and insulated cable. The cores which carry the current are insulated, and two or more are laid side by side with a bare core as an earth wire, and covered in a tough flexible sheath. This type of cable uses PVC for both insulation and sheathing, although it is still possible to obtain vulcanised rubber insulation and sheathing, and heat resistant variations of both insulation and sheathing for special situations.

Sub-mains and consumer control units

Starting at the beginning of any installation there is a heavy cable entering the property from the mains cable in the street. This terminates at a piece of equipment frequently known as a **cut-out** or **cable head**. This hard plastic box contains: one, two or three fuses, termed **service fuses**, a **neutral terminal block** and an **earth terminal block**. Our single phase supply will have one fuse, one neutral terminal and one earth terminal block. Current practice is to house the cut-out together with the supplier's meter in a cupboard accessible from the outside of the building.

Figure 13.2 is a photograph of a typical single phase cut-out and meter cupboard. There are many different patterns of cut-out and meter. The photograph is fairly self explanatory. The cupboard, a lidded box, is made of glass fibre and has a piece of 25 particle board fitted in the back onto which the equipment is

Fig. 13.2 Single phase cut-out and meter in cupboard.

screwed in place. The mains cable comes up a duct into the cupboard and is wired in to the cut-out. The main fuse and connection of neutral and earth wires can all be seen clearly. Note how both live and neutral wires go into the meter and then from the meter go through the wall into the house. The duct through the wall of the house is not very clear.

Inside the house, and ideally on the other side of the wall from the supplier's cut-out and meter, there will be a **consumer's control unit (CCU)**. A sleeve will be built through the wall to allow the easy passage of cables from the meter and earth terminal to the consumer unit. A cable should never be built solid into masonry as any movement could then damage the insulation or even break the cores. The cables from the cut-out to the CCU are known as **tails** and are generally single core insulated and sheathed cable. There is also an earth tail of insulated cable from the earth terminal in the box to the consumer unit. The tails should be kept as short as possible. Collectively these cables and the CCU are known as the **sub-mains circuits**. In large installations these sub-mains circuits can be complex, involving lots of switch gear, circuit breakers and cabling frequently in conduit and trunking[1].

Figure 13.3 is a photograph of a small selection of cables used in domestic work. The cables, starting from the top, are:

- 1.00 mm^2 twin and earth core cable (ECC) – note the solid cores
- 2.5 mm^2 twin and ECC – note the solid cores
- 6.00 mm^2 twin and ECC – note the twisted cores
- 6.00 mm^2 insulated earth cable – note the twisted core and the anti-clockwise twist
- 10 mm^2 sheathed and insulated cable as used for tails to CCU; the seven strands of the core have been separated out for purposes of illustration.

---

[1] Conduit is a metal tube used to give mechanical protection to insulated-only cables. In this context trunking is a rectilinear section of metal boxing with a detachable lid also used for mechanical protection. It can carry more cable than conduit and can be subdivided inside. Trunking is also available in plastic in a wide variety of forms and finishes.

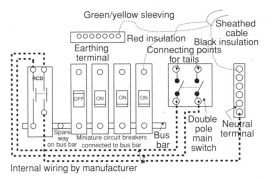

Fig. 13.4 Schematic of consumer control unit.

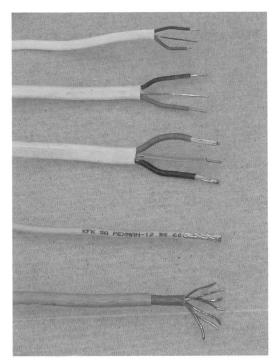

Fig. 13.3 Cables.

The consumer unit has a number of devices built into it:

□ A switch which will break the continuity of both the current carrying cable (**line cable** or simply **line**) and the **neutral cable** or simply **neutral**. A switch of this type is called a **double pole** switch. The earth wire is never broken by a switch or other device.

□ Devices to protect the cable(s) in the individual circuits which start in the consumer unit, such as those for lighting, cooker, water heating, small power, etc. They are known as **sub-circuits**. Each will have a cable of a size appropriate to the function (and therefore current expected) to be carried out. Lighting circuits which carry only 1 or 2 amperes (a measure of current) will not require cable quite so large as, say, the cooker circuit, which might be serving a double oven and grill and draw up to 12–15 amperes at any one time

□ A neutral terminal block

□ An earth terminal block

□ A **residual current circuit breaker (RCCB)** (also known by the now obsolete name **earth leakage circuit breaker (ELCB)**,

current operated type). See notes on RCCBs later in this chapter. The device is a double pole switch activated when there is a fault in the sub-circuits and generally in addition to the double pole manually operated mains switch. A schematic for an RCCB is shown later in this chapter.

Figure 13.4 is a schematic of a consumer control unit (CCU). It shows only the working parts. These would be mounted in an enclosure of either metal or plastic, the choice being made on whether or not the CCU might be subject to mechanical damage, and to what extent. The enclosure would cover all the terminals and exposed parts of the bus bar, and the internal cabling and sub-circuit wires entering the CCU. The lid on the CCU would cover everything except the mains switch. Note that the diagram shows all the sub-circuits protected by the RCCB.

Figure 13.5 is a photograph of a CCU. It has a plastic enclosure, and the flip-down lid covers the mains switch. Note that only four out of the nine sub-circuits are protected by the RCCB. It is important to emphasise that the fuse or MCB is provided to protect the wiring of the sub-circuit from a current which would damage it by overheating the core(s). It is not there primarily to protect anyone from electrical shock, although should a core come into contact with metal work in the house, the current would immediately rise, the fuse would blow or the MCB would trip and shock to the householder would be prevented, i.e. if the householder then went on to touch that metalwork no current would now be passing through it.

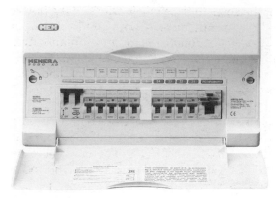

Fig. 13.5 Consumer control unit.

Personal protection is provided by the inclusion of an RCCB or ELCB in the CCU. The devices used to protect sub-circuits are of three kinds:

☐ Rewirable or semi-enclosed fuses
☐ HBC (high breaking capacity) cartridge fuses
☐ Miniature circuit breakers **(MCBs)**.

New installations now only use MCBs.

Figure 13.6 shows a photograph of rewirable fuses, and on the left is an MCB. The upper face in the photograph is what the consumer sees when the cover is opened. The screw holes for the electrical connections are at each end, and in the middle of the T-bar is the switch, tripped when the MBC is overloaded and which is depressed to reconnect the sub-circuit. The device clips into the CCU enclosure on what is called a DIN rail, and the bus bar is fixed into one end and the sub-circuit wires into the other end, tightening the screws

Fig. 13.6 Fuses and a miniature circuit breaker.

provided. On the right of the photograph are a pair of the now obsolete but widely used rewireable fuses. The upper one is of porcelain and the lower of plastic. The terminals for fixing the fuse wire can be seen on the lower fuse carrier.

Sub-circuits

Sheathed and insulated cable was shown in Figure 13.3, and two types were illustrated:

☐ Twin and earth core cable
☐ Single core cable.

Twin and earth core cable has a core with red insulation, a core with black insulation and a bare earth core embedded in the sheath. Alternatively (and fairly rarely) both insulated cores are coloured red and the cable is used solely for switch wiring. This then is the basic cable used to bring electricity out of the CCU to fitting(s) and appliances.

Three core and earth core cable has three cores with red, blue and yellow insulation and a bare earth core embedded in the sheath. In domestic work this cable has always been used for two-way switching of lights on stairs or in hallways; however, two-way switching of central lights in bedrooms is becoming more common. Using this cable it is possible to have any number of switches in the circuit controlling just one light. For example, in a bedroom a central light can be controlled by a switch at the door and two other switches, one either side of the bed position. In a circuit with multiple switches only two are **two-way switches**, the rest being **intermediate switches**.

With these cables it is possible to wire up all the mains sub-circuits necessary for the modern house. Circuits themselves can be either **radial** circuits or **ring main** circuits. Each circuit is connected to the CCU, the red core to the fuse protecting the circuit, the black core to the neutral terminal and the bare earth wire to the earth terminal.

In ring main circuits the cable is led from the CCU round the fittings being served by the ring main and then back into the CCU. Both red cores are connected to the MCB protecting the circuit, both black cores are connected to the neutral terminal, and both earth cores to the earth terminal. When the sheath is stripped back to get cable into CCUs or other fittings, the bare earth wire is fitted with a short length of green/yellow striped sleeve to prevent it from coming into contact with any parts carrying electricity.

Examples of radial circuits protected by a MCB would be:

- All the lights on one floor or half the floor area of a single-storey house
- A cooker – hob and oven
- An electric immersion heater or circulator
- An outlet point for supplying power to the central heating control system. This is not electrically powered central heating
- A dedicated circuit for telephony and/or computing.

Examples of ring main circuits protected by a single MCB would be:

- Small power supplying 13 ampere socket outlets. If the floor area of the house is less than $100\,\text{m}^2$, an unlimited number of socket outlets can be attached to the circuit; however, it is prudent to allow two ring mains in the average house, especially with the ever increasing number of appliances which can be plugged into these socket outlets. One ring main should serve the kitchen and any utility room(s) in larger houses.
- Power supply to a series of electric heating appliances such as storage heaters. Whether one or more ring mains would be necessary for this has to be determined by the circumstances but it does mean more ring mains, not heavier cable and larger fuses or MCBs.

Each of these circuits will expect to have a very definite amount of current supplied, and as the amount of current passing in a cable determines how hot the cable becomes[2], the size of the cores in the cable is critical if the house is not to burn down. Core size is given as the cross-sectional area in square millimetres. Looking at the circuits mentioned above, we can use the core sizes from Table 13.1 and connect them to protective devices with the noted current ratings.

Cables with 1.00, 1.50 and $2.50\,\text{mm}^2$ cores have single solid cores. Cables over these sizes have stranded cores of seven wires with a very slow anti-clockwise twist. The most usual metal for the wires in the cores is copper, although copper covered aluminium has been used when copper was expensive[3]. Copper can now compete economically with aluminium.

Further notes on ring mains must mention that socket outlets can be disposed of in different ways on the ring main. The majority must be wired in **in parallel**, i.e. the cable coming into the socket outlet and the cable going out are wired in red and red, black and black and earth and earth. These socket outlets may be either one or two gang. In addition to this, a number of further sockets may be wired to these earlier sockets as **spurs**, i.e. a separate length of cable – $2.5\,\text{mm}^2$ – is connected to the terminals of a socket on the ring main and run out and wired to an outlying socket. Furthermore, it need not be socket outlets which can be wired to the ring main or as spurs. **Cable outlet plates** can be wired in as well as, or totally in place of, socket outlets.

Cable outlet plates are used where there is an appliance – such as a space heater or water heater – to which power must be supplied. A cable – usually a three-core flexible cord – is wired from the plate to the appliance, sometimes hidden in conduit or ducting or simply set in the voids of a framed structure, and on other occasions it comes out of a hole in the face of the plate and out to the appliance.

When attaching anything to a ring main, either through a plug top in a socket outlet or

---

[2] Several factors affect the heating effect of a current passing through a cable, e.g. how cables are disposed, in conduit, bundled or bunched, clipped to surfaces, buried in walls or floors, under insulation, etc.

[3] The reason for not using plain aluminium was that it would suffer from corrosion problems. However, even after covering it with copper it was found to suffer fractures when put under pressure at screw down terminals. It would literally harden and break off.

Table 13.1 Circuits and cables.

Circuit	Core area in mm^2	Current carrying capacity of cable enclosed in a wall	Protective device rating (amperes)
Lighting	1.00 and 1.50	11 and 14	5 and 10
Oven and hob*	6 or 10 or 16	32 or 43 to 52 or 57	30 or 45
Immersion heater or circulator	2.50	18.5	15
Central heating	1.00	11	5
Telephone/computing	1.00	11	5
Small power ring main	2.50	18.5	30
Electrical heating ring main	2.50	18.5	30
Supply to garage**	2.50	18.5	15
Supply to garage†	10	43 to 52	45

*Depending on the total rating of the appliances or whether it is only an oven, and a gas hob has been fitted.
**This would only allow the fitting of a light and a double socket outlet in the garage.
†This would allow a lighting circuit with several lights and a ring main with several socket outlets to be installed.

by a cable outlet, each individual appliance must be fused with a 3, 5 or 13 ampere cartridge fuse. Plug tops all make provision for these fuses. Cable outlet plates have a fuse carrier built into them. Both socket outlet plates and cable outlet plates can have some optional extras: switches which are in the main single pole; neon indicators which glow when power is passing to the appliance connected; and RCCBs.

Work stages

Electrician's roughing

Like many trades, electrical work is carried out in a number of stages. The first stage, electrical roughing, is the installation of sub-circuit cable. First of all, the house has to be reasonably watertight overhead; the work is then carried out when the internal partitions have been put up but before any plaster or plasterboard is applied to masonry or timber framing.

Where there are masonry walls, the cable has to be let into a groove or **raggle** cut into the surface of the masonry. The cable is protected from mechanical damage by a short length of conduit fixed into the raggle. Current practice uses plastic conduit. It is not as resistant as steel conduit to nails being hammered in, but anyone with common sense does not hammer nails into a wall immediately above or below a switch or a socket outlet etc. So the plastic conduit is more about not embedding the cable solidly into the wall and also about being able to withdraw it if rewiring is to be undertaken.

Also, in masonry walls which are to be plastered, accessories such as switches and socket outlets will require a rectangular box to which they are fixed – a **back box**. The box is let into a recess cut into the masonry, with just the thickness of the plaster protruding. Boxes can be simply 'cemented' into position but it is more usual to have them plugged and screwed back to the masonry. The boxes are made of galvanised pressed steel and have **knock outs** for the insertion of the conduit and/or the passage of cable. The edges of the hole at the knock out can be quite sharp, and to protect any sheathed and insulated cable passing through, a rubber grommet is fitted in the hole. A knock out is formed in the metal of the box by pressing a die against the metal with sufficient force to weaken the circumference of the hole but not remove the metal. The removal is done by the electrician, and these pieces of metal can be poked out with the end of a screwdriver.

In timber framed housing or in stud partitions, the cable is simply drawn through holes drilled on the neutral axis of the framing timbers. Where the cable crosses or passes alongside roof timbers it is usual to fix it down with small plastic clips which have a hardened steel nail already in place. Various sizes of clip are available to accommodate different sizes of sheathed and insulated cable, and these are identified by referring to the core size of the cable, e.g. 1.00 mm^2, 2.50 mm^2 and so on. Cables passing down through stud partitions or frames in walls merely pass through holes drilled in the timbers.

An important point is that the cable is *not* pulled tight or fixed down under any kind of stress. To do so would stretch the insulation and the sheath, resulting in their becoming thinner and so less capable of protecting or insulating the core. In extreme cases the core itself would become stretched and therefore thinner and less able to carry the full current for which it was rated.

When the electrician finally connects up all the cores to appliances or accessories, he or she takes great care to ensure that the sheath of the cable is enclosed within the appliance or accessory, and always uses any cable clamps to grip the sheath and insulated cores together. Insulated cores are never left outside appliances or accessories or back boxes and are never clamped on their own. So when using steel back boxes the sheath is always drawn inside, but there is generally no clamping arrangement. Quite generous lengths of cable are left at the places where accessories such as switches and socket will be mounted.

Modern practice in plasterboard clad walls and partitions uses a box of plastic which is placed in a cut-out in the plasterboard and which clips to the plasterboard. The boxes are known as **dry lining boxes**. With the dry lining boxes, the entries into the boxes have a flap which allows the cable to pushed through and which springs back to grip the sheath, an improvement on the steel back box with nothing to grip the cable.

Earth bonding

Earth bonding is an extremely important part of the installation and there are two parts to it which must be carried out.

The first part is the connection of the earth core in the twin or three core cables to the earthing terminal in the CCU. Where cables reach appliances or accessories, the earth core must be secured to the earthing terminals supplied as part of the appliance or accessory. This earth core in sheathed cables has no insulation, and yet to reach the earth terminals it may have to be left quite long. To prevent this length of bare wire from touching the other terminals, a short length of green/yellow plastic sleeving is pushed over the wire. Any old length will not do; it has to be capable of insulating the whole of the earth wire protruding from the sheath.

The second part comprises a core with green/yellow insulation only installed from each metallic appliance or piece of equipment which is to be installed in the house, all the way back to the CCU position. Metallic objects would include any metal sink, pipework to sanitary appliances, metal cisterns, radiators or other emitters, gas or other pipework and so on. Patent clips are used for thin metal edges, and earthing clamps are used on pipework.

Accessories such as switches, socket outlet plates, etc. are fixed in position to boxes with threaded inserts for the screws. The boxes are made in different styles to suit different situations. Accessories are either surface mounted or flush mounted. For surface mounting, the box to which they are fixed is itself mounted on the surface of the wall and is fully exposed. For flush mounting, the box is sunk into the wall so that only the thin plate on the top of the accessory is visible. All these boxes are called back boxes.

Back boxes are made from steel (galvanised, zinc plated or black enamelled) and plastic, and different shapes and styles have evolved for surface and sunk finishing, but in the main all require to be fixed back to the wall, floor or ceiling with screws and plugs.

One special box for mounting within a plasterboarded wall has patent clips which allow it to be fixed flush with the surface held only by the clips which tighten on the plasterboard as the accessory plate is screwed into place. This particular style of box is commonly known as a **dry lining box**.

All the boxes are available as one or two gang or circular, and the surface mounting ones are also available in different depths. Surface mounting boxes in plastic are also sometimes called **pattress boxes**.

Steel back boxes and plastic dry lining boxes, like the accessories which are fixed to them, are either single or double (it is possible to have three or four but that is rare) and this is referred to as **one gang** or **two gang**. To make things more complicated it is possible to have a two or three gang light switch – i.e. two or three switches – in a plate which only requires a one gang back box. Figure 13.12 should make this clear.

Figure 13.7 shows a steel back box with connector for plastic conduit. It is a one gang galvanised steel back box of 25 depth, and in one of the knock outs a connector for plastic conduit has been fitted – a male connector with a back nut. Alongside it is another conduit connector, a female bushed connector with a male bush. In both cases the plastic conduit is solvent welded into the plain socket and the cable drawn down once the solvent weld

Fig. 13.8 Plastic dry lining boxes and cable clips.

cement has dried out. Note that the solvent in the cement would attack PVC insulation.

Figure 13.8 shows plastic dry lining boxes – one gang, two gang, and circular, and cable clips. All push through plasterboard or other sheet material with the side clips withdrawn inside the body of the box. Once in position, the side clips are pushed out behind the plasterboard and pulled up tight. There is usually a ratcheting action built into the plastic extrusions. Finally, when the accessory is fixed to the back box, the fixing points for the screws are located in the lugs holding the box tight to the plasterboard so that tightening up these screws serves only to fix the box in position more securely. The rectangular boxes are used for accessories with rectangular plates while the circular box. . .

In the bottom right of Figure 13.8 are a few cable clips of the type used to secure twin and three core sheathed cable. The nails supplied already fitted are of plated, hardened steel and will drive into masonry as well as timber backgrounds.

Figure 13.9 shows two water supply pipes to a wash basin, with earthing clamps fixed to them, and an earth bonding cable attached and running off to the left – hopefully to an earthing point.

Figure 13.10 shows a patent clip with an earth wire attached, which can be screwed firmly to the welded edge of something like

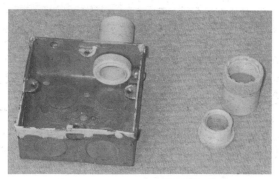

Fig. 13.7 Steel back box and plastic conduit connections.

Fig. 13.9 Earthing clamps on pipework.

a radiator, a metal sink or similar piece of equipment.

Final fix

Once the first rough is completed, the electrician must wait for wall, floor and ceiling finishes to be applied right up to the painting and decoration. When that is at least under way the electrician can return and fit the light switches, socket outlets, etc. to the back boxes or dry lining boxes, the ceiling roses for lights, etc. and make the electrical connections in each. The CCU is also mounted in a cupboard where the tails come through the wall from the meter, and all the connections are made in that. All the earth cabling from around the house must also be connected to an earthing terminal and this connected to the earthing terminal on the supply company's cut-out.

Fig. 13.10 Earthing clip on radiator.

Testing and certification

The installation must now be tested for continuity, polarity, insulation and earth resistance.

Continuity means testing to see that all cores are correctly connected and do not impede the flow of current to the appropriate accessories or appliances.

Polarity means testing to see that the cable (red insulation) connected to the phase wire is connected to the correct terminals of socket outlets and fixed appliances, and that it is the cable broken for switches in circuits with separate switches such as lighting.

Insulation means testing to see that there is no leakage of current from a core to earth or another core due to the insulation being cut or otherwise damaged.

Earth resistance means testing to see that the resistance to the flow of current in all earthing cores connected back to the earth terminal at the supply company's earth terminal is as near zero as possible. This includes earth cores included in twin and three core and earth cables, as well as separate earth cores leading from isolated metalwork.

Once this has been carried out satisfactorily, the electrician can sign a certificate that this has been done and that the installation complies with the IEE Regulations.

Now the supply company can be asked to come and connect the system to the mains. Their engineers no longer test installations, so all they have to do is open the cover on the cut-out, mount the meter in the box, and using the cable left in the tails make the connection from cut-out to meter to CCU. Putting the appropriate fuse in the cut-out and replacing and sealing the cover completes the whole installation.

More on protective devices

A further few points about RCCBs, MCBs and fuses.

Figure 13.11 shows a schematic of a current operated RCCB. It is essentially quite a simple idea. While current flows through both

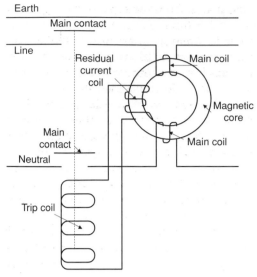

Fig. 13.11 Schematic of a residual current circuit breaker.

conductors or cores in the cabling, the current is in near perfect balance, but should a fault develop which causes the current to leak from any conductor to earth, then the imbalance pushes up the current flowing through the little coil, which in turn trips the spring-loaded double pole switch in a matter of a tiny fraction of a second. RCCBs generally have a test button which can be used not just to test the device but also to shut down all the sub-circuits which follow, thus obviating the need for a double pole switch. They should be tested regularly.

It was stated earlier that the present design of RCCB has rendered the ELCB obsolete. The ELCB was originally designed to operate in one of two ways: by reacting to an imbalance in current or to an imbalance in voltage between two conductors. Either type is now obsolete, and they should be replaced with RCCBs which only react to an imbalance in current.

MCBs are the preferred protective device for domestic sub-circuits, and all modern CCUs sold in the UK now make provision for these to be fitted as standard They are built with a whole range of ratings – the amount of current they carry before tripping – from 5 to 45 amperes. HBC or HRC (high rupturing capacity)

cartridge fuses, obtainable in ratings of 5, 10, 15, 20, 30 and 45 amperes, are termed Type 1 fuses and used in CCUs; and the 60, 80 and 100 amperes, available as Type 2 or 3 fuses, are used by the supply company in the cut-outs. These fuses should not be confused with the fuses used in 13 ampere plug tops and which are rated at 3, 5 or 13 amperes. These latter cannot be used in CCUs.

Last, and no longer installed in the UK, are rewirable or semi-enclosed fuses. There must be hundreds of millions if not billions of this type worldwide, but they are definitely frowned upon in the UK and EU now and should be replaced with MCBs if at all possible. They comprise a base fitted over the bus bars in the CCU into which a fuse wire carrier is plugged, two pins making contact with the bus bars. The two pins have screw terminals into which a length of plain or tinned copper fuse wire is fitted, passing through a hole in the carrier between the pins. The CCUs made with rewirable fuses at the time HBC fuses and MCBs were being introduced were generally of a pattern which would allow any of the three devices to be fitted. In some instances only the overall cover requires to be changed, making the conversion of an existing CCU to a more effective circuit protection unit much more feasible and less expensive than complete replacement.

There is no retrospective legislation to make consumers update their control units, but when premises are being rewired or largely altered, competent electrical engineers will always insist that MCBs are fitted into any CCUs.

Wiring diagrams

No text on electrical work seems to be complete without line drawings of how to wire up a light with a one-way switch, with two-way switches (there is seldom any mention of intermediate switches), ring mains and so on. This seems a pretty futile exercise as the drawings themselves are never colour coded and the insulation is colour coded so that correct polarity of the wiring can be observed. Also, the insulated cores in twin and three core cable are shown all over the place, while in reality they are never split out of the sheath or are otherwise running together in conduit. The final objection must be that it might provide the inexperienced with just enough information to make them think they can do their own wiring up with disastrous results. Besides which, the amateur will not have the equipment to test the installation correctly and so will never know if it has dangerous faults or not.

Leave this type of work to those with proper training employed by a reputable company of electrical engineers with the proper tools and test equipment.

Accessories

Accessories are such things as switch plates, socket outlets etc. The photographs which follow are of accessories and small parts used in a typical installation.

Figure 13.12 shows switch plates and a ceiling lighting pendant. On the left are three switch plates, from the top a one gang, a two gang and a three gang, and to the right of the last is a one gang architrave switch. The three square switch plates all fix back to a one gang back box. Architrave switches have special back boxes in galvanised steel only.

The other item in Figure 13.12 is a ceiling pendant. It comprises a number of components. Top left is a dark circular plate which is screwed back to the ceiling board direct or to a circular dry lining box. A curved terminal block can be seen on one edge, a length of twin core flexible cord is wired into two of the three terminals, and to the right-hand side there is a bright metal terminal, the earth terminal. The flexible cord passes through the white disk on the right. This is the cover for the back plate, and together they form a **ceiling rose**. Finally, the flexible cord is connected to an all insulated lamp holder. (Note, electricians don't talk about a *light bulb* – the proper term is **lamp**).

Flexible cord is different from the cable used to wire up the sub-circuits, and it is principally the cores which make that difference. Insulation is PVC just like the sub-circuit cable, but it may be flat or round and there may or may not be an earth core. Line and neutral cores are insulated brown and blue normally and the earth is always green/yellow. The cores are formed by twisted copper wires, anything from 19 strands upwards being used. The

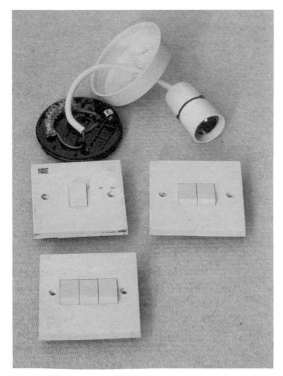

Fig. 13.12 Light switches.

Fig. 13.13 Socket outlets, plug top and cable outlet.

Fig. 13.14 Electrician's roughs – cable coming together at the point where the CCU will be sited.

particular example here has 19 strands of copper, blue and brown insulation and a white circular sheath. The more strands, the more flexible the cord. Note: **cord** is the correct term, although many tradesmen refer to it as **flex**.

Figure 13.13 shows socket outlets, plug top and cable outlet plate. On the left are a pair of two gang socket outlet plates, the upper one being a surface mounted plate with an enamelled steel surface back box. This type is used for commercial and industrial work as well as domestic garages etc. It is generally used with a conduit system, although conduit to drops is a viable alternative. The lower two gang socket outlet has a metal plate – brass – for mounting on a recessed back box. On the right at the top is a one gang plastic socket outlet plate and below it is a plastic switched fused cable outlet plate. The fuse carrier can be seen with the little screw holding it in place and the plate is switched but does not have a neon indicator lamp. At the bottom is a **plug top**. These should not be referred to as simply plugs. Plug tops have provision for fusing individual appliances.

Finally, Figure 13.14 is a photograph of the electrical cable in a single-storey house coming to a point in the floor where the CCU will be sited. Cable can be seen clipped to the sides of the ground floor joists, and some plumber's roughing can also be seen – the insulated pipework for central heating. None of the cables is identified but the size of cables for specific circuits can be readily identified on site, and if there are more than one or two, a meter can be used to determine what goes where.

APPENDIX A
Maps and Plans

Maps

Maps are the two dimensional representation of the three dimensional features of a landscape[1]. They are produced to large scales by Ordnance Survey: 1:50 000, 1:2500, 1:1250.

The name Ordnance Survey came about following the introduction of longer range, accurate artillery. To be able to hit a target some miles distant, the artilleryman must know where his gun is sited as accurately as possible. He must also know where the target is, relative to the gun. Guns of all kinds are collectively known as ordnance and so any cartography carried out to assist in their use became known as Ordnance Survey. The first survey of the whole of the UK was carried out with this in mind following the onset of the Napoleonic wars, and although the civilian population derive great benefit from these surveys, their military use has been the driving force behind all such surveys worldwide.

1:50 000 is the familiar Landranger scale. Maps to this scale are used by a variety of the public and professionals but are familiar to climbers, walkers and ramblers. Each sheet covers an area of approximately 1600 square kilometres, and has a grid of kilometre squares. Levels are shown by contours at 10 m intervals and spot heights on prominent high ground.

Individual large buildings in open country are distinguishable but not capable of being measured. Location is only possible to the nearest 100 metres by giving a six figure reference. By prefixing that reference with the letter reference for the square of 100 km side, a unique reference within the UK is obtained. A typical grid reference is given in Figure A.1 with reference to Grange Farm from OS Landranger sheet 66. Try it with a Landranger or map of similar scale which you have. The procedure is simple and is illustrated in Figure A.1.

Larger scale maps are of use to the builder or developer, utilities companies, etc; 1:2500 and 1:1250 maps show much more detail but cover a very much smaller area. On the 1:1250 and 1:2500 maps, all buildings are distinguishable and their size can be 'measured' to some degree. Levels are shown as 'spot levels' and benchmarks are shown. Even at these larger scales there is insufficient accuracy to allow us to plan the positioning of new buildings and also show all the detail of services and ancillary works.

Plans

A typical site plan is shown in Figure A.2. 1:500 is the largest workable scale and plans of this type (note change of terminology with change of scale) have not been available from OS, but this will change in the very near future. They are generally plotted out from a survey carried out by the building client's surveyors. Traditionally, civil engineers or surveyors would 'survey' the area and plot out the results to a scale of 1:500. A 1:1250 or 1:2500 map might be used as a basis for such a plot.

[1] Charts show underwater features, buoyage and lights off shore, land masses and prominent features on land masses such as major hills, church spires and tall buildings and including lights for navigation etc. and instead of contours show depths of water. They are prepared by the Admiralty Hydrographic Department.

Read letters giving 100 000 metre sided square reference:	NT		
Locate first vertical line to left of location.	22		
Enter two numbers. Now estimate tenths:	4		
Locate first horizontal line below location.		86	
Enter two numbers Now estimate tenths:		6	

Together these give a unique reference within the UK **NT 224 866**

Fig. A.1 How to determine a grid reference from a 1:50 000 map.

The survey was traditionally carried out using **chains**[2] or steel measuring tapes and a **dumpy level** or **tacheometer** (see Appendix B). Now sophisticated instruments are used which measure levels and distance as well as relative angles between straight lines joining all points. They employ EPDM – electronic position and distance measurement. The data is recorded in magnetic format and fed into a computer which does the plotting.

Surveyors must be able to measure out land and buildings and plot out all the salient features to a suitable scale as a site plan. The electronic distance measurement equipment is very expensive and is only warranted on the largest of surveys or where the survey has to tie in with work previously done and/or which will be processed on a computer. It is possible to hire specialists who do any size of survey – for a price – and who will use the latest equipment to cut down on the staff overhead. Larger developers use it where the initial ground survey is done, the 1:500 or other scale plans are plotted and then the same equipment is used to lay down a grid of reference points on the actual site. These are used by the trades foremen to set the buildings correctly on the site as well as the alignment and levels of roads, sewers, etc. Illustrations are included in Appendix B, Levelling.

If the need for survey work is infrequent or the size of the survey does not warrant the expense of hiring specialists, it will pay to have the capability of reverting to traditional methods. That means having the basic equipment to hand and keeping the necessary skills honed.

Carrying out a survey allows us to 'fix' the absolute positions of all existing features, levels, etc. To the drawing which we plot out from the survey we add the proposed works – buildings, roads, services, alterations to topography. There

may be so much to add that it is split up over a number of separate plans. Our site in Figure A.2, however, is not so large and is now shown as a proposed development in Figure A.3.

Scales of 1:200 and 1:100 are occasionally used for site plans where the 1:500 would not be capable of showing the required depth of detail and would not allow more accurate transfer of information to the physical site: for example, a block plan of a large building to be altered or extended. The overall size and shape of each building which is shown on the site plan is taken from the figured dimensions shown on the building plans. These dimensions should never be scaled off. The position of the building should be marked with written dimensions on the site plan.

Plans of buildings

When a building is planned, drawings of it are prepared to a variety of scales depending on the detail required at each stage in the design. Plans drawn to scales of 1:50 or 1:100 are prepared to show foundations, accommodation, roof, etc., as shown in Figure A.4.

'Sections' are also prepared. These are views of a 'slice' across the building. Scales used are 1:50, 1:20, 1:10, as shown in Figure A.5.

Finally, 'details' are prepared. These show how materials and components fit together. Scales include 1:1, 1:5, 1:10 and 1:20, as shown in Figure A.6.

☐ Scales of 1:500, 1:1250 and 1:2500 are known as engineer's scales
☐ Scales of 1:1, 1:5, 1:10, 1:20, 1:50, 1:100 and 1:200 are known as architectural scales.

Only use these standard scales.

If the drawing paper or film cannot accommodate the full building use a smaller scale or break up the building into different parts with a 'key' drawing to a smaller scale. *Never* use nonstandard scales. These can be misinterpreted by someone, with disastrous results. For example, a designer was employed in the Far East and had a team of Philippine draughtsman to assist him. They were American trained and when asked to prepare full size details of window and door sections, drew them up 1:2 – half size, marked the drawings full size and didn't put a scale on the drawings. 1:2 is a common American scale. The first that was known about all this was when flimsy windows started to arrive on site!

---

[2] There were two kinds of chain in use, the Gunter chain and the engineer's chain. The Gunter chain was 66 ft long (22 yd or one tenth of a furlong). The engineer's chain was 100 ft long.

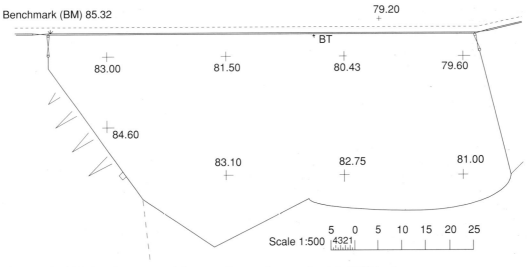

Benchmark (BM) 85.32

79.20

BT

83.00   81.50   80.43   79.60

84.60

83.10   82.75   81.00

Scale 1:500   5 0 5 10 15 20 25

Fig. A.2 A 1:500 site plan prepared for a small area of approximately 0.903 hectares.

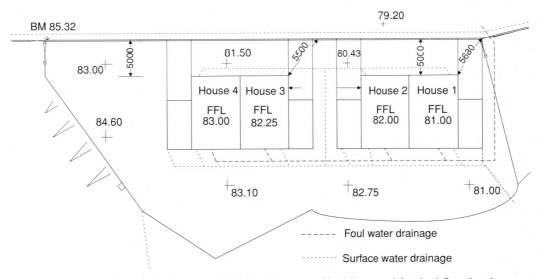

79.20

BM 85.32

5000

81.50   5500   80.43   50C0   5630

83.00

84.60

House 4 | House 3   House 2 | House 1

FFL 83.00 | FFL 82.25   FFL 82.00 | FFL 81.00

83.10   82.75   81.00

------ Foul water drainage

.... Surface water drainage

Fig. A.3 A completed site plan with details of the placement of buildings and finished floor levels.

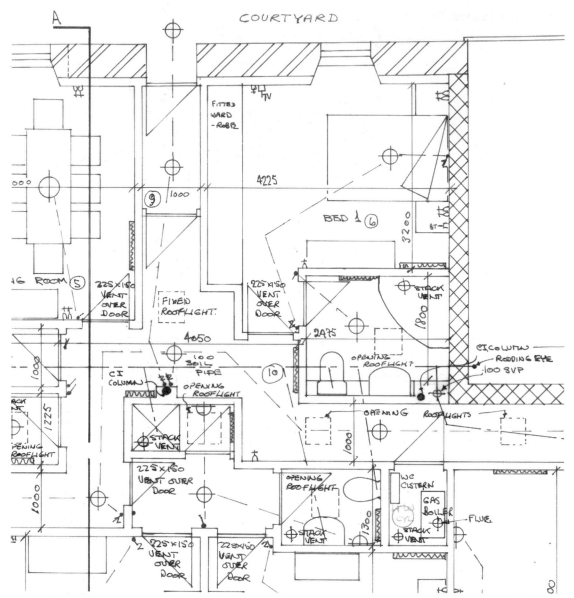

Fig. A.4 Part of the accommodation plan of Old Grange Farm to a scale of 1:50.

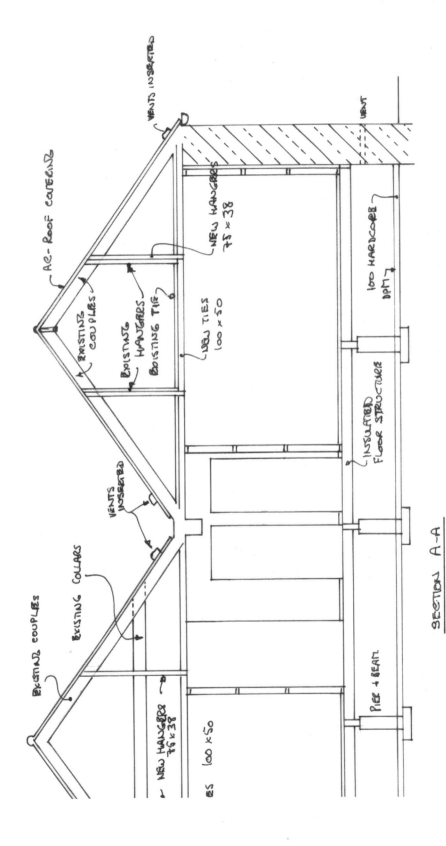

Fig. A.5 Part of a section through the redeveloped Old Grange Farm to a scale of 1:20.

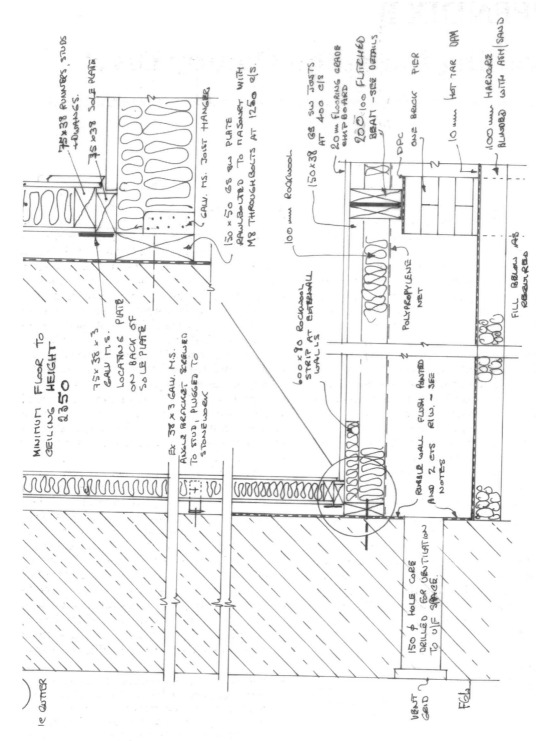

Fig. A.6 Part of a sheet of details showing wall, ground floor and underfloor construction with intermediate support to the floor.

APPENDIX B
Levelling Using the Dumpy Level

A variety of surveying instruments can be used on a building site to assist in the accurate setting out of a building on the ground and in setting the levels at which various parts of the building are to be built. The most basic of these instruments is a **level**, or **dumpy** as this type of instrument is often called. Many types of dumpy level have been 'invented' but many old models are still and will continue to be in regular use. These all use the principle of setting a telescope so that the line of sight is truly horizontal no matter which way the level is facing. A model popular with builders on smaller sites is one which uses a telescope but the line of sight is set horizontal automatically by a series of pivoted mirrors which are counterweighted. The whole instrument is contained in a small box which mounts on a tripod. It is not as accurate as the older style dumpies.

A more modern version uses a revolving laser light projected in a truly horizontal plane so that instead of viewing a measuring 'staff', the laser illuminates a staff as it comes round. This is more accurate than the little automatic box but not as accurate as the full size optical instrument. Our explanation will cover the use of a basic optical dumpy level[1]. Once the principles have been grasped, the use of slightly different instruments is only a matter of familiarisation with the particular product being used.

Levelling

Levelling is a technique used by surveyors and others in the construction industry to determine the height of land surface or objects in a landscape relative to a known fixed height. The known fixed height may be a permanent mark or a temporary mark. Such marks are known as **benchmarks** (BMs) and can be incisions on the masonry face of a building or pillar for permanent benchmarks, or simply a block of concrete, the corner of a manhole cover or even a peg driven into the ground. The whole point about a benchmark is that it has a known fixed height. This might be a true Ordnance Survey (see Appendix A) related level or an arbitrary value. Arbitrary values are frequently used where there is no need to relate to an Ordnance Survey (OS) value or where bringing in an OS level would be impractical. Arbitrary values are usually set at 100.00 metres. All other levels on the job are relative to this.

Figure B.1 is a photo montage of some benchmarks taken around the UK. It shows (from top to bottom): an OS triangulation (trig) point with a benchmark on its side; a detail of the benchmark on that trig point; the top of the trig point showing the fixing arrangements for the theodolite[2] used by the surveyors during the survey work; a benchmark set up by the then War Department as part of a coastal gun battery; and finally a typical benchmark inscribed on the wall of a building. The actual levels of the benchmarks are found by consulting a map with an appropriate scale, e.g. all maps will give the level

---

[1] For some general information on surveying instruments look on the Internet and visit such sites as Leica at www.leica-geosystems.com

[2] A theodolite is a more sophisticated instrument based on a telescope which swings in a horizontal plane *and* a vertical plane. The arc of the swing for any reading can be read off a pair of scales in degrees. In addition, the instrument will allow the surveyor to measure distance from the instrument to the staff. This technique is called tacheometry. There are now modern instruments which automate much of this type of work.

(a)   (b)

(c)

(d)

(e)

of the OS trig point but only 1:1250 or 1:2500 will give the level of the mark inscribed on the building.

From the relative heights we can calculate and show as spot heights or contours the 'level' of the land on which we propose to build. The technique uses a dumpy level and an analogue measuring device or **staff**, shown in Figure B.2, and the usual notebooks, pencils, etc. Note that the staff is of sectional telescopic construction, having five sections and extending to a height of 5 m.

Basic dumpy level

The basic dumpy level comprises:

☐ A telescope with a sighting device or **stadia wires**
☐ A bubble level on the telescope
☐ A horizontal mounting which will allow the telescope to rotate through 360°
☐ A tripod on which to mount the instrument in a stable state.

The sketch in Figure B.3 shows the basic operational parts of only one variety of dumpy level but will serve to illustrate what they are about. Starting at the centre top and working down the sketch from left to right: the accurate bubble level is used to finally set the telescope truly horizontal just prior to each reading taken on the staff; sometimes there is a mirror fitted over it or there is a small eyepiece which through a

Fig. B.1 Some benchmarks and an OS triangulation point. (a) is a photograph of an OS triangulation (trig) point, a tapering concrete pillar of square cross-section. The face of the pillar bears a benchmark which is cast in bronze and then cast into the concrete. The casting incorporates a recess into which a bearer for a levelling staff can be placed. (b) is an enlarged view of the benchmark. The hole under the benchmark is a drainage hole for all the recesses in the top of the trig point. (c) is a detail of the top of the pillar showing the arrangement for securing instruments or sighting targets. Trig points are only found on the tops of high hills and mountains where they were used in the base survey work. (d) is a typical OS related benchmark carved into a house wall. No indication of the level is given. This has to be obtained by reference to a 1:1250 or 1:2500 map of the area. (e) is a benchmark erected by the military next to a World War II shore battery overlooking Scapa Flow in Orkney.

Fig. B.2 Typical dumpy level and staff.

complex optical system allows the surveyor to view both ends of the bubble, and when these are in alignment, the instrument is truly horizontal. The object lens is the large light-gathering lens facing the staff; the larger it is the more light there will be and the further out the surveyor will be able to use the instrument. The trunnions either side of the telescope allow it to pivot up and down on its optical axis in bearings set in the mounting for the instrument. The focusing eye-piece allows the surveyor to focus on the sighting and reading marks inside the telescope – see stadia wires below. The levelling screw is used to bring the telescope from nearly level to truly level as indicated by the large bubble level. The bearing shown is the pivot on which the whole of the telescope and its mounting revolves in a horizontal plane. The index mark allows the scale below to be read. The transverse locking screw clamps the bearing so that the telescope is pointing in a 'fixed' direction. The scale of degrees on the lower fixed part of the instrument mounting is not much used in general levelling but can be useful in fixing the horizontal angle of various sights or readings relative to a fixed point in the survey. The coarse levelling screws are used to set the mounting to an approximate level as indicated by the bubble level between them – what is not shown is that there are three coarse levelling screws. Finally, the bottom plate of the mounting has a large threaded hole for securing the instrument to the top of the tripod.

The instrument works on the principle that if the telescope is set so that the line of sight is horizontal, then measurements can be taken below (and above) that line of sight or **collimation**. These measurements can be used to determine relative height. If a known fixed height is included as one of the measurements then all others can be calculated relative to it, i.e. a benchmark either OS or temporary.

Using the dumpy level

Figure B.4 illustrates the dumpy level in use. To begin, the dumpy level has to be set up on its tripod so that the telescope's line of sight is truly horizontal. The tripod has to be set on firm ground and the spikes on each leg pushed firmly into the ground or otherwise stabilised so that they don't slip. The table on which the dumpy

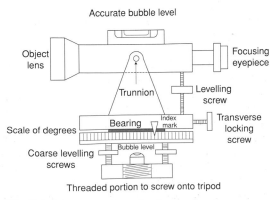

Fig. B.3 Schematic view of a dumpy level.

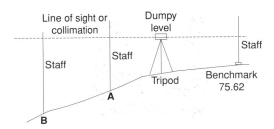

Fig. B.4 Schematic of dumpy level in use.

Table B.1 Noting the readings in a level book

BS	IS	FS	Collimation	Level	Notes
1.45			77.07	75.62	*Benchmark*
	2.30			74.77	*A*
		2.96		74.11	*B*

level is mounted must be set fairly level – within the limits of the small bubble level(s) mounted on the table. Finally, the telescope itself is either levelled for any angle of viewing (360° in a horizontal plane) or brought level as each reading is taken; it all depends on the type of dumpy level being used. The dumpy illustrated in the sketch is of the latter type.

Measurements with the instrument are taken by placing the staff on the selected point and sighting on it with the telescope. The telescope is focused on the staff and measurements on the benchmark and points A and B are taken in turn. The first reading on the BM is termed a **back sight**. All others except the last are **intermediate sights**. The last is a **fore sight**. A is therefore an intermediate sight and B a foresight. It is not by chance that the readings are termed back and fore sights, remembering that the work of the OS was first carried out for men who fired guns.

Table B.1 shows how the readings are written down in the level book and how the levels are worked out.

Stadia wires

A stadia wire is a device placed inside telescopes used for aiming, such as telescopic gun sights, periscopes in submarines, and here in a dumpy level. They are also used in other surveying instruments such as theodolites.

When first looking through the telescope to take readings, the surveyor focuses the eyepiece of the telescope on the stadia wires. Once this is done it generally needs no further attention, the stadia wires remaining in focus no matter how the focus on the staff is changed. Various styles of stadia wires are made but the one in Figure B.5 is a fairly common pattern.

The complete telescope is in turn focused on to the staff and the reading taken as shown in Figure B.6. Note that the staff appears upside down. This is the most common scenario, the reason being that the additional lenses required to turn the image right way up would reduce

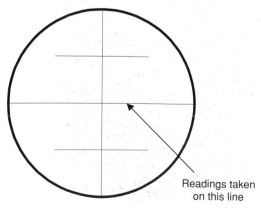

Fig. B.5 Common form of stadia wire.

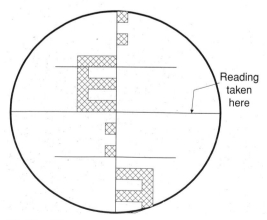

Fig. B.6 How the staff appears to the surveyor when viewed through the telescope.

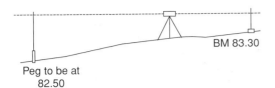

BM 83.30

Peg to be at
82.50

Fig. B.7 Setting to a 'level'.

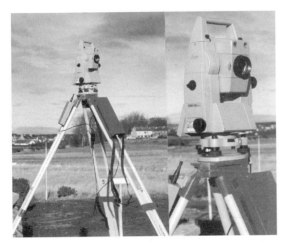

Fig. B.8 An EPDM instrument on its tripod and in close-up.

Fig. B.9 The peg and the nail under the EPDM instrument.

the amount of light reaching the surveyor's eye. This is not a problem for work on a building site where distances would be relatively short, but in the wider surveying field could be quite critical.

Patterns of markings on staffs vary.

Setting to a 'level'

The instrument can also be used to assist in the setting of objects to a given height in the landscape. Refer to Figure B.7. The process commences with setting the telescope level and

Fig. B.10 The target.

sighting on the staff over a benchmark, thus obtaining the line of collimation. If the reading at the BM is 2.33, then the collimation is 85.63. To obtain the level on top of the peg, the reading there must be $85.63 - 82.50 = 3.13$. The peg is driven in until the reading is obtained, and the peg is then at the correct height.

This technique is frequently used on building and construction sites to set the levels of foundations, floors, beams, column bases etc.

On the subject of modern instruments, Figure B.8 shows an EPDM instrument on its tripod with its associated electronic gadgetry and power pack. It is set up over that most basic indicator of position, a wooden peg with a nail in it (Figure B.9). In Figure B.10 is the target on which the telescope of the instrument is aligned and which also houses its own set of electronic gadgetry, the two sets communicating with each other.

The complete set was being used by a squad of bricklayers under the supervision of the site engineer to set the position of the substructures being built. Only the peg with the nail had been set up – again with the EPDM equipment – by the engineer. There were no profiles on the site for these buildings. Position and levels were all transferred by the equipment. The engineer did admit to checking the setting out with a good steel tape!

APPENDIX C
Timber, Stress Grading, Jointing, Floor Boarding

This is not meant to be a comprehensive treatise on timber, stress grading or timber joints and jointing. Several textbooks have been written on the subject already. What we are trying to give in this appendix is an overview of the more basic features of timber as applied to the limited construction forms and types in this text; a brief overview of stress grading – what it is and where it is used; and an explanation of the more common joints and jointing methods used in modern domestic construction.

The structure of timber will not be dealt with in this text. The reader is referred to any good materials textbook or one dealing only with timber technology.

Sawn and regularised timber for carpentry work

In general timber can be divided into two categories:

- **Softwoods**, obtained from resinous, cone bearing evergreen trees (gymnosperms) which have needle-shaped leaves – firs, pines, spruces, hemlock, etc. The only deciduous tree from which commercial softwood is obtained is larch.
- **Hardwoods** are obtained from deciduous, broad leaved trees (angiosperms) – oak, ash, beech, mahogany, teak, etc.

In the UK the bulk of timber used in construction is softwood and is mainly imported either from Russia and Scandinavia – the European softwoods – or from the USA and Canada. A small, but ever increasing, amount of home grown timber is used. This discussion will concentrate on the use of softwood.

Timber is obtained principally from the trunks of trees. A log is a tree trunk with the branches removed. Logs are rounded in cross-section – the timbers we require are generally rectilinear.

The heavy line in Figure C.1 is the outline of the log and the thinner, straight lines indicate saw cuts. The log in Figure C.1 has been **slab sawn**. Note that the first and last slabs have only one flat face. Slabs have a rough edge called **wany edge** or **wane**. Cutting up the log is known as **conversion**.

Besides slab sawing, there are other ways of converting timber (Figure C.2). The upper illustration shows a typical log from a European source and would be of 300–450 mm overall diameter. The lower illustration is a North American sourced log and would be typically around 700–1200 mm diameter.

Naturally, pieces of rectilinear timber have acquired names over the many years of use:

- A **baulk** is the largest square sided timber which can be obtained from a log.
- **Planks** are from 50 to 150 thick, 275 to 450 wide and 2.40 m and upwards in length.
- **Deals** are 50 to 100 thick and 225 wide.
- **Battens** are 50 to 100 thick and 125 to 175 wide.
- **Boards** are of any width and length but not exceeding 50 thick.
- **Scantlings** are resawn timbers 100×100, 100×75, 100×50, 75×75, 75×50, etc.

The third common way to cut softwood logs is quarter sawing (Figure C.3). Other cuts are used on hardwood logs to enhance the figure or grain of the timbers. To get full use of all available sections, quarter sawing can be combined with slab sawing.

A wide range of cross-sectional sizes are available from sawmills. Those from BS 4471 are listed

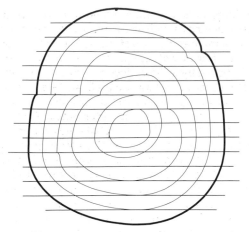

Fig. C.1 Log, slab sawn.

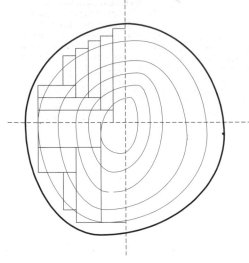

Fig. C.3 Log, quarter sawn.

in the précis of that BS in Appendix L. The shortest length available is 1.80 metres and stock lengths rise in increments of 300 mm. Lengths are therefore: 1.80, 2.10, 2.40, 2.70, 3.00, 3.30, 3.60, 3.90, 4.20, 4.50, 4.80, 5.10, 5.40, 5.70, 6.00, 6.30, 6.60, 6.90 and 7.20 m.

Lengths and cross-sections available obviously depend on the availability of suitable trees for conversion. As the larger trees are felled or become the subject of conservation, longer lengths and wider and thicker timbers become scarce and more expensive. Timbers over 6.00 metres long are more expensive, as are timbers with one side over 200 mm wide. Combine

excessive length with excessive width and even more expense is incurred. The précis of BS 4471 in Appendix L gives some indication of the current situation regarding stocking at mills, resawing, availability and relative economics of the different lengths and sections.

Much of the timber imported now comes in a complete variety of sizes as given in the British Standard but some will inevitably be **resawn** in the UK from larger boards or deals shipped in. Resawing is done with either circular saws or band saws. Band saws have thinner blades and thus waste less timber.

Timbers produced and used directly from the saw are described as **sawn**. BS 4471 lays down tolerances for finished dimensions of sawn timbers:

- −1 to +3 mm on timber thicknesses and widths not exceeding 100 mm
- −2 to +6 mm on timber thicknesses and widths exceeding 100 m
- On length −0 and unlimited excess length.

There are two further possibilities:

- Sawn timbers can be **regularised**, i.e. brought to a common overall size. This can be achieved by accurate sawing to closer tolerances or by machining.
- Sawn timbers can be **planed**, **dressed** or **wrot**, i.e. brought to a common overall size and given a smooth finish all round. This can only be achieved by machining.

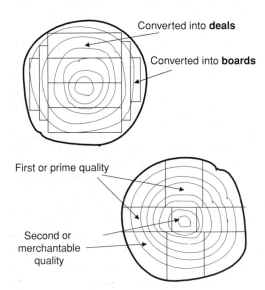

Converted into **deals**

Converted into **boards**

First or prime quality

Second or merchantable quality

Fig. C.2 Logs, alternative sawings.

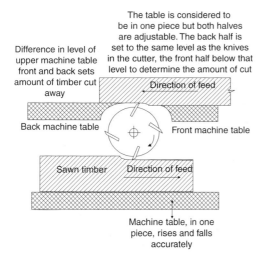

Difference in level of upper machine table front and back sets amount of timber cut away

The table is considered to be in one piece but both halves are adjustable. The back half is set to the same level as the knives in the cutter, the front half below that level to determine the amount of cut

Direction of feed

Back machine table   Front machine table

Sawn timber   Direction of feed

Machine table, in one piece, rises and falls accurately

Fig. C.4 Schematic of planing and thicknessing machine.

In machining, sawn timbers are passed through a machine with from one to six sets of rotating cylinders, each cylinder having two, three or four sharp blades and each cylinder presented to a separate face of the timber. The timber runs along a flat surface, a table, or in large machines a horizontal roller bed, to be presented to the cylinder(s). In large machines timber is fed by powered rollers and in small machines is pushed through by the operator.

Figure C.4 shows a schematic of a combined planing and thicknessing machine, generally referred to as a thicknesser, with a four-cutter cylinder. The individual cutters are termed 'knives'. In the figure a four-cutter cylinder is shown. On planing machines and thicknessing machines with one and two tables respectively this is the norm, and a single four-knife cylinder thicknesser is the most common machine of this type found in small mills and joinery workshops. To obtain a wrot or dressed surface on a piece of wood the timber is passed over the top of the cylinder using the split table, which in turn determines how much is being machined off, i.e. the difference in level between the front and back parts of the table. The front half is lower than the top of the cutter knives; the back part is the same level as the cutter knives.

To obtain a piece of timber of known and accurate thickness it must first be wrot on one face and then passed between the cylinder and the lower table, the gap between them being set to a predetermined distance.

In large mills it is more usual to have a **four sider** machine as well as a **thicknesser**. The four sider as the name suggest cuts all four faces of a piece of timber at one pass. Each cutter cylinder is driven by its own electric motor. However, these faces need not just be plain surfaces. They can be any shape which can be ground onto the knives in the appropriate cutter block. A window section could have a rebate and rounded groove cut on one face for the glass; grooves cut on a second face for a pressure break and a draught excluder/weather seal; a splay and a rounded arris on a third face; and a splay, rebate and groove cut on the fourth face. For each face the knives could be ground to cut different parts of the timber, e.g. on the third face the splay could be set on two of the knives and the rounded arris on the third (and the fourth knife set for the splay), it taking less effort to cut the rounded arris than the splay.

Setting up these machines is a highly skilled job. It is very expensive to grind the knives and set them up, including test runs and the waste that involves. So mills tend to stick to stock sections which they run and keep in stock, or for very large orders they will grind and set up cutters, but *large* orders only are economical. Even if a client was willing to pay the premium, mills are not easily persuaded to take on short runs of special shapes as it disturbs the general running of the organisation.

In Figure C.5 the sketch shows a schematic layout of a 'four sider' with six cylinder cutters and a roller bed with power feed rollers at the start and end of the machine. Having two cylinders to serve the upper surface of any timber presented means that that surface can have deeper and/or more complex shapes made, the high load that these shapes put on the machine being spread over two separate motors. There is one cylinder for the underside and one for each side, plus a beading cutter on the underside. Although it is shown on the underside in our schematic, these cylinders are frequently capable of being set up to cut on any of the surfaces. They are useful for cutting small quirk beads, small splays and chamfers, small cavettos or roundings or small grooves. These are frequently to a similar size no matter what else is being done to the timber, so the little cutters can be reused time and again and they save a lot of money setting up special large cutters on the full size cylinders for a single job or run.

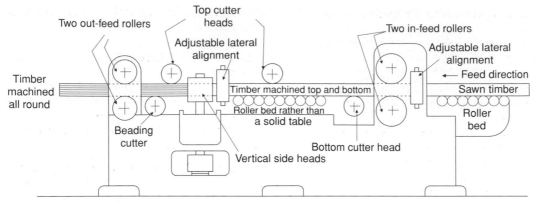

Fig. C.5 Schematic of four sider machine.

Regularised timbers are sawn or machined to give a close tolerance on width and thickness. BS 4471 requires a tolerance of −0 +2 mm. The maximum permitted reductions off sawn sizes of timber for regularising are 3 mm off any dimension up to 150 mm, and 5 mm off any dimension over 150 mm.

Wrot or **dressed** timbers are machined to closer tolerances but also to a finer surface finish. BS 4471 requires tolerances of −0 +1 mm. The actual amount of reduction from the basic sawn sizes depends on two factors:

☐ The application for which the dressed timber will be used
☐ The basic size ranges.

The précis of BS 4471 gives details.

Regularised timbers are used for the manufacture of structural carpentry work or components such as trusses for roofs or beams for floors – anywhere the timber is 'engineered' and regular sizes are required for jointing. Engineered components include trussed rafters where the timbers are joined at the nodes by pressing toothed steel plates into opposing faces of the node. The teeth on the plates are relatively short so the timbers must be of a good, even thickness, otherwise they will prevent the teeth from penetrating fully on one side of the joint.

In carpentry work, wrot or dressed timber would be used, for example, to finish round the edges of a roof structure where the timbers were exposed, or anywhere the timber has to have a good appearance either for painting, staining or varnishing, for example, internal finishing timbers such as skirtings, architraves, door frames, etc.

North American timber

Timber obtained from North America will normally follow the CLS or ALS sizing rules (Canadian Lumber Standard or American Lumber Standard). Surfaced softwood will be available 38 mm thick and in the following widths: 63, 89, 114, 140, 184, 235 and 285 mm. All arrises are rounded, not exceeding 3 mm radius.

Measuring moisture content

Timber changes shape as its moisture content varies. It is, therefore, important to measure timbers at a constant moisture content (**MC**).

BS 4471 stipulates that all measurements are taken, and given, at an MC of 20%. For an MC lower or higher to a maximum of 30%, the size will be adjusted by 1% for every 5% difference in MC.

Moisture content and seasoning of timbers

An important feature of timber is its ability to absorb moisture and dry out, cyclically. This affects its use in a number of ways:

☐ Absorbing moisture or drying out affects the size and shape of the timbers
☐ Logs when first cut have a high MC in the form of sap

- There are optimal MCs for specific end uses of timber
- Too high an MC renders the timber liable to insect or fungal attack.

MC should be held below 20% to avoid fungal or insect attack. Some insects will attack timbers at far lower MCs than 20%.

MC is measured as the percentage mass of water against the oven dry mass of timber. MC can be measured using electronic instruments or by taking samples and weighing them 'wet' and then oven dry, calculating mass of water and thus MC.

MCs for various end uses are:

- Ground floor joists – 18%
- Upper floor joists and roof timbers – 15%
- Framing and sheathing – 16% (timber frame construction of walls)
- Windows and external doors – 17%
- General internal joinery finishings – 8–12% depending on heating system.

Movement of timber due to varying moisture content can cause problems unless this movement takes place *before* the timber is fixed in place and, in the case of dressed or specially shaped timber, before it has been machined. Timber components such as doors and windows or made up joinery should have a much reduced MC before manufacture.

Figure C.6 shows the distortion of timber when cut from various parts of the log and in different ways. Starting at the top and working clockwise:

- The slab sawn plank will cup and bow as it dries out
- The radial cut section gives least distortion
- The quarter cut timber turned into a cylinder becomes oval after drying out
- The quarter cut timber left rectilinear 'diamonds' after drying out.

Reducing the MC to acceptable levels is called **seasoning**. Unless the sap and moisture is removed from timbers, they will warp and shrink between the time of building into position and natural drying taking place, with the following results:

- Joints will open up
- Joints will be weakened
- Timbers will split
- Timbers will shrink – usually unevenly
- Timbers will warp
- Various types of shakes will develop
- Timbers are weakened
- Timbers are extremely susceptible to fungal and insect attack before seasoning.

Figure C.7 illustrates some natural defects in timber.

Seasoning can be done by **natural seasoning**:

- By stacking smaller section or thin timbers in the open air with spacers between each section and a waterproof roof over the stack
- By immersing logs or baulks in water for a few weeks before stacking.

Seasoning can be done artificially by subjecting the stack of timber to a warm air stream in a **kiln**. Steam is sometimes employed in the initial part of the process or the whole process for small batch work. The kiln can be a simple box or a barrel for small timbers. At the other extreme large purpose-built buildings are used, with complex controls for temperature of air and steam and final MC of the timber.

Natural seasoning is a long slow process and is expensive as one has a resource – the timber – which costs money to grow and procure, tied up in a pile somewhere for anything from 18 months to two years. Kilning is also costly – the capital cost of the kiln, its maintenance and the energy cost to run it. However, the timber is seasoned in a matter of a few days or weeks rather than

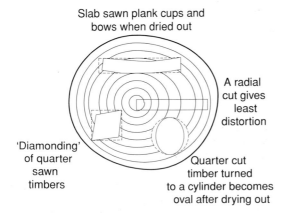

Fig. C.6 Distortion of timber sawn from various parts of a log.

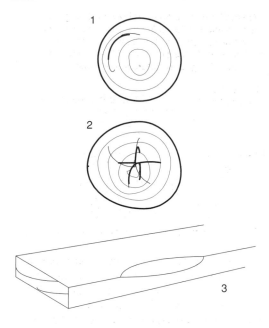

1   An example of a 'cup shake' where the annular
    rings of the timber separate. This can be
    exaggerated during seasoning.

2   A 'star' or 'heart shake' where the fibres of the
    timber separate along the medullary rays.
    This can be exaggerated during seasoning.

3   'Wany edge' caused by cutting the balk or
    board too near the outside face of the log.

Fig. C.7 Natural defects in timber.

years, and with sufficient throughput the process is very economical.

Before using naturally seasoned timber, it must be brought indoors to allow the MC to adjust down to the ambient MC before it can be worked. Kilning can produce timber with just the right MC for the job in hand. Naturally seasoned timber can be subject to insect and fungal attack. Sawmill yards can be a prime source of fungal spores and insects.

Inexpert seasoning can produce a number of faults:

☐ Timbers can be warped, twisted, sprung, bowed or split
☐ Too high a temperature in a kiln can 'case harden' the timber
☐ Steam seasoning can alter the colour of the timber so that it is no use for joinery timbers.

Figure C.8 illustrates some seasoning faults.

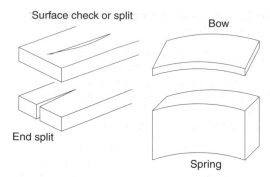

Fig. C.8 Defects caused by seasoning.

Timber should not be fixed in a position within the structure of a building unless it is protected from excess moisture. This is generally done by building in DPCs and by providing ventilation. The MC of structural timber can then be kept below 20%.

Timber can be wetted for prolonged periods when leaks occur in roof coverings, overflowing gutters or leaking pipework. To give protection during such accidents it is common practice to 'treat' timber with chemicals which are fungicides and/or insecticides.

Preservative and other treatments for timber

BS 1282, *Classification of wood preservatives and their methods of application,* lists three classifications of preservative:

☐ TO – Tar-oil types
☐ OS – Organic solvent types
☐ WB – Water borne types.

Tar-oil type is better known as creosote, of which there are two kinds recognised by the BSI:

☐ Coal tar creosote for pressure impregnation
☐ Coal tar creosote for brush application.

Creosote is not favoured for internal use even in building structures as it has a strong odour, contains phenols, bleeds out through paint etc. finishes and can be harmful to people and animals long after it has supposedly dried out.

Organic solvent types include solutions of chlorinated naphthalenes and other chlorinated hydrocarbons, copper naphthenate, zinc naphthenate, pentachlorophenol and tri-butyl-tin oxide. Of these:

☐ All can be applied by pressure or double vacuum impregnation (see below)
☐ Most are suitable for paint finishing
☐ Some are clear and all can be tinted either for decorative effect or to indicate extent of treatment, depth of penetration, etc.

Water borne types include solutions of copper-chrome, copper/chrome/arsenic, fluor/chrome/arsenic/dinitrophenol, copper sulphate, sodium fluoride, sodium pentachlorophenate, zinc chloride and organic mercurial derivatives.

☐ These are suitable for painting over as they do not bleed
☐ All but a few are non-corrosive
☐ Those made of two or more chemicals will not leach out as they 'lock' chemically to the wood tissue.

Application of preservatives can be by:

☐ Brush
☐ Dipping
☐ Immersion
☐ Pressure treatment
☐ Vacuum treatment
☐ Combination of pressure and vacuum.

Brushing and dipping are self-explanatory. Immersion is similar to dipping except that the timber is left to 'soak' for a period of time. None of these treatments allows penetration of the preservative into the timber for any appreciable depth and are therefore unsuitable on their own. Immersion or dipping can be used on cut ends of timbers which are treated before cutting to length. Brushing such ends is frequently specified but is a poor treatment.

Pressure treatment involves immersing the timber in preservative inside a closed vessel and applying pressure to hopefully force the fluid into the timber. This is only partially effective as Figure C.9 shows.

Vacuum treatment involves a sealed cylinder into which timber is placed; the air is evacuated and preservative introduced. The fluid hopefully occupies the voids left by the extraction of the air from the cells of the timber. Because it is not commercially viable to create a perfect vacuum and because the vacuum causes the fluid to vaporise, the cells are not filled with fluid. So

No matter how far down or how hard the jar is pushed into the liquid, the air trapped will never be displaced. The cells in timber are in a similar state. As the outer cells fill, the inner cells containing air are merely pressurised.

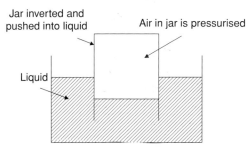

Fig. C.9 Demonstration of air entrapment in cells.

pressure treatment and vacuum treatment are not capable of giving complete penetration of the timber cells by the preservative. Figure C.10 shows a pressure cylinder treatment.

There is another bad side effect. A lot of preservative is left in the timber and must be allowed to dry off. It is dangerous to handle timber wet with preservative. Only the water or the solvent dries off, leaving excessive and expensive quantities of chemical behind. These can still be dangerous.

The double vacuum or vac-vac process avoids much of that trouble:

☐ Timber is placed in a sealed cylinder
☐ The air is evacuated as far as is feasible
☐ Preservative is introduced
☐ Pressure is applied
☐ Preservative is drawn off.

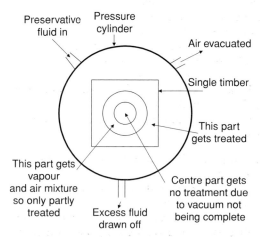

Fig. C.10 Schematic of preservative treatment pressure cylinder.

A vacuum is created and the excess preservative from the timber is drawn off for reuse. The timber is recovered from the cylinder. The timber will have just the right amount of preservative chemical deposited in the cells. Because of the pressure applied, the preservative is driven deeper into the timber. It will never penetrate 100% in very large section timbers but is very effective in the range used for general carpentry and joinery. There is no wasted material. The timber is almost dry when taken out of the cylinder.

The double vacuum process is licensed by a small number of preservative manufacturers to a large number of UK sawmills. Some trade names are **Tanalith** and **Celcure**. There are others. Tanalised timber has a greenish tint. Celcure can be tinted or clear. The latter is preferable for exposed timber which has been wrot or dressed and will be varnished.

Fire resistance

Besides treatment against rot or insect attack, timber can be treated to improve its resistance to fire. Timber has a natural resistance to fire which may seem a contradiction in terms when it is used as fuel by the bulk of the world's population. This natural resistance is due the build up of **char**, the carbon on the surface of timber exposed to heat or flame.

When considering fire in buildings it is well understood that it is impossible to build a structure which cannot possibly be damaged by fire. Buildings or parts of buildings are constructed to resist fire in three ways for set periods of time. The periods of time vary according to the type of building, its use and number of occupants. The best that can be done with fire prevention is:

□ To build in such a way that the building will remain standing long enough for occupants to escape and to allow fire fighters to enter to control or extinguish the fire
□ To prevent one part of the building radiating heat onto another part, thus setting the latter alight or harming occupants while they escape
□ To prevent the passage of smoke, flame and hot gases through a part of the structure, thus setting another part alight or harming occupants while they escape.

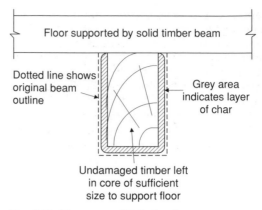

Fig. C.11 Char on a solid timber beam.

Despite being a flammable material, timber has a high natural resistance to attack by fire. To fully utilise this resistance, exposed timbers used in a structure should be oversized by the amount of **char** a fire would produce in a given period of time for which resistance is required – see Figure C.11. That way there will be sufficient undamaged material in the centre of the timber(s) to support the building.

Treatments for improving fire resistance were originally based on strong solutions of zinc chloride. Solutions of boracic acid salts are now preferred:

□ They can be much more dilute
□ They are more effective
□ They are more economical.

How to join timbers in carpentry and joinery work

This section deals with the joining of natural timbers in carpentry and joinery work. Many of the joints can be made in both sawn and wrot timbers. All of them can be made using hand tools but obviously are more easily and accurately accomplished with machines, both stationary and portable. With the development of more powerful and sophisticated portable power tools, many of these joints can now be made on site or even in-situ. How they are made is less important to the student at present, but knowledge of the variety of joints and their use in specific circumstances is vital.

All joints depend on cutting the timber into a variety of shapes or forming holes of various shapes – 'operations'. These 'operations' all have names and to fully appreciate jointing and other work on timber these operations must first be understood. Neither the list of operations nor that of joints is exhaustive but it is fairly representative of what is found in modern domestic construction.

Joints have been developed over the centuries to exploit the natural strengths and obviate the natural weaknesses of timber. Timber is generally cut so that the grain runs parallel to the long side of any plank or board. Loads applied at right angles to the grain can be resisted most easily. Loads applied parallel to the grain cause shear or splitting. Splitting is the most common mode of failure in joints and is usually due to improper use or production of the joint.

Of course there is an unceasing quest to make things 'easier', and pressed steel fittings have been developed over the past 40 years to allow complex joints to be made simply by cutting the ends of timber square and applying the appropriately shaped metal fitting between them. The fittings are normally **hot dipped galvanised**[1] or **sherardised**[2] and are pre-holed for small diameter nails. These fittings will be described later in association with Figure C.46.

Nails should be galvanised wire nails and their length should be at least 40 mm – sometimes less if the timber is thinner. All holes in the fitting should have a nail, otherwise the fitting will not develop its full strength. Forcing very large diameter nails through the fitting will end up with the timbers splitting.

Before looking at operations a brief explanation of the two most common power tools used in joinery work is necessary – the **circular saw** and the **router**.

[1] Hot dip galvanising is done to protect steel from corrosion, principally rusting. It means thoroughly cleaning the steel and immersing it in a bath of molten zinc. The layer of zinc, if of adequate thickness, is of a crystalline structure which if scratched will self-heal – the crystals grow towards each other and seal the surface damage. There is a British Standard for galvanising fittings used in construction.

[2] Sherardising is another method of coating steel with zinc metal. Again the steel is thoroughly cleaned but the zinc is spread over the surface in the form of a fine powder. Steel and powder are heated to a temperature for a period during which the zinc is absorbed into the surface of the steel.

The circular saw

Figure C.12 shows a 250 mm portable saw. Saws can be fixed, in which case the timber is passed over or under the saw, or they can be portable and are passed over the timber. In either case the saw is fitted with an adjustable guide or 'fence' to maintain the width of cut. The blade retracts into the 'table' or sole plate to adjust the depth of cut. The sole plate can usually be tilted to allow bevelled cuts to be made.

Portable saw blades are generally no more than 300 mm in diameter. Fixed saw blades can be up to 2.50 metres in diameter, although most are around 300–450.

The circular saw illustrated in Figure C.13 is an overarm cross-cut saw. Timber is placed on the table against the fence and the saw is pulled across, making a cut through or to any depth in the timber. Multiple passes at less than the timber thickness can make a wide slot across the timber (see trenching in a later section). The blade and motor can be swivelled, and if set with the blade parallel to the fence, timber can be passed under the blade to cut it lengthways to any depth. A router can be substituted for the blade and its motor, and routing carried out across or along timbers.

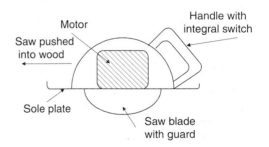

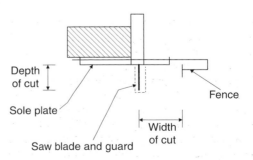

Fig. C.12 Portable circular saw – schematic.

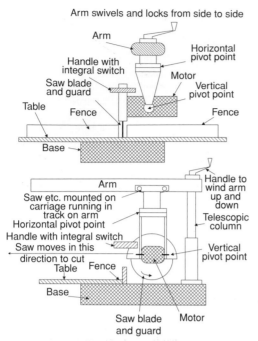

Fig. C.13 Overarm, cross-cut circular saw – schematic.

The router

The router, illustrated in Figure C.14, was originally conceived as a portable tool (passing over

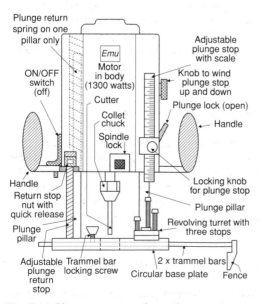

Fig. C.14 Plunge router – schematic.

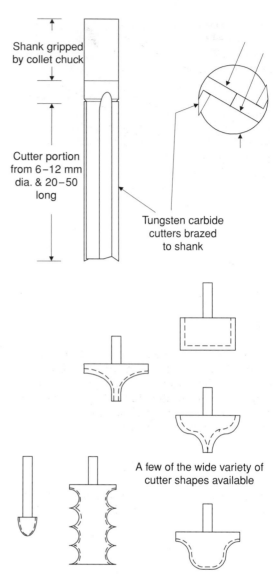

A few of the wide variety of cutter shapes available

Fig. C.15 Selection of router cutters.

timber) but it can be adapted to fit into fixed mounts (the timber is passed over *or* under the tool). A wide variety of cutter shapes is available and some are shown in Figure C.15. The most common is the straight fluted cutter which can cut rebates, grooves and shape timber and board by following a template.

Depth of cut is variable on the router. A fence can be fitted on **trammel bars** to vary width of cut. The fence can be replaced with a 'point' to allow circular cutting. The trammel bars can be almost any practical length.

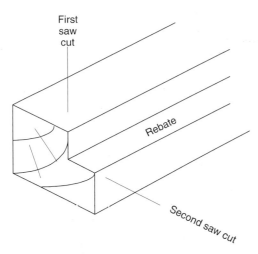

Fig. C.16 A rebate.

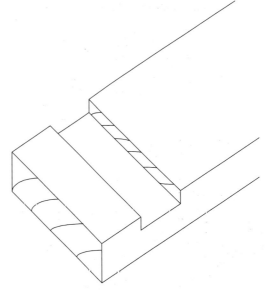

Fig. C.18 A trenching.

Operations

Rebating or rabbeting (Figure C.16)

Cutting out a rectilinear cross-sectioned slice of timber from an edge of a plank or board can be done with two passes of a power saw or a single pass of a router, depending on the size of rebate and the capacity of the machine. It is most frequently done on wrot timber.

Grooving (Figure C.17)

Cutting a rectilinear cross-sectioned piece of timber from the body of a plank or board and parallel to the grain can be done by taking closely spaced, repeated cuts with a power saw or passes of a router. It is most frequently done on wrot timber.

Trenching (Figure C.18)

Trenching is the same operation as grooving but at an angle to the grain – usually 90°. It is done on sawn or wrot timber with a portable saw or router against a timber clamped to the board, or with a cross-cut saw or router on a cross-cut saw arm.

Tonguing (Figure C.19)

Cutting away a rectilinear cross-sectioned slice of timber from the corners of one edge of a timber can be done with two passes of a saw or router but most frequently and more accurately is done in a stationary machine with multiple cutters in a single 'head'. It is mostly done on wrot timber.

Notching or halving (Figure C.20)

This involves cutting away a portion of the cross-sectional area of a timber, i.e. across the grain.

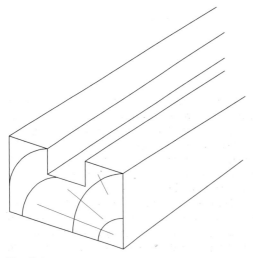

Fig. C.17 A groove.

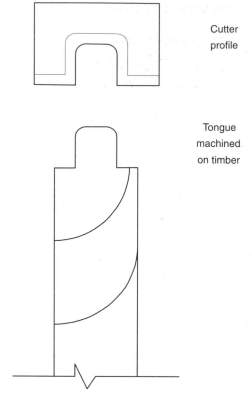

Cutter
profile

Tongue
machined
on timber

Fig. C.19 A tongue.

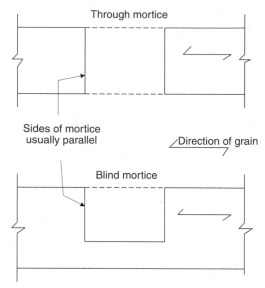

Through mortice

Sides of mortice
usually parallel

Direction of grain

Blind mortice

Fig. C.21 A mortice.

Anything less than half the thickness or width of the timber is a notching; cutting away half the timber is a halving; cutting more than half the timber is never done. Notching or halving are done with multiple passes of a power saw or router, or multiple saw cuts and hand chiseling, on wrot and sawn timbers.

Morticing (Figure C.21)

Morticing is cutting a rectangular-shaped hole in the face or edge of a timber at right angles to the grain. A through mortice extends the full thickness of the timber, i.e. a hole is visible on opposite faces. A blind mortice is a hole which stops short of the opposite face. If a tenon is fitted into the mortice, the short sides are tapered so that the expanded tenon is jammed into place when wedges are driven into place.

Morticing is most common in wrot timber and is done using a purpose-built machine or is drilled and chiselled out by hand.

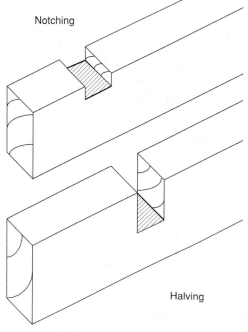

Notching

Halving

Fig. C.20 Notching or halving.

Tenoning (Figure C.22)

Tenoning is reducing the cross-section of the end of a piece of timber, usually to fit into a matching mortice. The reduction can be made on one, two, three or four faces of the timber. Variants

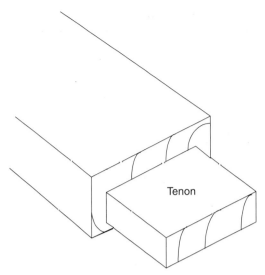

Fig. C.22 A tenon.

are shown in the later illustrations on mortice and tenoned joints. It is most common on wrot timber and is done using multiple passes of a fixed saw or router, or a purpose-built machine.

Splaying (Figure C.23)

Splaying is cutting away the corner of a piece of timber along the grain, usually at an angle of 45°. It is done using single or multiple passes of a power planer or router, or a single pass of a power saw. The cutters used for tonguing and grooving are modified. Splaying is frequently used to disguise a joint, and on both sawn and wrot timber.

Moulding (Figure C.24)

Moulding involves cutting a shape into the regular edge of a timber, usually parallel to the grain but it can be done across the grain, most usually to match the moulding parallel to the grain. It is done for decorative effect but can be combined with a jointing technique to disguise a joint. Normally done on wrot timber, a router is used in either mode of operation with suitable cutter(s).

Dovetailing (Figure C.25)

This operation involves cutting one timber with a splayed projection or 'pin' and the other with a

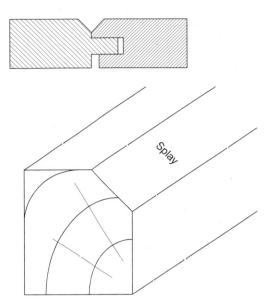

Joint hidden by two 'splays'

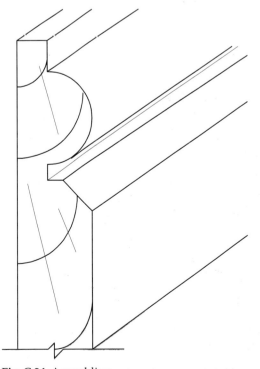

Fig. C.23 A splay.

Fig. C.24 A moulding.

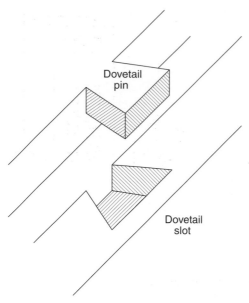

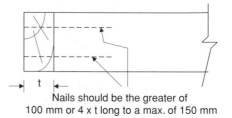

Nails should be the greater of
100 mm or 4 x t long to a max. of 150 mm

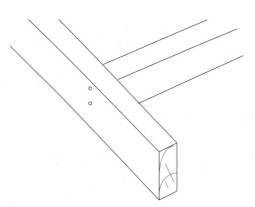

Fig. C.25 A dovetail.

matching splayed slot or groove. There are many variants from single dovetails to multiple dovetails, used in furniture and cabinet making. It used to be cut by hand but is now done using a special jig and a router. It is most common in wrot timbers.

Joints

Now that we have considered the tools and operations, we can look at how these are combined to give simple joints. The list is not exhaustive, many more joints being permutations of several techniques used in several simple forms of joint.

Butt joint (Figure C.26)

This is not really a good joint on its own. The timbers are cut square. The end grain of one is placed against the face of another and nails or screws are driven through into the end grain of the first piece. Gluing is a waste of time and the joint relies for its strength entirely on the nails or occasionally screws. Improved nails give a better grip in the end grain. This joint is satisfactory for rough framing and panels covered with boarding.

Fig. C.26 A butt joint.

Half checked or halved joint (Figure C.27)

The timbers can be joined to give:

- A single continuous length
- A corner joint
- An intersection

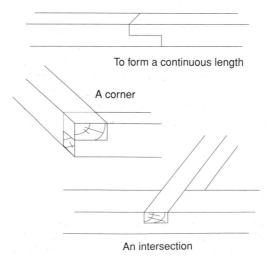

To form a continuous length

A corner

An intersection

Fig. C.27 A half checked or halved joint.

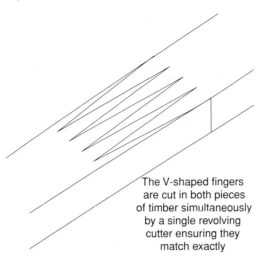

Fig. C.28 A finger joint (1).

The V-shaped fingers are cut in both pieces of timber simultaneously by a single revolving cutter ensuring they match exactly

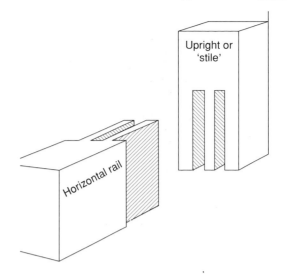

Fig. C.29 A finger joint (2).

In carpentry work the timbers are nailed but in joinery work will be pinned and glued or glued and screwed from the hidden face.

Finger joint (1) (Figure C.28)

This variant of the finger joint is cut by a special machine and can be used to join any regularised or dressed timbers, even structural timber. Only glue is used in the joint, generally a waterproof resin-based type cured by **RF** (radio frequency) radiation. The joint is frequently stronger than the timber itself.

Finger joint (2) (Figure C.29)

This variant is used to join the corners of frames such as door and window frames. Generally only glue is used in the joint, a waterproof glue allowed to set naturally or perhaps at a slightly raised temperature. A wooden or metal (star) dowel can be driven through the centre of the joint. The joint is frequently used as a corner joint for framed material such as window sashes and light doors.

Plain housing (Figure C.30)

This is frequently used to join the uprights or jambs of a frame to the upper or lower horizontal portion of the frame. It is also used to join studs to the head and sole plates of timber framed construction. It is most frequently used for simple door frames. No glue is used, the pieces being simply nailed together. Two nails are a minimum.

Shouldered housing (Figure C.31)

This is used to join wider timbers or sheet materials at right angles. The extra width allows the edge of the joint to show a plain 'butt' joint while

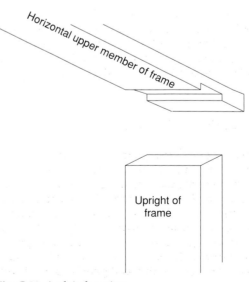

Fig. C.30 A plain housing.

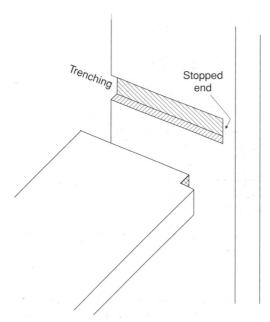

Fig. C.31 A shouldered housing.

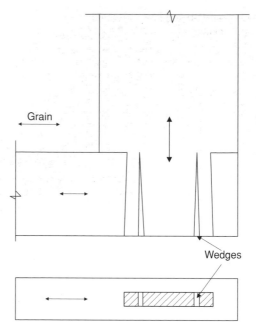

Fig. C.32 Plain morticed and tenoned joint.

the housed portion adds the necessary strength and rigidity. It can be glued or glued and nailed but is often just nailed.

Plain morticed and tenoned joint (Figure C.32)

The illustration shows the most simple of these joints with the tenon being made with an equal reduction in cross-section all round. Note the use of wedges to expand the tenon in the mortice and so wedge it into place. Also note that the pressure from this expansion is exerted *along* the grain of the timber with the mortice. This joint is sometimes glued; *no* nails or screws.

Shouldered morticed and tenoned joint (Figure C.33)

One of the problems with a plain m&t joint, shown in Figure C.32, is that on narrow timbers the tenon is not wide enough to prevent twisting. By adding a shoulder to the joint we are in effect 'housing' one timber to the other for their full width but need only a small portion of the width for the mortice.

Barefaced morticed and tenoned joint (Figure C.34)

This variation of the m&t joint is used for the bottom and mid rails of bound lining doors – see Chapter 8 on Doors. The illustration shows the bottom rail with the tenon cut, the stile with

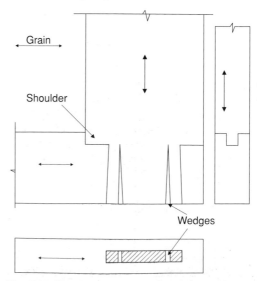

Fig. C.33 Shouldered morticed and tenoned joint.

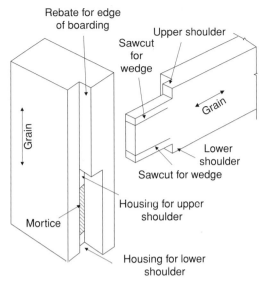

Fig. C.34 Barefaced morticed and tenoned joint.

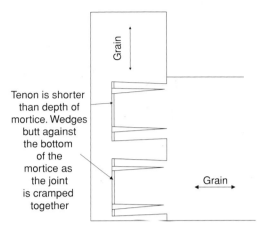

Fig. C.36 Blind morticed and tenoned joint with fox-tail wedges (2).

a mortice and the space left for the thickness of the boards covering the door.

Blind morticed and tenoned joint with foxtail wedges (Figures C.35 and C.36)

This variation is used to join the rails to the stiles of a door in high quality work. The mortices are 'blind', i.e. they do not show on the exposed edges of the doors stiles. Note that the tenon is

now made up of four separate pieces all worked on the end of the rail. This is described as a split, double tenon. Cramping the stiles close together on opposite sides of the door forces the wedges into the saw cuts in the tenons. Wedges are dipped in glue just prior to cramping up.

Tusk tenoned joint (Figure C.37)

This is used to join beams and joists when framing up large openings in timber floors etc. The wedge draws the joint tightly closed while the main part of the tenon provides a bearing in the larger part of the mortice.

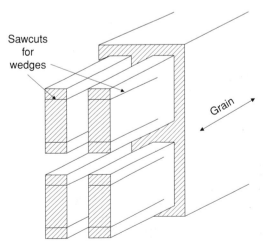

Fig. C.35 Blind morticed and tenoned joint with fox-tail wedges (1).

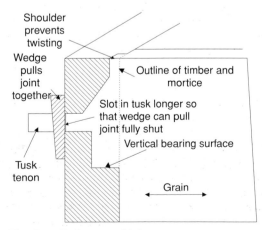

Fig. C.37 Tusk tenoned joint.

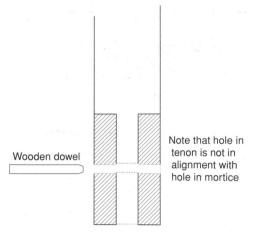

Wooden dowel

Note that hole in tenon is not in alignment with hole in mortice

Fig. C.38 Draw boring morticed and tenoned joints.

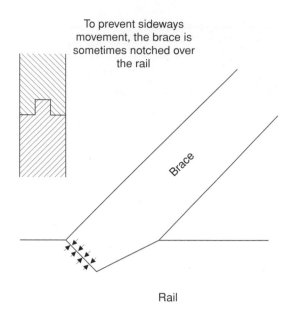

To prevent sideways movement, the brace is sometimes notched over the rail

Brace

Rail

Draw boring morticed and tenoned joints (Figure C.38)

M&t joints are always wedged but to keep the joint firmly closed the timber frame must be restrained with sash cramps until the wedges are driven up tight. On frames which are subject to a lot of vibration etc. the wedges might not be strong enough to maintain the joint firmly and so a timber dowel is often passed through both timbers to coincide with the centre of the tenon. The holes are bored such that in passing through the hole in the tenon, the dowel pulls the tenon more firmly into the mortice.

Toe joint (Figure C.39)

This joint is frequently used for struts when in compression and for diagonal bracing on simple ledged and braced doors. Any compression forces on the strut are transferred to the main timber through the flat angled face in the matching angled notch.

Scarfed joint (Figure C.40)

The simple scarfs shown here join two carpentry timbers together as a continuous length. Note the difference between a **splayed** and a **plain** scarf. Scarfs are only suitable for timbers in tension. The joints are nailed together; glue is not used.

Fig. C.39 Toe joint.

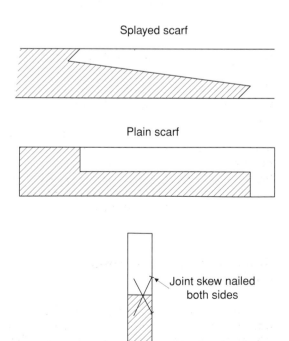

Splayed scarf

Plain scarf

Joint skew nailed both sides

Fig. C.40 Scarfed joint.

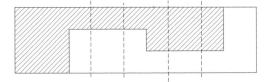

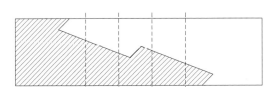

Fig. C.11 Scarf joints – variations (1).

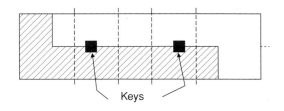

Keys

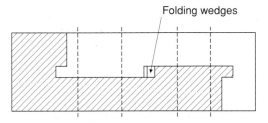

Folding wedges

Fig. C.42 Scarf joints – variations (2).

Scarf joint variations (1) (Figure C.41)

This variation shows the joints made with a 'joggle' which if properly fitted should prevent the timbers pulling apart. The scarfs shown are termed **tabled** scarfs and **splayed indented** scarfs. Note that the joints are bolted together with coach bolts, nuts and washers.

Scarf joint variations (2) (Figure C.42)

The joints here have either 'keys' or pairs of wedges inserted into square mortices cut across the diagonal cut of the joint. The 'keys' are simply square section pieces of hardwood set into the mortices which prevent the joint from pulling apart. Wedges inserted in the mortices first of all pull the joint together and go on to prevent pulling apart. Once the wedges are tightened, the joint is bolted together.

Single and double notching
(Figure C.43)

The upper timber is bearing down on the lower timber and the notchings provide resistance to lateral movement of one or both timbers. This joint is used for joists over beams etc. and is nailed but not glued.

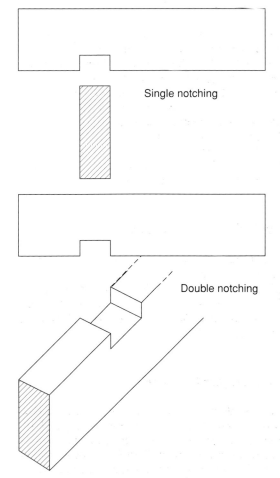

Single notching

Double notching

Fig. C.43 Single and double notchings.

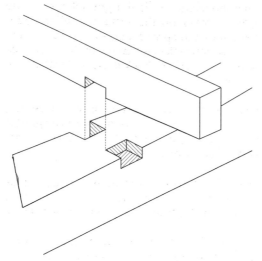

Fig. C.44 Cogging.

Cogging (Figure C.44)

This is similar to double notching and is used for jointing rafters over purlins, and similar purposes.

Dovetailing (Figure C.45)

Dovetailing is more common in cabinet work and the illustration is of a dovetailed joint for

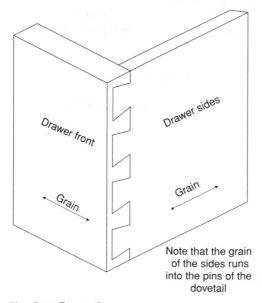

Fig. C.45 Dovetailing.

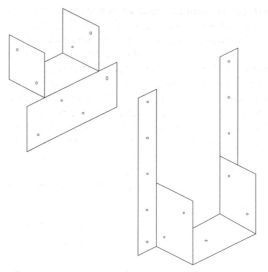

Fig. C.46 Sheet metal connectors.

a drawer side to a drawer front. The dovetail shown is a blind dovetail, i.e. it does not show on the drawer face.

Using a sheet metal connector (Figure C.46)

The connectors illustrated here can be used to 'bolster up' a plain butt joint or can be used to join two timbers crossing one another at right angles. Note that they are stamped and folded from one piece of sheet metal.

Stress grading

The Building Regulations now require all timber which is used for structural purposes to be stress graded, so it is important that what is meant by the term is understood even if the intricacies of the technique etc. are not.

Timber being a natural material can vary in any of its qualities, even within the same planks sawn from the same log or even from plank to plank. Compare this with steel products such as beams columns and joists. These are all man-ufactured to known sizes and shapes from a material whose properties are closely controlled and thus entirely predictable. We can confi-dently assume that when we specify Grade 43 steel, the ultimate fibre stress of that steel will be the same as that of every other piece of Grade 43 steel produced. So it is that a structural engineer can calculate exactly what sizes of steel to use when building anything from a block of flats to a chemical engineering plant. Prior to stress grading, it was difficult to design timber structures without providing much greater cross-sectional sizes, just to be sure that the structure would remain standing. The timber engineer had to be very conservative with the stresses he assumed the timber would bear.

The first attempts to overcome this problem looked at the defects which could weaken timber:

- Knots. Particularly the size and condition of the knots – small and tightly fixed in the timber were not a great problem but large and loose or even missing were serious defects. Also, the position of the knot in the timber cross-section was important. Even a small tight knot at the edge or on the arris of the timber would 'pop' if overstressed and so weaken the timber.
- Shakes and splits. These gaps running with the grain of the timber mean that the fibres are no longer in contact with each other. This obviously weakens the timber but if they occur in the length this can mitigate their effects.
- Wane reduces the overall cross-section and thus weakens the timber.
- Insect and fungal attack would preclude the use of a piece of timber because although it might appear sound on the outside, with just a few pinholes or discolouring, no-one can see inside the plank or board. Insects can almost

take the inside of plank right out and hardly leave a mark on the surface. Similarly fungal attack can spread inside a timber, but the more serious effect is that the timber would be subject to further attack if the MC was to rise above 20%.

Stress grading using machines is now the norm, with visual checks for certain classes or grades and for certain purposes. The machines operate on one of two principles:

- Apply known load to a length of timber across a pair of rollers and measure the amount by which the timber deflects between those rollers
- Apply a load to deflect the timber by a set amount and record the load required.

Machines of the first type are used in the UK and those of the second type are used in North America.

Figure C.47 shows a schematic of a typical stress grading machine, which would have the timber power fed into it. Once the timber is in position the load is applied, the readings taken and the fibre stress of the timber is calculated, whereupon the machine stamps the grade on the timber. So simple when put this way, but the current standards recognise 15 discrete grades. In any length of timber therefore it is the lowest grade which that length can sustain which has to be stamped upon it.

There are at least three companies offering these machines in the UK market. All are

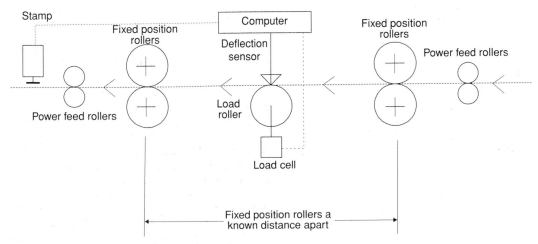

Fig. C.47 Stress grading machine – schematic.

different, all have advantages and disadvantages and only five of the machines are fully accepted as compliant with BS 4978.

The idea of permanently recording a grade on the timber is very important. It assures the design engineer that the correct timber has been used and assures the building contractor, the client and building control officer that the timber is what has been specified for the work. Figure C.48 shows two typical stamps for UK graded timber. It should be noted that to comply with BS 4978, the mark stamped on the timber should include the grade and the species.

Building Regulations now include load/span tables for structural timbers such as joists, roof timbers and structural partitions, so that to comply with these regulations the designer has only to match the expected load with the proposed span and select the appropriate timber size at a given grade.

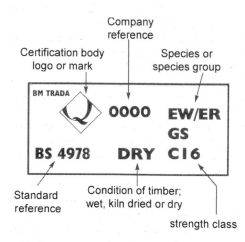

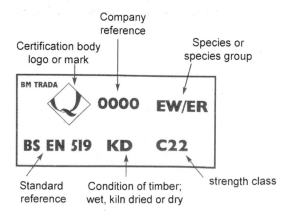

Fig. C.48 Stress grading marks.

Floor boarding

Traditional floor boarding is still available, the kind which is in long lengths and narrow widths with a tongue and a groove on opposite long edges. Figure C.49 shows a traditional flooring board machined from solid timber. Note in the top part of the figure that, as well as the tongue and groove, the board has a series of V shaped grooves on the underside. These ease the tensions in the board when drying out and help to reduce the tendency to cup and bow.

The thicknesses currently available vary from 16 to around 32 and can be supplied in almost any length. One cannot be dogmatic about availability, for the mills will tend to keep only a limited range of thicknesses and lengths generally called for in their area. One mill approached on the subject only kept two thicknesses in stock, 19 and 27, the first for new work and the second for refurbishment in old buildings in the area. Face widths available were 90 and 115 for the 19 boards and 140 for the 27 thick board. Lengths stocked were 4.20, 4.50 and 4.80 m.

Manufactured board materials are now widely used for flooring over joists or battens and also for providing floating floors over insulation where the sheets are glued together. Plywood was one of the first materials to be given a

Fig. C.49 Traditional flooring board machined from solid timber.

tongued and grooved joint on all four edges, and the standard size sheets (2440 × 1220) were the first to receive this treatment. These were of US manufacture and were too expensive to compete with traditional floor boards in the UK. They still reign supreme in North America. It wasn't until the introduction of home produced particle boards[3] that a product which could compete with floor boards was brought onto the market. This is now manufactured in a stock size of 2400 long × 600 wide and the most common type used has a tongue machined on two adjacent faces and a groove on the other adjacent faces. Thicknesses available are 19 and 22.

The material is made by taking debarked logs and shredding the timber into particles, which are bound with a resin and hydraulically pressed into a continuous sheet of material which is finally cut into suitable sizes as the penultimate act of the production line. The final act is the machining of the tongues and grooves. The laying down of the particles is no random act. The external faces of the sheet are made up of finer particles which press down to give a hard surface, while the core is of coarse particles giving a lighter centre to the sheet. The sheet can also be treated to render it moisture resistant (**MR**). This does not mean to say that the sheet can be immersed in water and not suffer damage – rather it is not damaged by superficial wetting as long as solid water is cleared from the surface reasonably quickly. Prolonged wetting can result in permanent swelling of the material, as we have already recounted.

Figure C.50 is a photograph of a corner of a sheet of flooring grade particle board, showing the upper surface.

Fixing down to joists or battens is done with improved nails, perhaps ring shank nails, the norm being four nails across the width of 600 for every joist. Alternatively the sheet can be screwed down, again four screws for every joist in the 600 width. Screwing of course would only be done using power screwdrivers and especially those which had an automatic feed for collated screws. Power nailers are used by large house builders/developers' workmen, and as for the screwdrivers, collated nails are used. All that the workman has to do is press the nose

Fig. C.50 Corner of a sheet of flooring grade particle board.

of the tool against the floor sheeting and this unlocks the safety mechanism. Pulling a trigger then drives either nail or screw through the sheet into the joist below – less than a second per nail or screw. The collated nails are of an improved pattern, usually square twisted shanks as they are tougher than the ring shank nails, which tend to be brittle. Figure C.51 shows a fixing with four nails in width across a joist or batten.

Nails, screws and other fasteners are discussed and illustrated in Appendix G.

The **t&g all round** facility means that the boarding can be laid across joists without having the ends fall on a joist, and thus reduces waste. It is much faster to lay and thus has almost ousted floor boards completely from the domestic market. Figure C.52 shows how it is laid, minimising waste, and with all the internal partitions built off its surface.

When laying these boards, the carpenter always starts at the LH (left-hand) corner, places the tongue of the sheet to the room side and fixes it down. On plan A, the carpenters have placed the first full sheet, FS, in the bottom LH corner of the building. They then cut a sheet so that the

---

[3] Particle board is the preferred term for board made from wood chippings or particles and bound with a resin and pressed into sheets of varying sizes.

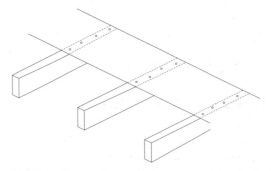

Fig. C.51 Four nails in width across a joist or batten.

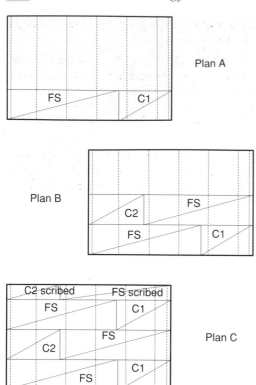

Fig. C.52 Laying t&g all round particle board.

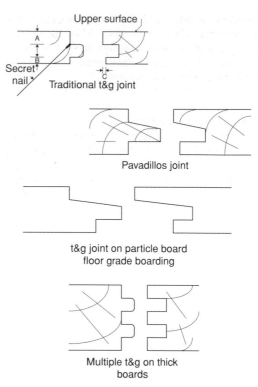

Fig. C.53 Tongue and groove forms and secret nailing of flooring boards.

LH end of the sheet fits into the space C1. It has a groove on the LH end which fits over the tongue on the RH end of the first FS. On plan B, the bit left from cutting C1 from a full sheet is placed on the LH side of the room (C2). The cut end is to the wall and the grooved end is ready to receive the next sheet, which is marked as a full sheet but might be a fraction less. If so, the narrow strip cut-off is discarded. On plan C, the laying continues with an FS laid on the LH end of the floor boards already laid, and a cut C1 is put in next, cut end to the wall on the right. C2 from that cutting is then cut down in width – scribed – but only the tongued edge should be removed as we need the groove to fit over the edge of the previous tongues. Finally, a full sheet is scribed to the wall and laid with the grooved edge over the previous tongues. Quite simple and logical once you work it out.

Tongue and groove boards have been the subject of innovation for as long as the joint has been around. Figure C.53 shows just a few of the forms adopted at one time or another, as well

as showing how secret nailing of tongued and grooved floor boards is carried out. The nails mentioned are illustrated in Appendix G.

The top illustration in Figure C.53 shows a slightly exaggerated form of the modern tongue on the left-hand side of the sketch; the tongued edge shows that there is more board above the tongue than there is below it. This is quite deliberate. Above the tongue is the wearing surface of the board, and should it be worn down at all, there is more timber above the tongue than below it, where no wear takes place. Also, on that side of the sketch a nail is shown hammered home in the angle of the tongue projection. There is no hammer made nor carpenter alive who can hammer a nail down into that position with the hammer alone. The nail is hammered close to that position and then a punch – a nail set – is placed over the head and the punch struck to drive the nail down so that the head is totally into the wood. Having A larger than B allows the nail some grip in the board and somewhere for the head to lodge. On the right-hand side the groove is shown with the underside piece cut back from the upper piece. This allows the board joint to close on the upper face without too much effort should dirt accumulate under the board joint over a joist.

Further information

The *Timber Designers' Manual* by Ozelton & Baird, now in its third edition and revised by Carl Ozelton, opens with a comprehensive chapter on timber, structural fasteners and structural glues. It is published by Blackwell Publishing.

BRE (Building Research Establishment) and TRADA (Timber Research and Development Association) are good sources of information on timber and stress grading. Some of their information is available in short leaflet format and is free.

A good website to visit is the Timber Trade Federation's site at http://www.ttf.co.uk, which has lots of good information and lots of good links – to BRE and TRADA especially.

APPENDIX D
Plain and Reinforced In-situ Concrete

Concrete has been mentioned in the text so often that there must be an early pause in our study to consider some of the more important features of the material before we proceed further. The study of concrete is vast. Books have been written about some of the features of the material, and there is no single comprehensive book – the subject is just so large.

So, how does the student go about finding out about the material and how to use it? Thankfully, it is possible to function quite well in the industry with a simple, basic knowledge appropriate to the type of work with which one is associated. If all you do is build one- or two-storey housing then you hardly need the knowledge required to build a multi-storey, prestressed, precast concrete car park.

We will proceed at a simple level appropriate to the type of construction we deal with in the text as a whole.

Concrete

Concrete is a plastic mixture of aggregates, OPC and water. Occasionally additives are included which have much the same effect as those put into mortar. The plastic mixture is poured into **moulds** or directly into the ground and allowed to set. Reinforcement of steel rod or bar can be fixed in place before pouring in the plastic mixture. Such concrete would be termed **reinforced** as opposed to **plain concrete**. The concrete can be placed in its plastic condition in the position it is meant to occupy in the building – **in-situ concrete**. If cast in moulds away from its final position, it is **precast concrete**.

Moulds

When concrete is placed in moulds, these are referred to as **formwork** or **shuttering**. Formwork is frequently made of **timber** as well as **plywoods**, various **fibreboards** and other timber products. Steel sheet, both plain and perforated, together with angles and other sections are also used. Formwork is frequently made to be used to **cast** the same shape over and over again. This makes it economical to make in relatively expensive materials and maybe introduce certain adjustable features so that different sizes of concrete products can be cast.

Concrete aggregates

Aggregates can be:

☐ **Coarse** – retained on a 6 mm sieve
☐ **Fine** – passing through a 6 mm sieve.

Except for the manufacture of **no-fines concrete** (see later in this appendix), aggregates are **graded**. Grading ensures that, when mixed, all voids between the particles are filled and all particles are evenly coated with cement. Aggregate size is given as the largest mesh size used for grading, e.g. 25 mm aggregate would have all particles passing a 25 mm sieve.

The sketches in Figure D.1 show a *theoretical view* of a graded coarse aggregate and a single size coarse aggregate. Note how the graded aggregate has the smaller particles filling the voids between the large particles.

Sources of aggregate are:

☐ Naturally occurring gravels from pit or river
☐ Crushed stone

Single size aggregate

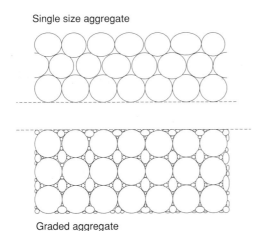

Graded aggregate

Fig. D.1 Grading of aggregate.

- Crushed brick
- Crushed concrete
- Furnace ash or clinker, pulverised fuel ash (PFA)
- Crushed slags – a residue left when smelting metals from ores
- Lytag – expanded clay pellets now frequently seen topping plant pots because of their ability to hold large quantities of water.
- Vermiculite and perlite; both natural minerals closely related to mica which have been soaked in water and subjected to a sudden rise in temperature which causes the leaves to expand or 'exfoliate'.

Crushed or irregularly shaped aggregates can bed down more tightly together than rounded aggregates. The latter generally require a higher proportion of fine aggregate to fill up the gaps. It is possible to purchase aggregates which fulfil the function of coarse and fine combined. They are known as **all-in aggregates** and are sized by the largest particle 'down', e.g. 40 mm down. Voids in coarse aggregates represent a proportion from 20% to 45% by volume.

Fine aggregate, also known as **concreting sand**, must also be graded and has a wider range of particle sizes than building sand. Due to a relatively cheap and plentiful supply, concreting sand is almost entirely sourced from pit or river sources.

Selection of aggregate is critical when the strength of the final concrete is important. Aggregates which contain soft or friable material cannot be used in high strength concretes but

might be all right in concrete used as a filling. We will see examples of this later in this appendix.

Water

As with mortars, water is required to react with the OPC to obtain a set. Water is also important in two important properties of concrete:

- In wet concrete, a high proportion of water allows easy placing of concrete in confined areas and round heavy reinforcement, but too much water weakens concrete and may also lead to water voids being formed where compaction of the concrete is inadequate.
- A low proportion of water is desirable where higher strengths are required.

Water/cement ratio

- Unlike mortars, the proportion of water used in concrete mixes is carefully controlled:
  - Enough to allow easy placement
  - Enough to give a 'set'
  - Just enough to avoid water pockets in the concrete.

It is given in litres per 50 kg of OPC.

When using low water/cement ratios a number of techniques can be adopted to ease placement:

- Use of vibrators
- Addition of air entraining chemicals to the mix – for every 1% by volume of air a 6% loss in strength is incurred
- Addition of water retaining or wetting agents
- Use of a smaller coarse aggregate.

Traditionally, concrete mixes for general building work were specified by volume. A 1:2:4 mix contained one part by volume of OPC, two parts fine aggregate and four parts coarse aggregate. A 1:2:4 mix was generally conceded to be a strong concrete, a 1:3:6 mix a medium strength mix.

Concrete proportions

For large works designed by an engineer, mixes used to be specified by mass, the aggregate being proportioned against 100 kg of OPC. Modern practice provides two options for specification:

☐ Designed mixes
☐ Prescribed mixes.

With **designed mixes** the purchaser is responsible for specifying the performance; the provider is responsible for selecting the mix proportions, aggregate size etc. to meet that performance. Performance is generally a minimum compressive strength in newtons/mm^2. For example, a purchaser might require a concrete which has a compressive strength at 28 days of 35 newtons and will be poured into column and beam moulds which have a high proportion of reinforcement.

The addition of a timescale takes account of the fact that setting OPC takes place over a period of time. Although the initial set of the OPC may occur in two or three hours, the concrete will not gain any appreciable strength for several days. The period of 28 days is generally recognised to be the minimum time for the concrete to gain its working strength.

There are two kinds of **prescribed mix**, both the subject of BS 5328:

☐ **Ordinary** – A list of grades of concrete is given in the British Standard, each with a full specification.
☐ **Special** – Special requirements not covered above. A full specification would have to be given in any specification issued to the contractor.

Reinforcement

Two types of steel reinforcement are available:

☐ As bar, which may be round, square twisted or deformed in cross-section
☐ As fabric manufactured in sheets or strips and made up from bar which is welded at all the junctions.

The steel may be mild steel or high yield steel, hot rolled or cold rolled, and may also be hot dipped galvanised. Figure D.2 illustrates:

☐ Round bar, also known as rebar (shortened form of reinforcement bar)
☐ Square twisted
☐ Deformed – many cross-sectional shapes are available.

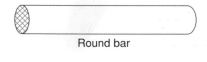
Round bar

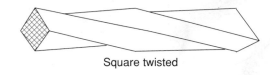
Square twisted

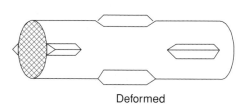
Deformed

Fig. D.2 Reinforcement bar shapes.

Reinforcement can be of **bar** or **mesh**. Both are made of steel. Bar reinforcement is covered by two British Standards:

☐ BS 4449 *Hot rolled steel bars*
☐ BS 4461 *Cold rolled steel bars.*

Steels used are:

☐ Mild steel 250 grade (tensile strength is 250 N/mm^2)
☐ 450/425 high yield steel (tensile strength is 450–425 N/mm^2).

Mild steel is only ever rolled into round section bar. High yield steels are available in round, round deformed and square twisted bars.

Mesh reinforcement is the subject of BS 4483. Fabric reinforcement is manufactured to comply with BS 4483. Fabrics are made in two types:

☐ Square mesh
☐ Rectangular mesh.

They are formed by welding smaller section rebar at intersections. Both types can be of mild steel rebar or high yield steel rebar. Only round or square twisted rebar is used. Rebar meshes are shown in Figure D.3.

Meshes are available as square and rectangular hole shapes, made by welding reinforcement bars at right angles to each other. Varying wire sizes are used to make stronger or weaker meshes. They are available as sheets or strips.

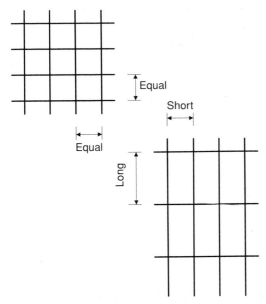

Fig. D.3 Reinforcement meshes.

Reinforcement must be firmly 'fixed' in the concrete. It must not move. Concrete, when setting, shrinks and so grips the bar. The grip can be measured by casting a bar into a block of concrete and placing in a testing machine which pushes against the protruding end of the bar. When the grip gives out, the load applied is recorded by the machine. A prepared sample for testing is shown in Figure D.4, and the difference between holed and bent ends on rebar is shown in Figure D.5.

Ways to improve the 'grip' are:

□ Bending or hooking the ends of the rebar
□ Using deformed or square twisted bar
□ Using both the above.

No-fines concrete mentioned above was a specialist material used extensively for local authority housing from around 1942 until the early 1970s. The technique was originally from Germany and was used there for the first time in the 1920s, being picked up and the subject of experimental housing during World War II by the then Scottish Special Housing Association (SSHA). Following several years of testing various techniques for casting the material in-situ, a technique was evolved where two-storey house walls were cast – for the complete house – in one fell swoop. The shuttering was erected on a prepared substructure and the whole of the

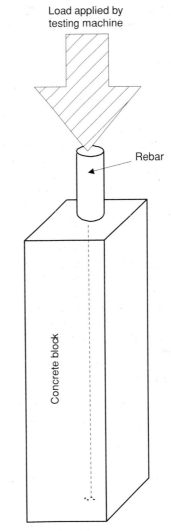

Fig. D.4 Testing for grip of rebar in concrete.

wall was cast in place with all the door and window openings, local reinforcement over and under the openings, vent holes, pipe holes, etc. A few days later the shuttering was stripped and a prefabricated trussed roof was hoisted into place and made watertight. Whole terraces of houses were frequently cast in lots of four or five houses at a time. The system was used by the Direct Labour Organisation (DLO) of the SSHA and one other contractor, Wimpy.

There are thousands of no-fines houses in Scotland, from single-storey bungalows to ten-storey

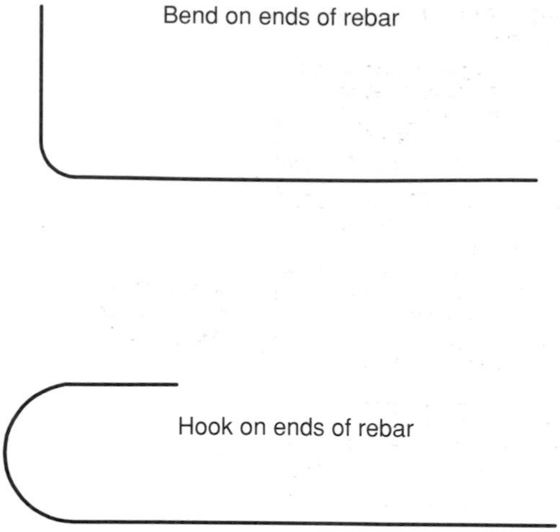

Bend on ends of rebar

Hook on ends of rebar

Fig. D.5 Bent and hooked ends of rebar.

blocks of flats. Both the SSHA and Wimpy developed their own detailed techniques and house designs, but there was general collaboration on the techniques and the mixes etc.

The mix was composed of one part OPC to 8 or 10 parts aggregate, plus a measured amount of water. The aggregate had to be a natural river washed aggregate of one size, as uniform as possible and with a good crushing strength appropriate to its use in the final wall. The mixing was critical, as what was required was an even coating of cement paste over the whole of the aggregate.

Weigh batchers were used for the mixing. A weigh batcher is a concrete mixer with some important attachments. The first is a mechanical shovel to load aggregate and OPC into the mixer drum. OPC was loaded by the bag, large batchers using several bags at a time. There was a water tank fed by the mains over the drum and it would deliver a measured amount of water at the appropriate time in the mixing cycle. The amount of aggregate being loaded was constantly weighed and shown on a dial just above the drum. The correct mass was shown on the dial using an adjustable needle of a contrasting colour.

The whole operation had to be capable of outputting an accurate and massive amount of concrete in as short a time as possible. The wet mix was poured out into a skip hoisted by crane to the top of the shutter, which incorporated a walkway for workmen who emptied the skip into the shuttering until full. Careful vibration

was used to ease the concrete around door and window frames etc. The wall thickness was usually 10 in (254 mm) for up to four-storey work, with slightly thicker lower storeys in high rise blocks and perhaps stronger mixes.

The concrete itself was almost the antithesis of conventional dense concrete. We talked there about getting an aggregate graded and combining coarse and fine aggregates to fill all the voids, so giving us an extremely strong product – well in excess of the compressive strength of the bricks we would normally expect to build with. Well, no-fines was not like that. Its crushing strength was much lower but then it seldom got down to anything like that of the bricks with which the industry is all too familiar. No-fines used only a well rounded aggregate and coated that with cement paste so that when the whole mixture was emptied into the shuttering, the individual stones stuck to their neighbours but there was a lot of empty space as well. This empty space had two advantages:

☐ It reduced the overall mass of the wall and therefore the load on the foundations. Ordinary strip foundations were the norm on all but the most difficult sites, and the walls were built in common brick, the earlier houses using selected common bricks, weather jointed where the wall was exposed.
☐ Any rain penetrating the outer covering on the no-fines would drain down the wall and run out of the bottom into the cavity of the substructure wall.

Reinforcing steel does not generally rust when embedded in dense concrete because of the alkaline environment it is in. However, rebar in no-fines concrete is badly exposed and it was the accepted practice to coat it with a rubber latex/OPC mixture before fixing it in the shuttering. Modern practice would use hot dipped galvanised steel.

The outer covering of the wall was a two-coat wet or dry dash render, the first or straightening coat being OPC and sand and the top or rendering coat being a cement/lime/sand mix. The top coat either incorporates whinstone chippings in it before application (a wet dash) or was dashed with coloured chippings after the coat was applied and still not set. White OPC and silver sand were used in backgrounds for light coloured chippings such

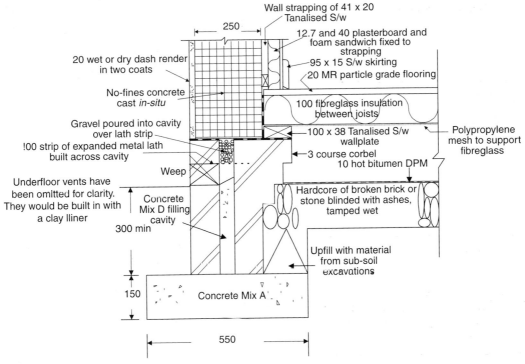

Fig. D.6 No-fines concrete – typical substructure detail.

as Skye Marble, which gave a dazzling white surface. The inside of the wall was finished with plasterboard, either nailed to treated wooden strapping or to vertical stripes of 'nailable' plaster applied to the no-fines. Fixing to the no-fines was simple. **Cut nails**[1] could be driven through any timber and into the concrete. When the point of a nail hit a hard stone it simply bent round the stone. Once fully in, it was almost impossible to get them back out.

Figure D.6 is a detail of a no-fines substructure. There are a few points to note about it. It is not exactly as the substructures were built back in the 1970s. The Building Regulations have moved on since then, the most significant change being the need to provide much better insulation in the walls and the ground floors. Note the gravel in a section of the cavity. This was placed there to prevent the wet no-fines concrete from actually

filling the cavity. That two course depth was required to act as a drain for the whole two-storey wall height. Note also the double layer of DPC over the inner leaf of brickwork, the upper layer turned up the inside face of the no-fines wall between it and the layer of insulation. This and the wall strapping holding the insulation clear of the wall contributes to keeping any water running into the cavity below and out of the weeps.

We will not pursue no-fines concrete any further in this text. It seems to be a forgotten technique now and there is no current literature generally available. One text, which the author had in his possession at one time, was *The No-fines Story* written by Ronald MacIntosh, who was, for many years, Manager of the SSHA's DLO. It might be possible to find a copy in some of our National Libraries or obtain it on inter-library loan.

A good website to visit belongs to the British Cement Association at http://www.bca.org.uk

[1] Cut nails are discussed in the main text and in Appendix G.

APPENDIX E

Mortar and Fine Concrete Screeds laid over Concrete Sub-floors or Structures

A screed may be described as a layer of mortar or a small aggregate, fine concrete laid in a smooth layer over a concrete subfloor (in-situ or formed with precast components) to give a smooth finish in itself or as a substrate for a sheet or tile finish such as vinyl, linoleum, thermoplastic material, mastic asphalt, carpet, clay or ceramic tile.

There are three ways to lay a screed:

☐ Monolithic
☐ Bonded
☐ Unbonded.

The following factors are interrelated and choice of one affects choice of another:

☐ Method of laying
☐ Thickness
☐ Mortar or concrete materials
☐ Mortar or concrete mix
☐ Self-finish
☐ Applied finish.

There is a choice of three basic materials:

☐ Cement/sand mix no stronger than 1:3
☐ Cement/granite dust/sand mix usually 1:1:2 and usually self-finished
☐ Cement/whinstone dust/sand mix usually 1:1:2 and usually self-finished.

Granite dust and whinstone dust are 'manufactured' aggregates, being quarried granite or whinstone (black basalt) passed through a crushing machine and graded through a set of sieves to give a fine aggregate, i.e. up to 6 mm. The mix containing granite dust is known as granolithic and is widely used as a self-finished layer for hard wearing paving and flooring.

Screeds using whinstone dust are also hard wearing but not as hard as granolithic.

Table E.1 summarises thicknesses of screeds by material and method of laying.

Laying monolithic screeds

Monolithic screeds are cast onto the concrete base within three hours of laying the concrete base. The concrete in the base is still what is termed 'green' and the fresh cementitious material laid on it mixes at the point of contact, ensuring a secure bond.

Laying bonded screeds (1)

Bonded screeds are cast on the concrete base more than three hours after the base has been laid, frequently having been left overnight. The steps in laying it are:

(1) Keep base damp with a spray or by covering with damp hessian until about to lay screed
(2) Clean surface of base and remove **laitance**
(3) Spread **thin neat cement grout** and lay screed within 20 minutes of laying grout.

Some terminology

Laitance: When beds of concrete are laid the wet material is spread out and compacted to the required thickness. Working the material in this way brings the mixing water to the surface, and this water contains cement. When the water evaporates, a layer of cement powder is left on the surface. This layer sets but does not adhere properly to the surface and must therefore be removed before a screeding layer is added

Table E.1 Thickness of screed by method of laying and type of material.

Method of laying	Sand/cement screed	Granolithic or whinstone/cement screed
Mono	12–15 maximum	10–25 maximum
Bonded	40 minimum unless in small areas – 25	40 minimum unless in stairs – 15–25
Unbonded	Minimum 50 – unheated	50–75
	Minimum 60 – heated	

Thin neat cement grout. Grout is basically a mixture of cement powder in water. A bonding agent may be added to the water. *Neat* refers to the need to have only cement powder in the grout without any fillers or extenders such as PFA. Aggregates are never included as it would then be a mortar. *Thin* grout has a 'creamy' consistency

Laying bonded screeds (2)

The alternative method for laying bonded screeds utilises a bonding agent instead of a cement grout. The screed is laid more than three hours after the base has been laid:

1. Remove laitance from the base
2. Spread bonding agent over surface of base
3. Lay screed within drying time of the bonding agent.

Note that the base should be covered with wet hessian between laying and screeding to prevent drying out, but there is no need to keep it 'wet'.

Bonding agents

Bonding agents are generally based on polyvinyl acetate – PVA, or ethylenevinyl acetate – EVA. Proprietary names include FEBOND and TRETOBOND – look at the literature from the manufacturer or visit their websites: www.feb.co.uk

It is usual to add some bonding agent to the mixing water used for the screed. These same agents are useful in plastering and rendering on 'difficult' surfaces.

Screeds containing PVA based bonding agents are *not* suitable for use in areas which are continuously wetted or damp, such as shower areas, swimming pools, dairies, etc. where EVA

would be better. Rates of application should follow the manufacturers' recommendations, which vary depending on the precise formulation of the agent.

Laying unbonded screeds

Brush loose material from surface of base. The mortar or fine concrete layer is laid direct onto the dry, set concrete base. No attempt is made to bond both layers together. The screed can also be physically isolated from the base to prevent a mechanical key by first laying a separating layer – bituminous felt, building paper or a layer of insulation.

Unbonded screeds can fail in principally two ways:

☐ They can 'curl' – just like a dried out sandwich – even when quite thick, and then the raised areas break up under the imposed floor load.
☐ Unless reinforced, they can crack badly.

To prevent cracking, or at least control it, a simple layer of galvanised chicken wire was often included, but fine wire welded meshes are now available which are more effective. Even when reinforced, the screed can still 'curl'.

Large areas should be laid in bays. Bay sizes should be confined to areas of approximately 50 m^2. Unbonded screeds should be separated from surrounding masonry walls by a compressible strip, and bays should be divided by compressible strip. Suitable material for compressible strip includes bitumen impregnated fibre board (Flexell), cork strip, rigid foamed plastic, bitumen impregnated soft foam plastic, etc.

A silicone or polysulphide-based mastic is sometimes 'gunned' over the compressible strip to prevent occasional water, cleaning materials, etc. penetrating the joint.

Heated screeds

Reference to heated screeds was made in Table E.1. Electrical cables can be embedded in the screed or can be drawn into ducts cast into the screed, or plastic or metal pipes can be cast into the screed and fed with hot water. Microbore plastic pipes can also be drawn into ducts cast in the screed. Try the Polypipe website for further information on underfloor heating using plastics pipework buried in a screed – http://www.polypipe.co.uk

Any of the above will heat the floor and so heat the room above. Insulation under and around the edges of the screed is essential. The output of cables or pipes should not exceed 150 W/m^2, which is the upper limit for human comfort and will generally meet the needs for domestic heating.

APPENDIX F
Shoring, Strutting and Waling

This refers to what is commonly known in the industry as **timbering** – the provision of support to the sides of excavation works for whatever purpose. This is done for two basic reasons:

☐ To protect the workmen down in the hole from collapse of the ground into the excavation. This happens more frequently than the lay person can imagine but thankfully not always with fatal results.
☐ To obviate the necessity of re-excavation following a collapse. This can not only mean doing it all again but could also mean the excavation is oversize and could mean additional concrete to fill up a foundation or just concrete to infill what would not have been required in the first place. Either way it costs money for which the contractor will not be reimbursed.

The exact technique for timbering varies according to the type of soil and depth of excavation as well as the type of excavation. At present the text is concerned only with relatively shallow excavations for trenches into which concrete strip foundations will be poured or in which pipes and other services will be laid. Any trench fill work is usually done in a naturally cohesive soil which might not require timbering at all, or only a minimal amount.

The material traditionally used for this work is timber and breaks down into three main types of sizes/kinds of sawn timbers:

☐ **Poling boards** are relatively short but fairly wide planks which are placed vertically against the side of the excavated hole. Typical sizes are 200 × 38 to 250 × 50. They are put in, in pairs, and each is propped one against the other with a stout timber called strutting.

☐ **Strutting** is usually square section timbers from 75 × 75 to 100 × 100 wedged between the poling boards, sometimes simply cut as a tight fit and more unusually fitted with folding wedges for adjustment. Telescopic, tubular, metal props are frequently used rather than timber struts, the base plates being drilled for nails to hold the struts in position.
☐ **Walings** are broad and relatively thin and quite long timbers, from 200 × 38 to 250 × 50, which are used to support poling boards where these are placed closely together and which would result in the use of too many struts. Struts are therefore spaced out along the lengths of the walings. If the excavation is relatively shallow there might only be one waling per face, otherwise two per face are put in.

A quick look at Figure F.1 will explain all of the above much more simply than a further thousand words.

Finally, we have to be able to deal with the fluid-like conditions of **running sand** and **silt.** The technique used is either close sheeting or sheet piling. Close sheeting is the technique of placing the poling boards close together edge to edge, braced with walings at least top and bottom and strutting the walings. It is suitable for dryish conditions where there is not a lot of ground water. Timber sheet piling or steel sheet piling has to be used where the ground water is high – and plentiful. This is necessary to reduce the pumping effort to keep the excavation dry and workable. Figure F.2 illustrates these techniques, the views being of one face of the excavation.

What is important to realise is that virtually every excavation of any depth and especially in

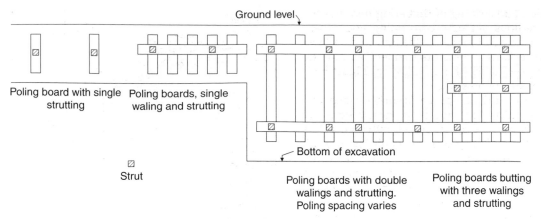

Fig. F.1 Shoring strutting and waling using timbers.

friable or 'fluid' soils requires shoring, strutting and waling to some degree or other, and all other work in the bottom of that excavation has to be carried out around the timbers or other material inserted.

Once the work in the excavation is complete, the timbering has to be taken down and removed. This operation is called **striking**. Where excavation of trenches, particularly for extended lengths of pipe or cable, is required, the frequent

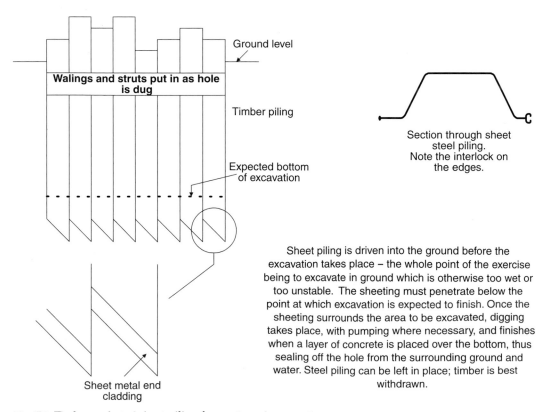

Sheet piling is driven into the ground before the excavation takes place – the whole point of the exercise being to excavate in ground which is otherwise too wet or too unstable. The sheeting must penetrate below the point at which excavation is expected to finish. Once the sheeting surrounds the area to be excavated, digging takes place, with pumping where necessary, and finishes when a layer of concrete is placed over the bottom, thus sealing off the hole from the surrounding ground and water. Steel piling can be left in place; timber is best withdrawn.

Fig. F.2 Timber and steel sheet piling for earthwork support.

erection and striking of timbering becomes not only tedious but time consuming and expensive. In an effort to reduce the time and cost factors, specialist contractors frequently make use of purpose-made all-metal support. Figure F.3 is a schematic of how this works.

The use of this equipment is fairly self explanatory – which does not mean to say there are not do's and don'ts associated with its use. It is generally used for pipeline or cable laying work where there is a need to excavate a trench continuously, digging shoring, laying striking and digging again to progress the pipe or cable laying swiftly. Note that although using steel sheet for the sides, it is not a watertight shoring and ground or surface water would have to be dealt with by pumping. For our purpose in this text, just remember that such equipment is available.

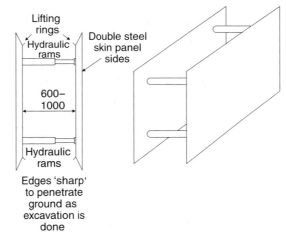

Fig. F.3 Steel trench support.

APPENDIX G

Nails, Screws, Bolts and Proprietary Fixings

This appendix describes a small selection from the vast numbers of fixings and fastenings available to the building industry. Figures throughout the appendix attempt to illustrate the range described, with photographs and sketches.

Nails are generally of two types:

☐ **Wire nails** (Figure G.1)
☐ **Cut nails** (Figure G.2).

These are available – although not in all sizes – in a variety of materials and forms. The basic material is steel for both types. For wire nails that steel can be round or oval in section. The former gives the name **round wire nails** to the most general of carpentry and joinery fastenings. They have a small flattened head, a sharp point and are available in lengths from 20 to 150. Anything larger than 150 which resembles a nail is more correctly termed a **spike**. Closely related are the nails made from oval wire – **oval brads**. They have a bullet head, a sharp point and are available in lengths from 12 to 65. Of course, just to confuse things, brads are also made in round wire in the same size range. Brads are used for finishing work such as the fixing of skirtings, architraves, window sill boards, etc. Very small gauge or short versions of the round brads are commonly known as **panel pins**, **glazing pins** (for fixing timber glazing beads – not the glass[1]) **hardboard nails**, etc.

The larger sizes of wire nails and brads are available in special finishes – hot dipped galvanised, sherardised and plated with zinc. Brads

can also be obtained in brass in the smaller sizes. Galvanised wire nails are used in situations where rusting would be a problem, such as exposed timbers in roofs, tile battens and sarking boards, and in external works such as fencing and screens, garden huts and garages.

A galvanised wire nail with an enlarged head, termed a **clout nail**, is used in a variety of lengths from 12 to 65. The shorter lengths from 12–25 are frequently referred to as **felt nails** owing to their use in fixing down bituminous felt roof coverings and similar sheet materials. Lengths of 50 to 65 are used to fix down slates and roof tiles and were referred to as **slate nails** or **tile nails**, but they are all simply clout nails.

It is for roofing that other materials for wire nails become common, the most common being aluminium and copper. These metals are generally quite soft, but for these nails a specially hardened version is used.

Cut nails are quite literally cut from steel sheet. Before the advent of cheap wire nails they were the first of the mechanically formed nails. Prior to that, all nails were hand forged, not in factories but as a cottage industry where individual smiths working from their own homes would contract to make nails for a larger company to sell on. Steam driven machinery allowed the use of large cropping machines which would slice a sliver of steel off the end or edge of a mild steel plate in a rough nail form. Some nails required to be stamped to form their head.

Cut nails are available hot dipped galvanised and in stainless steel to special order. They made a resurgence for a long period in the 1950s to 1970s when they were used in large quantities to fix timbers to no-fines concrete, the shank of the nail being soft enough to bend round the aggregate in the concrete, making them impossible

---

[1] The small steel pins used to fix the glass in place when glazing in putty are called glazing sprigs and were originally formed by cutting sharp slivers from sheet steel. They had no head so that they could be tapped into place close to the glass with the edge of a firmer chisel.

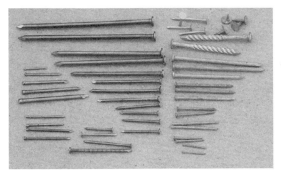

Fig. G.1 Nails. Top left, starting with 150 and 125 wire nails down to 25 wire nails; to the bottom of that group, a few 'improved nails', and to the left of that are two groups of oval and round brads. On the right, a few aluminium slate nails and galvanised felt nails, then two improved nails called **drive screws**. The latter are used to fix down corrugated metal sheet to timber. Below them, a range of wire nails which have all been galvanised.

to pull out. So strong was the grip that the timber being fastened would have to be split and the nail broken off if things went wrong. From the 1960s to the present day they have found ready use whenever aac concrete blocks have been used. Provided the carpenter doesn't hit the nail too often – in other words, only enough to get the head down to the surface of the timber – the cut nail will grip well in that concrete. As with the wire nails, the coating with hot zinc

is there to protect from excessive dampness, but the use of stainless steel sheet for their production was occasioned by a specialist treatment to add thermal insulation to the outside face of no-fines concrete walls. The head of the nail was specially formed to give a T-section, and they were then known in the trade as T-nails.

There is hardly a trade or industry which does not use nails of some kind but round wire, cut nails and brads are the mainstays of the building industry.

The introduction of the wire nail brought nails with sharp points to the industry, and with that came the problem that when nailing through thin finishing timbers these would often – and still do – split. This is because the sharp point pushes its way past the timber fibres, pushing them aside and so inducing a split. The old cut nail had a square, blunt end which actually cut the timber fibres and forced the cut ends to curl back; if the nail was driven with the long side of its section parallel to the grain, there was little chance of splitting. This is why we see joiners, when nailing up thin timbers with wire nails, blunting the point of the nail and then driving it through.

Occasionally copper boat nails (Figure G.3) are used for fixing timber cladding, especially cedar cladding. These nails are made from square section material with a small countersunk head stamped at one end and a point cut at the other. The point about using them is their high resistance to atmospheric attack, especially in coastal regions.

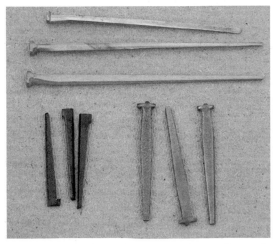

Fig. G.2 Cut nails. At the top are a 125 and two 150 stainless steel cut nails; bottom right, plain steel cut nails and, bottom left, some flooring brads.

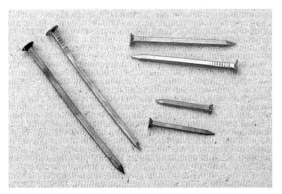

Fig. G.3 Copper boat nails 35, 50 and 100 long. Note the square shank and what is termed a **rose head**. Used for fixing weather boarding etc. in exposed, aggressive situations – near the sea etc.

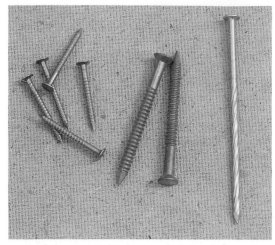

Fig. G.4 Improved nails – on the right the long nail has a square twisted shank and on the left are two groups of nails with ring barb shanks; the smaller nails are of plain steel and the larger are of phosphor bronze.

Fig. G.5 Collated nails for an electrically powered or pneumatic nailer. Both sets of nails are 75 mm long; one set is sherardised and each has a ring barb shank, and the other set is zinc plated and they have square twisted shanks.

Copper tacks also feature in plumbing work where lead sheet has to be secured along an edge. They are used in lengths from 12–25.

Improved nails are a feature of everyday work on today's building sites. These nails give a better grip in the timber and are used for fixing wall ties in timber frame construction, sheet floor boards, plasterboard, even structural timbers and pressed steel framing plates, anchors and hangers. The annular ring shank nail is one of the most popular; a twisted square shank nail is also widely used, plated versions of both being used for plasterboard work. Several improved nails are illustrated in Figure G.4.

Driving lots of nails with a hammer is a tiring and time consuming business and when time is money the use of power nailers can become an attractive option. They can be powered by compressed air or electricity, both at 110 and 230 volts a.c. Compressed air is most economical when a central compressor and pressure tank is set up, as in a factory situation. Electricity, particularly 110 volts, comes into its own on building sites. These nailers use collated nails, i.e. the nails are held in a plastic strip or gathered close together and bound with an adhesive tape (Figure G.5).

Screw nails are widely used; indeed they have supplanted ordinary nails in a lot of areas of structural timberwork. There is a huge range of thread forms available, from the traditional tapered thread on the first two-thirds of the length

to the multi-start forms on parallel shanks from tip to head. Figure G.6 shows a variety in a range of sizes.

Top left are three screw heads – from left to right a **slotted head**, a **Phillips head** and a **Pozidriv head**.

To the right, a group of 3 **bj** (black japanned), **rh** (round head) slotted screws; then 3 brass rh slotted screws. On the extreme right, a 3 inch and a 4 inch long screw, **csk** (countersunk) slotted head, and on their left a pair of brass csk slotted screws and, further left, 3 **zp** (zinc plated) steel screw nails with csk slotted heads. These six distinct sizes of screw nail are of a traditional pattern, with approximately two-thirds of their length having a tapered thread and with a relatively blunt point. The shank above the thread is always parallel and the full diameter of the screw.

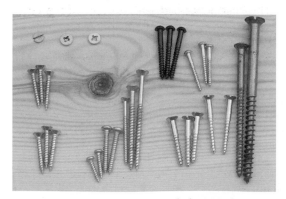

Fig. G.6 A selection of screw nails for timber

The remaining screws are of more modern design. They have either csk or **pan head** shapes – easily distinguishable – and have either Phillips or Pozidriv recesses for a screwdriver. Up to around 40 long, the thread is formed for the complete length of the shank. Over 40 long the proportion of thread to plain shank is the same as the traditional pattern, the real difference being that on many this plain portion of shank is smaller in diameter than the given size of the screw nail.

The threaded portion is parallel and there are actually two threads running one inside the other – **twin start** threads. Note how sharp the points are in comparison with the traditional screw nails on the right.

While it is perfectly possible to drive traditional screw nails with a power screwdriver, these tools were never meant to be used that way, and the plain flat blade required frequently 'cams out' of the slot and may score or otherwise damage the surrounding woodwork. The modern screw nail with a Phillips or Pozidriv head (less likely to cam out), the twin start thread and the sharp point are designed for fast power driving. This can be done one at a time or the screws can be mounted in a plastic strip or belt and fed automatically into a power screwdriver driven either electrically (battery or mains) or with compressed air. We have already seen examples of nails which are fixed this way, and the term **collated** can also be applied to screw nails.

Each screw nail is supposed to perform a particular function better than any other. Screw nails have very definite features which set them apart from other fastenings:

☐ They have a partly or fully threaded shank with a sharp point
☐ The point may be self-drilling
☐ The head is formed with a plain slot, a cruciform recess – Phillips or Pozidriv, a square recess (a very new idea), a Torx recess, or a multi-splined recess; the last two are virtually confined to the motor industry.

Besides the arrangements for driving the screws, the heads are obtainable in several different shapes – countersunk (csk), raised countersunk, round, pan and cheese.

Screw nails which the building industry might use are available in lengths from 10 to 100 and in diameters from 2.00 to 6.00. Diameter of screw nails used to be measured as a gauge, the smaller the number the smaller the diameter. For example, 8 gauge or No. 8 screws were obtainable in lengths from around 15 to 75 long. The shorter screws in that range were also obtainable in smaller diameters and the larger screws in larger diameters. 8 gauge was the equivalent of 4.5 millimetres. With metrication the same situation applies.

Screw nails are available in a wide range of finishes and materials, plain steel being available and also plated with chromium, brass, nickel and zinc. Black enamel is also used but only to finish round-headed screws. Hot dipped galvanised was available but was used mainly by the boat building industry. Other base materials used are brass, which may also be chromium plated, stainless steel, bronzes and aluminium, which may be left plain or anodised to match satin anodised ironmongery.

Screw nails are not available in every size, base material and finish permutation; only the more usual combinations are made, and not all are stocked by every supplier.

Drilling holes

Driving screws into softwood can often be achieved with nothing more in the way of preparatory work than making a small hole with a bradawl, especially if a power tool is used. However, with larger and thicker screws it would be helpful to drill a **clearance hole** in the first timber and then a **pilot hole** in the second timber. This is always done when fastening hardwoods. If the screw nail has a countersunk head, the equivalent hole – the countersink – must be made in the timber. The countersink can be drilled with accuracy by using a countersink block. This only works if the drill is in a vertical drill stand.

A countersink block is a steel or iron block with a selection of countersunk holes and through holes drilled in it. Take the screw which is to be used for the fastening and fit it to one of the holes and countersinks in the block. Mark that hole with chalk. Put the countersink drill bit in the drill chuck. Place the timber on the drill stand work table; place the countersink block on the timber; pull the countersink bit down until it fits neatly into the selected shape in the block and set the depth stop on the drill at that point. Now reverse the position of the timber and the block,

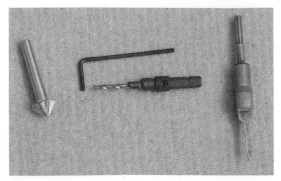

Fig. G.7 Cutters/bits for, left to right: a countersink; countersink and clearance hole; and countersink, clearance hole and pilot hole.

and countersink the timber. You will now find that the screw head fits neatly and as flush or as recessed as you set it in the block.

Drill bits are available which combine a drill to make a clearance hole in the timber with a countersink recess and even a pilot hole. Figure G.7 shows such drill bits, from left to right: a simple bit to cut a countersink shape in a previously drilled hole; a drill bit for a clearance hole with a countersink and counterbore attachment; and a pilot, clearance and countersink and counterbore bit all in one. Note that the portion of the bits which drills the countersink portion of the hole slides onto the hole drilling portion and is locked in place with a grub screw. This allows for different lengths of screw/thicknesses of timber.

Drilling the start of the hole through a timber as large as the diameter of the head allows the screw head to be sunk well below the surface of the timber. Forming this deep hole is called **counterboring** and the hole is a **counterbore**. It can be used to allow a shorter screw length to reach the second timber, but its real use is to allow the joiner to **pellet over** the head of the screw, thus hiding it from view. The pellet used is cut from a scrap of the same timber, fastened using a **plug cutter**. Plug cutters come in a range of standard diameters so that one can be selected to make a plug for the hole drilled or vice versa. The plugs are slightly tapered and are tapped into the hole with a little glue using a hammer. When the glue has set, the bit of the plug protruding can be left proud as a feature or more usually it is flushed off with the rest of the timbers. Plugs cut from the same timber and with decent grain matching and alignment can pretty well disappear – on other occasions

Fig. G.8 Plug cutter and plugs through various stages of pelleting over.

a contrasting timber can be used to decorative effect.

Figure G.8 shows a plug cutter, how the plugs are cut and broken out from a block of timber, a plug set and glued into a counterbore, and a plug planed down flush with the timber.

Coach screw

One screw which is used for heavy work is the **coach screw** (Figure G.9). The shank is tapered and the bottom two-thirds have a tapered thread with a relatively sharp point, but the head is either square or occasionally hexagonal, both

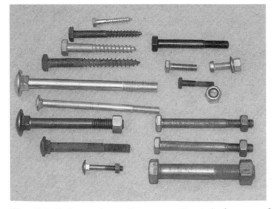

Fig. G.9 Bolts and coach screws. Top left, four coach screws, galvanised and black iron; below them, five coach bolts; on the right, a selection of hex bolts and a single nyloc nut.

to allow the use of a spanner. It is used to join heavy timbers together, particularly where only one side of the timbers can be accessed. It is also used under the same circumstances to join metal sections to timber.

Coach screws are made from steel and can be obtained either black enamelled or hot dipped galvanised. Care must be taken when tightening lest the metal thread strip out the timber. This is easily done if long-handled spanners are used or a spanner is given more leverage by slipping a pipe over the handle. Avoid these things if possible.

Bolts

Bolts (Figure G.9) are used for heavy work in carpentry and for light work in finishings, particularly fixing knobs and handles and sundry items of ironmongery.

Starting with the heavy stuff first, these bolts come as two types, **coach bolts** or **hex bolts** (short for hexagonal, which is descriptive of the head shape). They come in a range of lengths from about 25 upwards, the longest the author has used being hex bolts in excess of 350.

The coach bolt has a head shape which was often described as **cup/square** or **pan/square**. It was primarily designed for work with timber. The head starts with a large diameter (compared to shank diameter) flattened dome and then a short length of square section shank followed by a round shank, which has a length near the end which has a thread. The nut supplied can be either square or hexagonal shape; washers used should be large diameter.

All timbers to be joined should be bored the bare shank diameter. The bolt is hammered through the hole to a point where the squared part of the shank penetrates the timber. Care has to be taken with the hammering, and it is often better if the bolt is pressed home with a cramp or similar device to avoid splitting the timbers. Finally, a large diameter washer and a nut are run onto the threaded portion of the shank and tightened. The squared portion of the shank prevents the bolt from turning in the hole while the nut is tightened. Overtightening is always a fault with these bolts and they should only be tightened up to the point where the large diameter pan head *or* the large washer is being drawn into the timber. As soon as timber fibres start to break

at either end *stop* tightening up. These bolts can be used to bolt metal sections to timber but the cup/square head should always be on the timber side.

Hex bolts have a hexagonal head, a round shank part of which is threaded, washers and a hexagonal nut. Washers should be large diameter, and in certain circumstances 50 square steel plate washers are used. Holes through timbers should be bare shank diameter or 1–2 more where multiple timbers are being bolted. Large sizes are used in stand-alone mode, and smaller diameters and shorter lengths can be found in proprietary fastenings.

They can be used for all timber joints or metal/timber joints. They are used in preference to coach bolts for certain structural work requiring the use of **Bulldog connectors** (Figure G.10) or **shear plate connectors**. These connectors were developed to allow the designed joining of timbers in trusses – lattices of timber made up of several layers which are bolted at the nodes with the connectors between every timber. Both connectors were made double-sided – timber to timber – or single-sided, which allowed timber to metal joints.

The truss had to be made in a jig, and it was not uncommon in the 1950s to see a rough shelter built on a site and under it the carpenters cutting out and laying timbers in a jig, boring holes through them all and then fitting connectors in place over a high tensile steel nut and bolt with several large, well greased washers fitted. A long-handled ring spanner was used to tighten up, the teeth on the Bulldog connector or the ring on the shear connector biting into the timbers. If there were too many layers, the bolt had to be put into only a few at a time, and then more timbers and connectors added until the whole

Fig. G.10 Bulldog timber connector.

joint was properly bedded down. Once that was done, the high tensile bolt was taken out and replaced with an ordinary hex bolt, washers each end and hex nut tightened up, but only to start the washers pulling into the timbers.

Once the truss had been in place in a dry building for a few months these nuts were often found to be quite loose, as the timbers shrank when drying out, and old specifications stated that the contractor had to retighten after a certain period. These connectors were most often used to build roof trusses and a series of designs was done and promoted to the industry by TRADA, the Timber Research and Development Association, High Wycombe. Look at their website on http://www.asktrada.co.uk

The smaller bolts (diameters less than 6) can have hex heads or the same variety of heads as screw nails and can also find themselves incorporated in proprietary fastenings.

All threads on bolts now manufactured have metric threads, and the sizes of bolts are given, for example, as M10 × 65 – a metric threaded 10 diameter bolt, 65 long. The length is measured from the end of the threaded part to the underside of the head.

For all ordinary timberwork the bolts need be no better than mild steel and they are generally called **black bolts** since they come as forged, covered in the black scale of oxide caused by heating. Bolts the reader may be familiar with are somewhat polished and steely in colour. These are bright steel bolts and are used in the metalwork industries, the automotive industry, etc. Threads invariably run for the complete length of the shank. They are more commonly known as **machine screws**. They are rarely used in carpentry work but find their way into joinery in the small bolts used there for ironmongery. Only where fastenings have to be ultra strong would high tensile bolts be used.

Nuts on the other hand can be plain or self-locking. To ensure that plain nuts do not come loose due to vibration, a star washer can be placed under them before tightening. The teeth on the star washer grip into the nut and underlying material, preventing it from turning back. Self-locking nuts have a hard fibre or nylon ring set into the top of the nut, and when the nut is tightened down to that collar the thread on the bolt has to press out a corresponding shape in the fibre or nylon. This prevents the nut from shaking loose.

Special fasteners

Special fasteners have been invented and continue to be invented – every one a better mousetrap! They all purport to fulfil some special function in some special background material to which we wish to fasten timber or fittings or fixtures in every trade.

First, the special nail – most basic of all is the masonry nail (Figure G.11). This nail has a *precision ground point*, or so the literature tells us, and it is certainly smooth and curved and quite sharp. The nail is made of hardened steel with a bullet head and can be driven straight through wood into a brick or concrete block background with ease. Softer concretes present no problem but some very dense concretes can prove difficult. It is an effective means of fixing but not really suitable for finishing joinery work. The nails can be driven direct with an ordinary claw hammer.

The nails are also available with a steel washer jammed on the shank near the point. These nails are meant to be driven with the aid of a special hand tool which holds the nail, protects the operative's hand and allows him to use a much heavier hammer with fewer blows to drive the nail home. The washer has two functions: to align the nail in the tool and to prevent the bullet head from disappearing into the timber. This is even more important in the 'gun' used to drive these nails (Figure G.12). It is literally a gun, the explosive charge being a blank cartridge and the projectile the washered nail. Both nail

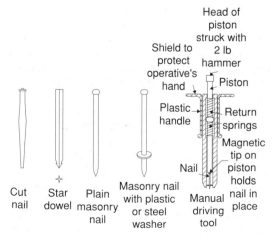

Fig. G.11 Some nails for masonry and the hand tool for masonry nails.

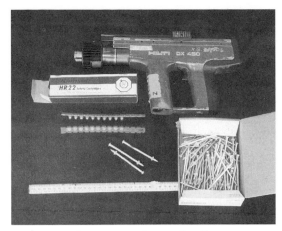

Fig. G.12 Hilti gun, cartridges and nails.

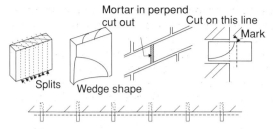

Dooks or plugs driven into wall at approximately 900 centres. Chalk line stretched over plugs on the line required and 'snapped' to mark top of plugs. Cut off plugs at mark. Nail timbers to plugs

Fig. G.13 A plug or dook shaped to wedge into a brickwork perpend.

and cartridge are loaded into the gun, and the nose of the barrel is depressed against the timber. This releases the safety catch and the trigger can be pulled. The cartridge fires and the nail is driven into the timber and the background. A range of cartridges is available which allows backgrounds of different types and densities to be penetrated. It is possible to nail timbers to the flanges of steel joist and beams. A Hilti gun, cartridges and nails features in the photograph in Figure G.12.

Fixing into masonry backgrounds has always been an area which has attracted much invention – ever since the first Rawlplugs came on the scene just over 80 years ago. The first wall plug was a brass strip pressed into shape and marked with a thread, which was folded, pushed into a pre-formed hole and the screw threaded into it. The name is so well known that even plugs not made by the company and 'invented' by others are known by the name. The company went on to give us Rawlbolts and a host of other devices whose designs have been imitated, tweaked, improved upon, etc.

At the most basic level there is the wall plug made from an offcut of timber and driven into a hole in the masonry. Nail(s) or screw nails can be used to fasten to this. Many plugs are simply roughly rounded in section to fit a similar hole but others can be split and shaped to fit into the perpend of brickwork or stonework (Figure G.13). These can be given a twist when they are cut and so wedge into the joint much more tightly. In Scotland the latter are not just plugs, they are **dooks**.

So, fixing with screws to masonry is perfectly possible (Figure G.14):

☐ With a wood screw and a *soft material* inserted into a pre-drilled hole; *soft materials* include preformed fixings in plastic, bituminous coated fibre and 'woven' metal wires such as silicone bronze. Other soft materials include amorphous mixtures of silica fibre and OPC supplied dry and applied as a plastic mixture with water.
☐ With a large coarse threaded screw (No. 12 and larger) directly into aac block.
☐ With a hardened steel screw directly into lightweight concretes and softer grades of bricks or a precisely drilled hole in denser concretes (Figure G.15).

The preformed soft material fastenings were first introduced by the original Rawlplug Co, now Artex-Rawlplug, and were simple tubes of

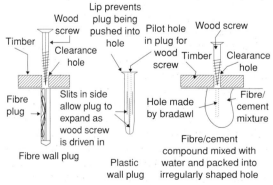

Fig. G.14 Fibre, plastic and amorphous plugging for fixing screw nails.

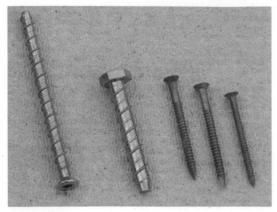

Fig. G.15 On the left are two bolts, one with a hex head and the longer one with a Torx head. These are designed to be screwed into concrete requiring a precisely drilled and clean pilot hole. On the right are improved shank bronze nails, again used to fix weather boarding etc. in aggressive situations such as near the sea.

fibre bound with bitumen. Various lengths and diameters were made to accommodate various sizes of wood screw. Later versions were made in bronze wire woven into tubes of various lengths and diameters. These could be used where heat would damage the fibre-based plug. The hole drilled in the timber allowed the wood screw to pass through, but a larger hole was drilled in the wall to take the plug.

Also introduced was a dry mixture of silica-based fibre (originally asbestos fibre) and OPC. This could be used to give a fastening in holes of, in theory, any shape or size. Mix sufficient of the material with water, pack the stiff paste into the hole, make a pilot hole in the damp mixture with a bradawl, and drive in a wood screw. Once the OPC had set you had a strong fixture.

A large number of companies produce 'wall plugs' and other patent fixings, and information is available by obtaining literature or visiting websites. Among those with good catalogues and literature are Rawlplug, Fischer, Hilti (who also make the cartridge guns etc.). Try the websites: http://www.artex-rawlplug.co.uk and http://www.hilti.co.uk

For heavier load applications, there are expanding bolts which fit into holes in the masonry and when tightened up expand a shell or a sleeve which grips the masonry (Figures G.16 and G.17). Some expand for the full depth of the hole, others only at the bottom. No matter what the situation, there will be an expanding bolt to suit your purpose and in a size to take the loads to be applied. Apart from the bolt or nut showing, there are also ring eyes and hooks which can be bolted into masonry.

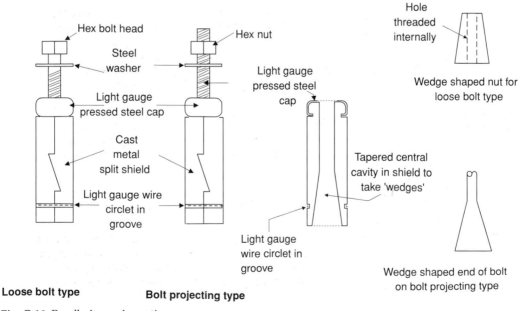

Fig. G.16 Rawlbolts – schematic.

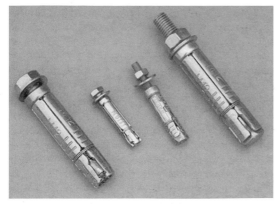

Fig. G.17 Rawlbolts M6 and M12 in bolt projecting and loose bolt types.

Fig. G.18 Spring toggle and expanding fasteners for blind fixing in hollow sections or panels.

Fixing back to a hollow background is another area where inventive minds have run riot. All the fixings manufactured depend on having a device which expands behind the skin or panel to which you are fastening – except one, which is the gravity toggle. This goes through the hole all aligned, then gravity takes over and part of it falls over and can be drawn up tight against the back of the panel. Figures G.18, G19 and G.20 show all this.

Frame fixings

Frame fixings (Figures G.21 and G.22) were specially developed to make the fixing of door and window frames into openings in masonry walls easier than with the traditional plug or dook. To use them, the frame has a clearance hole drilled for each fixing. Then a masonry drill is used through the clearance hole to drill corresponding holes in the masonry reveals. To do this

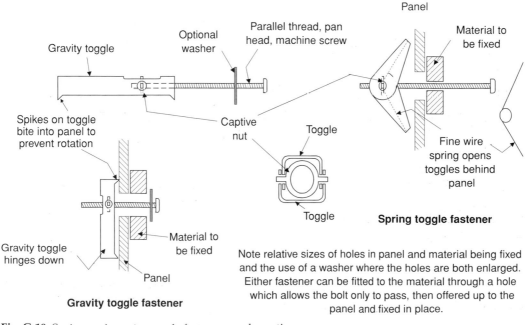

Fig. G.19 Spring and gravity toggle fasteners – schematic.

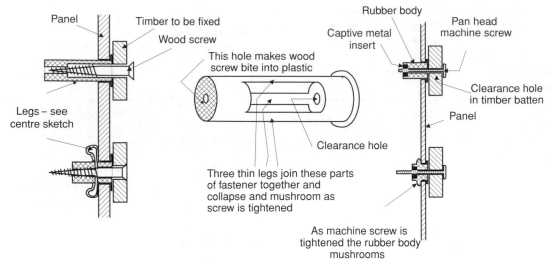

Fig. G.20 Rawlnuts and plastic mushroom fasteners – schematic.

successfully, the frame must be fastened securely in the opening so that the jambs of the frame are plumb and the sill and head horizontal – note that it would be easy to fix the jambs plumb and still have the sill and head sloping off, or vice versa. To fix the frames for drilling, packers are forced between frame and reveals, head and sill as close to the clearance holes as possible if not directly over them. Once the holes in the reveals have been drilled, each frame fixing is tapped into place and the screw for each is driven up tight with a hammer.

These devices are widely used in the industry but there appears to be little advantage over the traditional plug or dook. The plug or dook can be cut from waste timber, and waste from treated joists etc. can be used. Ordinary nails can be used with plugs or dooks, and for external work a galvanised nail is longer lasting than the plated screws used in frame fixings. With plugs or dooks there is no need for packers, as the frame is propped against the plugs, and these are marked for plumb and level before being cut back to size.

Fig. G.21 Frame fixings – schematic.

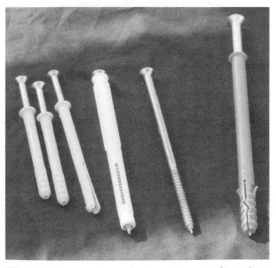

Fig. G.22 Frame fixings – the large one on the right is 150 long. Next to it is the drive screw which is used. Note the shape of the thread.

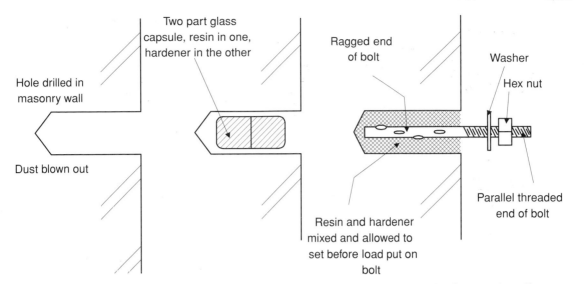

Fig. G.23 A chemical anchor schematic. A more up-to-date system uses a mastic gun and a cartridge of one part resin to fill the hole into which the anchor is set.

There are even proprietary brackets marketed in various thicknesses of plastic – more expense still. Where plastic packers are not used, cuttings of untreated plywood or even worse, hardboard, are used. Finally, there is always the chance that the packers are not packed tightly, and driving home the fixing will pull the jambs of the frame out of the plumb and distort the frame.

The last fixing is the 'chemical anchor' as illustrated in Figure G.23. A range of these are manufactured, and Artex-Rawlplug immediately comes to mind. They work on the principle of setting a length of threaded rod into a predrilled hole in masonry, into which has been placed a compound which sets due to chemical action. Once setting has taken place, timber or some other piece of construction can be placed over the threaded rod, and a washer and nut applied and tightened.

Early models used a two-part capsule, one part holding synthetic resin and the other the chemical hardener. Pushing the threaded rod into the hole broke the capsule placed there, and rotating the rod mixed the resin and hardener. Newer variations use a single-part resin, which hardens on contact with the air. The resin can be packaged in cartridges and injected into the hole in the masonry – no capsule to break, no chemicals to mix.

Figure G.24 shows a **skeleton gun** with a cartridge of silicon sealer in it. This is the tool used with the cartridges of chemical anchoring

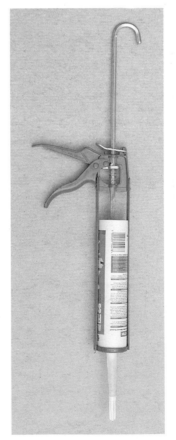

Fig. G.24 Skeleton gun for injecting compounds into holes, grooves or recesses in construction work.

compound. The skeleton gun shown can be used with any standard cartridge to place a wide variety of compounds, one of which is a glue which substitutes for mechanical fasteners for use on items such as skirtings, architraves and other trims. This particular compound is a form of contact adhesive with gap filling properties. This type must not be used for structural work of any kind.

It is used by simply applying a ribbon or series of spots of the compound on the substrate and then pushing the trim firmly against it. Temporary support may be required if the trim is heavy or overhead.

APPENDIX H
Gypsum Wall Board

Most lay people know what is meant by plasterboard – at least they think they do! It is a layer of white material with paper both sides. But that is too general a description of what is a truly versatile material and one which is used in a wide variety of forms in virtually all the construction work in the Western world and not a little elsewhere. So we need to have a general discussion of what the material really is, how it can be fixed to different backgrounds, and what it can do for the building we are constructing.

At its most basic, this sheet material is a sandwich of two layers of special paper with a layer of gypsum (calcium sulphate) between, which can be nailed or screwed to a framed timber or light steel section background. Thus it can be used to cover the underside of upper floor joists to provide a ceiling, or applied to walls and partitions to provide a smooth wall surface.

The website to visit is http://www.british-gypsum.com

There you will find that much of their publication, the **White Book**, is available and can be viewed and printed out with Adobe Acrobat. Another website to visit is the international company Lafarge, at http://www.Lafarge.com

Another company, Knauf, has very good on-line literature, particularly their 32 pages on the use of plasterboard in housing. Adobe Acrobat is needed to download and print this but it is very worthwhile. Find them at http://www.knauf.co.uk

All the companies manufacture a wide range of products for wall panelling or **dry lining** work, all with special features which make them suitable for the wide range of conditions met in any type of building work. Among the types of board made are those suitable for acoustic and thermal moderation, moisture resistant board, fire resistant board, flexible board, etc.

Concentrating on British Gypsum, in general the type of board the domestic market will use is plain plasterboard, Gyproc Wallboard. This type of board, the most basic, is perfectly adequate for domestic construction. It has good strength for covering walls and partitions where impact damage might occur, and by building up layers it can fulfil all the fire resistance requirements in low rise housing – protection of structural elements such as steel beams and columns, timber structural members and so on. Figure H.1 illustrates a double plasterboard layer.

The board is available in a variety of lengths – 1800, 2370, 2400, 2500 and 3000 – and in widths of 900 and 1200 plus the thicknesses 9.5, 12.5 and 15. Note that not every permutation of these sizes is manufactured. In addition to the range of dimensions, the boards can be **square edged** or **taper edged**. Again, not every board size manufactured is available in both types.

As the name suggests, square edged means that the board is an even thickness across its entire area. Taper edged board has the margin of the long edges reduced in thickness to accommodate tape and jointing compound or simply **jointing** in a dry lined wall or ceiling system, thus giving a flush finish which is ready for paint work as soon as the jointing is dry. Figure H.2 shows taping and jointing of both types of board.

The board is supplied with a different paper on either side. One side is ivory coloured and it is this side which has the taper edge. This paper is specially formulated to receive decoration direct without any further preparation. That doesn't stop people applying further layers of this or that in an attempt to improve. The reverse side is covered with a grey coloured paper which is

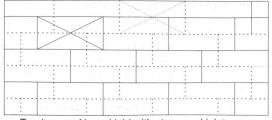

Two layers of board laid with staggered joints on background framed at 600 centres. Dotted lines indicate first layer, solid line the final layer. Diagonal lines indicate full sheet sizes.

Fig. H.1 Double plasterboard layer with board joints staggered.

specially formulated to receive a layer of plaster. The plaster would be a gypsum-based plaster about 3 thick – what is referred to as a **skim coat**. Skim coating does give a superior finish to any wall or ceiling but perhaps defeats the object of dry lining in the first instance, which is the elimination of a 'wet trade', plastering.

Concentrating on domestic application, the background to which plasterboard is fixed is usually timbers at centres with dwangs and noggings at intervals of around two per 2400 long board. Pressed galvanised steel framing is occasionally used but is more common in commercial work and in alterations work. Fixing back to timber can be done with nails – **plasterboard nails**, which are galvanised or plated and generally have some form of improved shank, a jagged shank being common.

Alternatively, fixing can be with screw nails – **dry lining screws** – which are made of hardened steel and have a specially shaped head with a Phillips or Pozidriv slot, a thread which has a fast drive into the timber, and a point which is self-drilling. The latter is of greater importance when fixing to the pressed steel framing. The shape of the head where it penetrates the plasterboard is important as what is required is to depress the paper surface of the board and allow the head of the screw to penetrate just below the general surface and still prevent the paper from tearing away. This is also helped by the use of power screwdrivers which have an adjustable clutch that slips at a predetermined loading and stops turning the screw into the plasterboard and timber or metal backing. Another device which can be used is a special screwdriver bit which has an outer annular ring to press on the board surface when the screw is at the correct depth, and which then makes the bit 'cam out' of the slot. Figures H.3 and H.4 show dry lining screws and bits.

For small jobs, hammering in nails with a joiner's hammer is not too tedious but to speed up the process and reduce the physical effort fixing with screws is extremely good. Battery powered tools have changed attitudes in the industry, and the use of collated dry lining screws in a portable tool makes fixing large sheets of

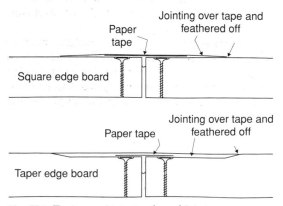

Fig. H.2 Taping and jointing board joints.

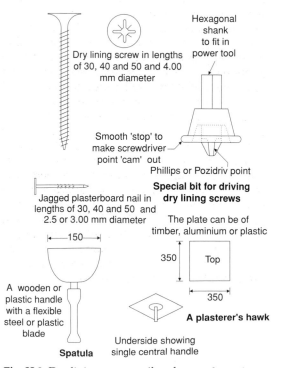

Fig. H.3 Dry lining screw, nail and screwdriver bit.

Fig. H.4 Dry lining screws and driver bit.

plasterboard the work of a few minutes and by *one* operative, 20–25 m² per man/hour being the norm for plain wall and partition work.

The frequency of the fastenings, whether screws or nails, is usually about 150–200 apart to every timber or framing member behind the board. The framing needs to be in rectangles not exceeding 600 × 400 to maintain a flat surface, particularly for ceiling boarding. Fixing back with plaster dabs or stripes to a masonry wall has been used a lot in the past in domestic work but now rarely finds a use except in refurbishment work. All the manufacturers mentioned above give details of the technique(s) in their literature.

Finishing off the board ready for decoration means applying jointing compound to all the joints, bedding in a paper tape or scrim, and then applying a final layer of jointing compound, all in a thin layer and with the edges feathered to make the joints disappear. The paper tape is supplied in rolls of around 300 m and is 50 wide with a centre crease to facilitate folding into corner joints. A reinforced paper tape is made which can be applied to external corners. The reinforcement is two strips of sherardised steel strip. Scrim is a woven tape of either cotton or jute fibre, the cotton being the finer texture. It is made in 50, 75 and 100 widths. Some decorators apply a coat of the jointing well watered down to the whole of the boards and then decorate, but this is not really necessary. Emulsion paint, oil paints, wallpaper, etc. can all be applied direct to the board. Some decorators will apply an oil undercoat before wall papering to facilitate removal when redecoration takes place others even go to the extent of using aluminium paint, although to what purpose is not clear.

The **taping and jointing** was originally carried out by both plasterers or painters and decorators but is now carried out by specialist **tapers and jointers**, recognised as such in most Working Rule agreements. These men will do the work with hand tools such as spatulas and hawks (see figure H.3), finishing off with sponges etc. to give the smooth finish desired. The bulk of their work is carried out using more mechanical means, even if still powered by human muscle. The process is then known as Ames Taping and comprises a number of tools with which to spread an even amount of jointing compound into straight and corner joints, and a tool which holds a roll of paper joint tape and applies it into the straight and corner joints, and then finally spreads the jointing compound across the joint, feathering down the edges.

Fire protection using wallboard is basically about putting a layer of a fire resistant material between the source of the fire and the material to be protected. Generally one or two layers of 12.5 thick board are sufficient. In Chapter 4, Floors, reference is made to ceilings with two layers of plasterboard when the floor divides flatted housing. The combination of plasterboard and sound deadening materials gives a fire resistance in excess of half an hour. Also in that chapter, literally wrapping a steel beam with two layers of plasterboard has the same effect. Loadbearing timber or pressed steel stud partitions always have any faces next to halls and staircases clad in one layer of 12.5 plasterboard to protect the framing.

In all these examples thick or multiple layers of plasterboard can slow down penetration of the partition, ceiling, etc. by fire or hot gases to give protection to occupants using the hall or staircase to escape or who live in the neighbouring flat.

APPENDIX I

DPCs, DPMs, Ventilation of Ground Floor Voids, Weeps

This is a somewhat mixed bag of an appendix but is an attempt to take out a lot of repetitive detail from the earlier chapters and explain some basic materials and techniques common across a number of chapters.

DPCs and DPMs

Damp proof courses and damp proof membranes are layers of material, impervious to water, which are placed in the construction to prevent penetration by dampness, rain, etc. DPCs are built horizontally into walls near ground level, over lintels and under sills and vertically at door and window jambs. DPMs are laid under floors or on or within the vertical thickness of walls. We will examine these situations and determine what is the best from a wide variety of materials to be used in the circumstances.

The conventional symbol for damp proof layers is shown in Figure I.1. Depending on the base material, these layers can vary from 150 mm thick down to a fraction of a millimetre. For those of a few millimetres and less it is impossible to show them to scale or in proportion.

The answer to this problem is quite simple. These layers are very important in the construction, therefore the symbolism must be clear on the drawings. The layers are shown larger than scale or proportion allows and the additional space required is taken off the adjacent layers – but only from one side. Dimensions are given to *one* face or side of the damp proof layer where the adjacent material is shown full size.

The detail shown in Figure I.2 was studied in connection with solid concrete floors (Chapter 2). Note the placement of DPC and DPM.

Note that neither is given a thickness. The DPC would be contained entirely *within* the mortar bed of the brickwork. The Visqueen sheet used for the DPM is only 250 microns thick.

The annotation on Figure I.2 states that the DPC and DPM overlap. Though not absolutely necessary, it is best if they overlap in the vertical plane alongside the wall. The DPM is laid first and turned up at the edges. The DPC is laid next and is turned down over the DPM. With the joint horizontal, any settlement of the floor slab neatly slices the join like a pair of scissors. With the lap vertical, the joint slides apart with the settlement, especially when edge insulation is in place. These points are illustrated in Figure I.3.

Only DPCs are built using thicker materials such as bricks, tiles and slates. In the case of these materials, they can be shown to proper scale as shown in Figure I.4.

Materials for DPCs and DPMs are the subject of several British Standards:

- BS 743 – *Materials for damp proof courses*
- BS 6398 – *Bitumen damp proof courses for masonry*
- BS 6515 – *Polyethylene damp proof courses for masonry.*

See also the précis of British Standards in Appendix L.

CP 102 is a Code of Practice, also published by the BSI, and is entitled *Code of Practice for protection of buildings against water from the ground.* BS 8215 is also a code of practice entitled *Design and installation of damp proof courses in masonry construction,* which deals with damp proof courses in the superstructure of the building such as around door and window openings.

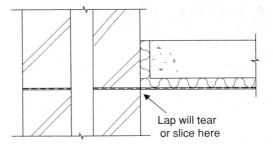

Fig. I.1 The conventional symbol for damp proof layers.

DPC and DPM materials, all listed in BS 743, can be classified as:

□ Bitumen sheet material
  ▪ Plain – BS 6398
□ Plastic sheet material – BS 6515
□ Metals
□ Mastic asphalt
  ▪ Applied hot
□ Masonry materials.

In addition to the materials listed in BS 743 there are:

□ Self adhesive bitumen sheet materials – not studied here
□ Cold applied liquid materials
□ Tiles of concrete or fired clay.

Bitumen sheet DPCs are generally classified by the 'base' materials used in their manufacture, and are detailed in BS 6398. Base materials available are:

□ Hessian – a single layer of plain woven jute yarn
□ Fibre – one or more absorbent sheets of felt made from a mixture of animal and vegetable fibres
□ Asbestos – absorbent sheet made from 80% asbestos fibres
□ Lead – sheet or strip milled lead with soldered joints.

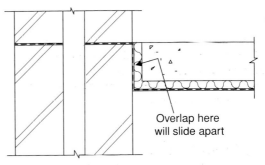

Fig. I.3 Laying and overlapping of DPC and DPM.

Table 1 of BS 6398 gives the Class and Description of each type of bitumen DPC (an example is shown in Table I.1) together with a full listing of the mass/m^2 of materials used in their manufacture. Classes are:

A Hessian base
B Fibre base
C Asbestos base
D Hessian base laminated with lead (also known as 'lead core DPC')
E Fibre base laminated with lead
F Asbestos base laminated with lead

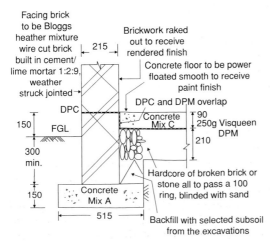

Fig. I.2 Placement and thickness of DPC and DPM in a detail drawing.

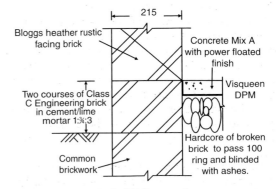

Fig. I.4 Brick and tile DPCs to scale.

Table I.1 Class and description of each type of bitumen DPX from Table 1 of BS 6398

Type	Application	Specification	Detail	Use
Sheet	Built in	Bitumen impregnated	Fibre bases with and without metal cores	Generally DPCs, rarely DPMs
		Plastic	Polythene and polypropylene	DPCs and DPMs
		Metal	Lead and copper	DPCs only
Liquid	Cold applied	Bituminous emulsions	Proprietary types	DPMs only
	Hot applied	Mastic asphalt	Natural or limestone aggregate	DPCs only and 'tanking'
		Coal tar pitch		DPMs only
Others	Built in two courses in 1:3 cement mortar	Brick	As British Standard 3291 for DPC brick	DPCs only
		Slate and tile	To appropriate BS for roofing material	DPCs only

Table 2 of BS 6398 lists suitable locations for the use of these DPCs, with the recommendations based on the following:

☐ Compressive loads
☐ Flexural loads
☐ Shear loads
☐ Water movement – up, down and horizontally.

While polyethylene DPC is listed in BS 743, the detailed standard is laid out in BS 6515. The material to be used is described as follows: black, low density, polythene sheet with a mass in the range 0.425 to 0.60 kg/m². The composition to include 2% evenly distributed carbon black and no more than 5% other material both by mass. No air bubbles and no visible pin holes. Table 1 in Appendix D of the Standard lists uses etc. using the same criteria as bitumen DPCs.

This material is supplied in rolls of 10 m² and in widths which are generally multiples of half a brick plus 200, 300 and at least 1000. There may be others – it depends on the manufacturer. This material is used in walls at DPC level, round door and window openings, etc.

Much of the polythene seen used as DPM material on building sites today is coloured blue. Figure I.5 is a photograph of a substructure and shows a polythene DPM laid over a solum. On the left-hand side it is ready to receive a concrete garage floor, which is reinforced with the steel fabric lying there. On the right-hand side the polythene has been laid and weighed down

with a layer of concrete. This will be under the timber hung ground floor of the house.

For more information try the website http://www.visqueenbuilding.co.uk

Other materials are specified in detail in BS 743:

☐ Lead should be Code 4 (i.e. 1.80 mm thick or 4 lb/ft²) and should be milled sheet or strip lead for building purposes complying with BS 1178.

Fig. I.5 Polythene DPM.

▢ Copper should be 0.25 mm thick complying with BS 2870 sections 2 and 7, Temper Grade O, **dead soft** or fully **annealed**.

▢ Mastic asphalt should comply with either

  ▪ BS 1097, Mastic asphalt for tanking and damp proofing (limestone aggregate), or

  ▪ BS 1418, Mastic asphalt for tanking and damp proofing (natural rock asphalt aggregate).

Whichever asphalt is used it is applied hot – generally around 230°C – in one or two layers from 12 to 20 mm total thickness. It can be applied to horizontal and vertical surfaces. It can be applied to the workman's side of a face or to the opposite side – **overhand**. It cannot be applied to overhead surfaces. It requires specially skilled labour and special plant for successful application.

Tanking is a technique applied to underground basements and is beyond the scope of this text. Materials other than mastic asphalt have been used for tanking but this is the best although the most expensive in capital outlay terms – what it would cost to redo the waterproofing of a basement wall carried out with inferior materials doesn't bear thinking about.

Slates should be not less than 230 mm long or less than 4 mm thick. They should comply with BS 5642 with regard to the wetting and drying test and resistance to sulphuric acid. A minimum of two courses should be laid to break joint in a mortar of $1:0-\frac{1}{4}:3$ cement/lime/sand.

Bricks should be of fired clay or brick earth. They should comply with BS 3921, *DPC bricks*. A minimum of two courses should be laid to break joint in a mortar of $1:0-\frac{1}{4}:3$ cement/lime/sand.

Other DPM material

Liquid, cold applied, bitumen-based materials are frequently used to provide damp proof membranes. They are principally 'emulsions' – a bitumen/rubber latex which while drying by evaporation of water give a waterproof seal. They are applied to a concrete or masonry surface in two coats by brush, the second coat brushed on at right angles to the first to cover pinholes.

They are frequently used in sandwich floor construction and on concrete bridge abutments. They can also combine the properties of an adhesive for block or parquetry flooring. Application rate is generally around 2.5 litres/m². RIW, Synthaprufe and FEB Hyprufe are three well known trade names.

Two websites to try:

http://www.ruberoid.ie
http://www.degussa-cc.co.uk

Hot bitumen is frequently specified as a DPM material. Originally this was 'coal tar pitch' obtained from the 'gas works' as a by-product of the destructive distillation of coal. It was delivered to site in insulated and heated tanker lorries and the hot pitch was spread over a solum from a watering can with a spout shaped like a fan, which would pour a wide ribbon of the hot liquid. The thickness of the ribbon could be set by varying the rate of pour.

'Bitumen' is now obtained from the distillation or 'cracking' of crude oil and this bitumen can be used in exactly the same way as the coal tar pitch. Delivery to the site is still in insulated and heated tanker lorries. Spreading is still carried out with the special watering cans.

Hot applied bitumen DPMs are generally laid over a **blinded hardcore bed** under hung ground floors. The blinding is of sand or preferably furnace ash, which is wetted and beaten down to a smooth hard surface. The ash being slightly pozzolanic (setting like a cement in the presence of water) gives the better finish on which to work.

A 10 mm thickness of pitch or bitumen is generally satisfactory. When the workman reaches a wall it is usual to allow the bitumen to splash up the wall, thus reducing the area of wall from which moisture can evaporate.

Plain clay or concrete roofing tiles can be used for damp proof courses in ancillary construction works such as screen walls and garden walls. A minimum of two courses should be laid to break joint in a mortar of $1:0-\frac{1}{4}:3$ cement/lime/sand. Materials used for mortar – cement(s), lime and sand – should comply with their respective and appropriate British Standards.

Table I.1 classifies the information on DPCs and DPMs by basic material.

Weeps

Weeps were originally used to allow the drainage of water trapped in a cavity in wall construction. This can occur in several areas of a building:

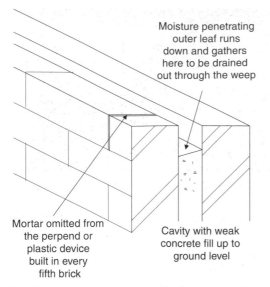

Fig. I.6 Schematic of a weep formed at ground level in a masonry wall.

Labels in figure:
- Moisture penetrating outer leaf runs down and gathers here to be drained out through the weep
- Mortar omitted from the perpend or plastic device built in every fifth brick
- Cavity with weak concrete fill up to ground level

Fig. I.7 A proprietary plastic weep.

☐ The substructure walls, just above finished ground level
☐ Above the DPC laid over a lintel
☐ At string courses
☐ At fire stops in timber frame construction.

Weeps can be formed by simply omitting the mortar from the perpend or by inserting a purpose made plastic weep.

Figure I.6 is a schematic of a weep formed at ground level in a masonry wall. Note that it is a basic weep with mortar left out of the perpend. Note the slope outwards of the cavity filling, which is standard on every detail you look at but must be well nigh impossible to form. The author has looked down a few cavities in his time and has yet to see this being done. Mostly the concrete is left quite rough; occasionally there is an attempt to level it off.

Figure I.7 is a photograph of a proprietary plastic weep merely set in place in some loose bricks to show how it is positioned. There must be at least a dozen manufacturers making their own version. The point about using these weeps is to see that the perpend does not get blocked with mortar as building goes on round about it. The internal part of the weep is specially designed to take water out and not let it be blown back in again. Figure I.8 is one of these weeps opened up to show the baffles and water channels. The bottom of the weep is at the bottom

Fig. I.8 Weep opened up to show the baffles and water channels.

of the picture with the gridded edge on the left showing on the face of the wall. Note that this particular weep has a drip built into it to throw water clear of the wall surface.

Weeps are made for building into walls which will eventually be rendered. The outer face has a plastic cover over the outlet. The weep is built into the brickwork, projecting the thickness of the render. The render is applied and, when set, the plastic cover is peeled off the weep. The result is a clean, unclogged weep and a neat finish.

Weeps are also used in timber frame construction, not just to take water away but also to provide a minimum amount of ventilation of the cavity between masonry cladding and timber panels. This has already been covered in Chapter 3, Walls. There it was stated that weeps had to be built in above and below the horizontal fire stopping at first floor levels. That fire stopping has to be the full cavity thickness and covered with a DPC to stop water getting across the cavity. Small particles of dust, waste material and dead insects and spider webs can and do build up in these cavities and especially over the horizontal fire stops. This waste material can block the flow of water down the inner surface of the masonry cladding and can even seal off the pores in the masonry. So water can build up and can run back into the timber frame. This is where the weeps above the fire stop act not just as ventilators but as weeps!

Ventilators

The ventilators being discussed here are exclusively those built into substructure walls to ventilate the void under a hung floor. They must let a minimum amount of air pass and this is covered in the British Standard. They must exclude vermin, birds and pets or other domestic animals and even small children. They must exclude water, so the liner should be built in with a slight slope to the outside and the ventilators themselves have the holes or slots arranged to shed water to the outside

Figure I.9 is a photograph of ventilators and liners made in terracotta, 215 × 65 and 215 × 140.

BS 493 lists two categories of air brick or ventilator, and the reader is referred to the précis of this Standard in Appendix L. The sizes of air bricks and ventilation gratings for use externally

Fig. I.9 Ventilators and liners made in terracotta.

always equate to the work size of full stretcher face multiples, the most common sizes being:

- 215 × 65 and
- 215 × 140

and so on to suit different brick sizes.

Cavities must be closed to duct ventilating air through the wall. This was done traditionally by closing all four sides with cuttings of roofing slate, as shown in Figure I.10. They must *not* ventilate the cavity between the masonry skins, so now **liners** are built in, as shown in the accompanying figures. This prevents the spread of fire from timber structures built into the cavity, where the cavity would act as a 'chimney', spreading the fire up to the next floor(s) and finally the roof.

Liners for use with external airbricks and gratings are made from clay or plastics. Off-cuts of clay or plastic drainage piping are sometimes used but a purpose-made clay liner is to be preferred.

One of the many gadgets made for the industry is a telescopic liner for underfloor ventilators. They can be used to duct air across and up or down a cavity where there is a non-conventional arrangement of ground and floor levels. For example, a hung floor might be built which is

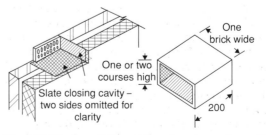

Fig. I.10 Closing all four sides of vent opening with cuttings of roofing slate.

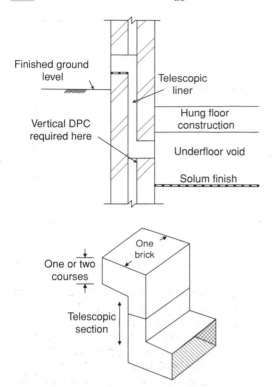

Fig. I.12 Terracotta liner built into a blockwork substructure wall.

Fig. I.11 Telescopic liner for under floor ventilators.

below ground level, and a straight through ventilator above ground level would ventilate the room, not the void under the floor. Figure I.11 shows how this might work. It is used in refurbishment work where there is no way of altering the relative floor and ground levels, but one wonders what is done about the lack of a DPC vertically in the walls joining the underfloor DPC to the wall DPC. It can hardly be a procedure to be advocated for new build work, not least because the unit is made of plastic and so is not fireproof as a clay liner would be.

Figure I.12 is a photograph of a terracotta liner built into a blockwork substructure wall. There is another course of blockwork for the outer leaf of the wall to be built, and then a timber frame kit is to be erected above this. Note that the liner is sealed to the blockwork with mortar. This will be repeated when the outer leaf course is built. The large clay pipe immediately below the vent liner is a duct for service pipework. In this case it is a 100 nominal soil and waste pipe.

Figure I.13 is a photograph of a similar liner built into a cavity wall in the wrong place. It is built two brick courses too high and is ac-

tually opposite the ground floor construction. Once the insulation for the ground floor is put in place, the ventilator will be completely blocked. Note also that the liner is not bedded all round with mortar. It should be 'sealed' to both leaves of masonry. The other vents in this house were in a similar situation, and unless the work was seen at this particular stage, no-one would ever know – until the floor started to rot.

Figure I.14 shows a typical vent in a cavity wall with a 100 thick outer leaf of artificial stone. The air brick is of clay.

Fig. I.13 A similar liner built into a cavity wall in the wrong place.

Fig. I.14 A typical vent in a cavity wall with a 100 thick outer leaf of artificial stone.

How much ventilation?

Although all regulations require ventilation of the underfloor space in hung floor construction, it would appear that only the Scottish Regulations give precise figures for its calculation, although these figures give some peculiar results, as we shall demonstrate.

Each ventilator built in will admit a given amount of air, and the amount required is calculated on allowing a free air flow through a given area per metre of wall – 1500 mm^2 per metre run; or on a given area of ventilator per square metre of floor area – 500 mm^2 per m^2 of floor area.

Note: the area of the ventilator in this part of the exercise is assessed on the size of the *hole* in the wall and not on the clear opening size of any grid or air-brick placed over the hole or any liner placed in the hole. This restriction is allowed for by requiring grids or air bricks to comply with BS 493.

Let's look at the problems involved in ventilating the underfloor void of a building with a ground floor area of approximately 45 m^2. A hole made by omitting two courses high and one brick wide has an area of 225 × 150, i.e. 33 750 mm^2. Divide this by 1500 and we arrive at a spacing of 22.5 metres along the wall between ventilators; or a typical ground floor area would be 45 m^2, which multiplied by 500 mm^2 gives 22 500. Divide by 33 750 and we have the equivalent of two-thirds of an air brick.

So for a building of approximately 7 m × 6.5 m on plan, using the first method we calculate that it will require one and a bit ventilators in the total run of the walls, and only part of a vent using the second method. Clearly impractical. Most buildings end up with a far greater number

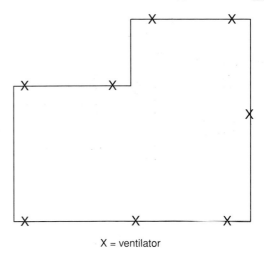

X = ventilator

Fig. I.15 Typical house plan with the position of vents.

of vents, so what are the criteria for deciding on how many and where to put them?

The placing of a single vent would give rise to 'dead spots' – areas of the underfloor space where the air would be stagnant and, if moist, would raise the **moisture content** of the building fabric. Figure I.15 shows a typical house plan with the position of vents marked as X. Ideally the vents should be placed within a metre of corners and should not be more than 4–5 m apart. It is best to place the ventilators in opposite walls and start and finish as close to the corners as possible. Observing these simple rules will avoid **dead spots**, i.e. parts where the void does not get proper ventilation.

Figure I.16 shows a house with part of the floor of solid concrete on upfill construction. This

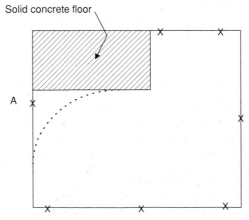

Fig. I.16 A house with part of the floor of solid concrete on upfill construction.

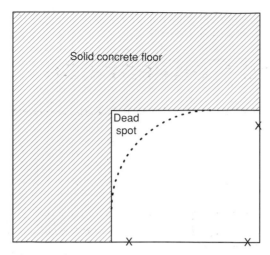

Fig. I.17 A house with a small area of hung floor – inadequate ventilation.

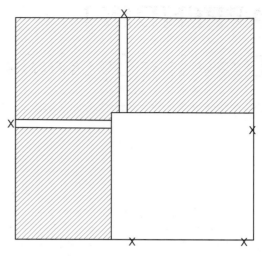

Fig. I.18 A house with a small area of hung floor – adequate ventilation.

effectively blocks off the possibility of ventilating a corner, so we break the rule about ventilators being opposite each other, and ventilate the odd corner and take out a potential dead spot.

Figure I.17 shows a house with a small area of hung floor, with two sides bounded by a solid floor construction. Three vents should be adequate following the rules of the Regulations but this leaves a large dead spot. The only way round it is to adopt the idea in Figure I.18, where ducts are placed under the solid floor (in the upfill) and begin with air bricks in the outer wall. The ducts are usually made from 150 diameter plastic or clay drainage pipe (the pipes must be strong enough not to collapse under the weight of the floor structure). The reason for putting in two pipes is that it will help the air pressure under the floor to balance when strong winds are blowing. Evening out the pressure in this way prevents any uneven pressure on the floor structure causing unwelcome draughts in the house above.

APPENDIX J
Drawing Symbols and Conventions

As one would expect, the drawing of architectural details requires a common methodology for the representation of building materials

Break line

Common brickwork

Facing brickwork

Natural stonework

Concrete blockwork

No-fines concrete

In-situ dense concrete

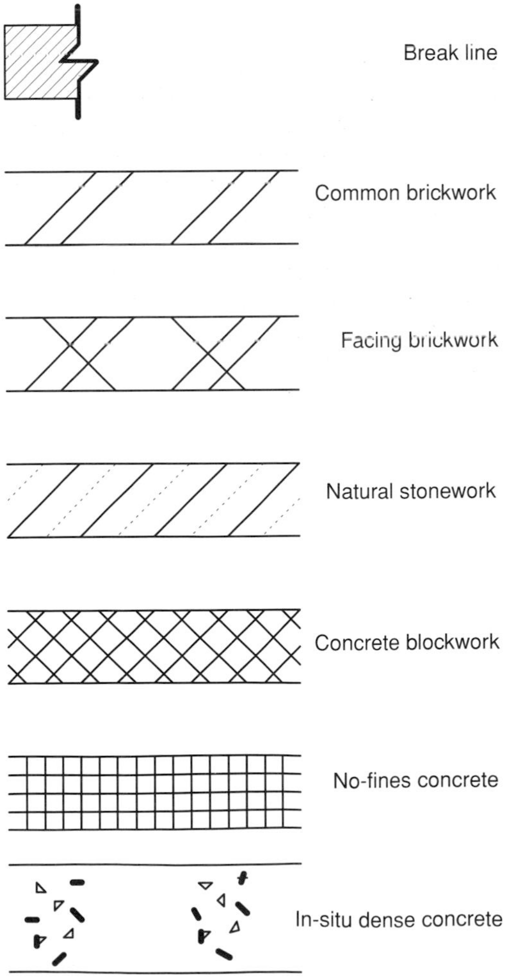

Fig. J.1 Architectural drawing symbols (1).

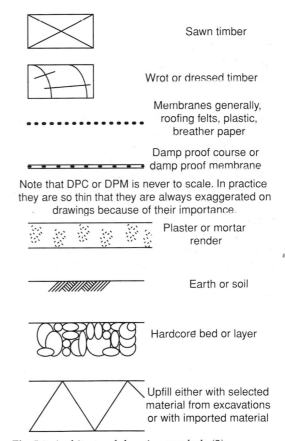

Sawn timber

Wrot or dressed timber

Membranes generally, roofing felts, plastic, breather paper

Damp proof course or damp proof membrane

Note that DPC or DPM is never to scale. In practice they are so thin that they are always exaggerated on drawings because of their importance.

Plaster or mortar render

Earth or soil

Hardcore bed or layer

Upfill either with selected material from excavations or with imported material

Fig. J.2 Architectural drawing symbols (2).

and construction forms. The samples given in this appendix are those used in this text; some may not be 'official' or their use may be slightly unorthodox. The author apologises if that upsets anyone.

While DPCs and DPMs are always shown to an exaggerated scale, the problem also arises with membranes. Where plastic, paper or felt layers are used, they are generally shown adjacent to the layer on which they are applied.

The British Standards Institution did publish a document some years ago which dealt exclusively with architectural drawings – everything from paper sizes and folding drawings to titles, line thickness and conventional symbols. Alas, this seems to have been withdrawn. There are, however, several good textbooks on the market which cover the topic. One which can be recommended is *An Introduction to Construction Drawing* by Arthur Thompson, published by Arnolds (ISBN 0-340-56823-2).

APPENDIX K
Conservation of Energy

Heat loss calculation using a spreadsheet

Conservation of energy has long been a requirement of Building Regulations although many argue that even the current requirements don't go far enough. Whether one agrees with that or not, the fact is that calculations have to be done to assess the energy efficiency of the building proposed.

One of the features of these calculations is the need to calculate how much heat will pass through a piece of construction – a floor, a wall or a roof, a door or a window. Then the areas of those features with a high heat loss have to be balanced out and items which cause cold bridging have to be assessed. Areas where condensation might occur within the building structure have to be identified and eliminated or prevented. The whole exercise is beyond the scope of this simple text but the student would be advised to come to terms with the one or two simple ideas put forward in this appendix, with a view to a more advanced understanding later in the course he or she is pursuing.

We must start with a few definitions. **Energy** is measured in watts. **Unit area** is one square metre. **Unit thickness** is one metre. **Unit temperature difference** is one degree kelvin.

Emissivity is the ability of a surface to pass or receive energy to or from the atmosphere in which it exists in response to a unit difference in temperature. It varies with the exposure of the surface, its texture and colour. Texture and colour have so little effect that rough/smooth and light/dark are sufficient to describe it.

Exposure, on the other hand, has a far greater effect. Three categories are generally used, sheltered, normal and severe, severe exposure being at least three times more emissive than one categorised as sheltered.

Thermal conductivity is the amount of energy which will pass through unit area, across unit thickness of a material in response to a unit difference in temperature, and is represented by the letter k.

Thermal resistance, r, is the reciprocal of thermal conductivity, i.e. $1/k$, and is the resistance of unit thickness of unit area of the material.

Thermal conductivity of a material of thickness l is given by dividing l by the conductivity k. l is always *used* in metres but may be given in millimetres when taken from an architectural drawing where only millimetres are supposed to be used. Take care, then, when transferring data from detail sheets to calculations.

U-value, or **thermal transmittance**, is the amount of energy which will pass through unit area of a piece of construction of given thickness and composition in response to unit difference in temperature, so it is the reciprocal of the sum of all the individual thermal resistances plus external and internal emissivities. It is measured as $W/m^2 K$:

$$U\text{-value} = \cfrac{1}{\begin{array}{c}\text{external} \\ \text{surface} \\ \text{resistance}\end{array} + R_1 + R_2 + R_3 + R_4 + R_n + \begin{array}{c}\text{internal} \\ \text{surface} \\ \text{resistance}\end{array}}$$

An Excel spreadsheet has been saved on this book's web page, www.thatconstructionsite.com, in 'Sample pages', where you can download it and use it. If you find it useful please feel free to use it; the only condition is an acknowledgement of the source and the author as printed on the first sheet of the worksheet WALUVAL2.

APPENDIX L

Short Précis of Selected British Standards

General	A brief summary of the history and working of the British Standards Institution
BS 187	Calcium silicate bricks
BS 493	Air bricks and gratings for wall ventilation
BS 743	Materials for damp-proof courses
BS 1199	With BS 1200 Sands for mortars, plasters and renders
BS 1282	Wood preservatives
BS 3921	Clay bricks
BS 4449	Carbon steel rods and bars for reinforcing concrete
BS 4471	Sizes of sawn and processed timber
BS 4483	Welded steel fabric reinforcement
BS 6398	Bituminous damp-proof courses for masonry
BS 6515	Polyethylene damp-proof courses for masonry

British Standards

General

The British Standards Institution was founded in 1901 and incorporated by Royal Charter in 1929. The objectives set out in its charter include:

- co-ordination of the efforts of producers and users for the improvement, standardisation and simplification of, in our field, construction materials and components
- to simplify production and distribution
- to eliminate the unnecessary production of a great variety of sizes of products and components
- to set up standards for quality and dimensions (in many cases these are generally considered to be minimum standards of quality, nor do the standards preclude manufacture of materials or components to other dimensions)
- to promote the adoption of British Standards.

The Institution is a non-profit making organisation funded principally by government grant, sale of publications and subscriptions from members. Any company, organisation or individual can be a member.

Individual British Standards are under constant review, and the pace of this accelerates with the introduction of new materials and techniques and, of recent years, the harmonisation with European Standards.

This constant updating has an important ramification for the designer of any building and other colleagues in the team of professionals running the contract to build it. When letting a contract to build, the team must agree a point in time past which no further amendments to Standards or new Standards will affect the contract. It is usual to state in the contract documents (in the Preambles) that reference to British Standards will include all amendments to the Standards promulgated before a date some time before the due date for receipt of tenders – usually at least

4 months. As the contract proceeds there will be amendments to some of the Standards but these cannot be introduced on the contract. The design team must stick with the amendments up to the date before tendering *or* they must negotiate with the contractor to have the new Standard introduced.

British Standard 187

Calcium silicate (sand-lime and flint-lime) bricks

This précis is intended only as an introduction to a particular British Standard to place particular information in the correct context within this text. The précis, therefore, does not include reference to the entire technical content of the Standard. Tables included in the précis are NOT the tables from the British Standard but may follow the same general pattern, including additional or excluding superfluous information as is thought appropriate for this text. At the appropriate stage in any course, students will be referred to the full Standard. Other readers should have recourse to their local public library or technical college/university library. Additional comment is given in italics.

Note that this Standard is no longer current but is included here as it is referred to in current editions of the Building Regulations.

This Standard is divided into two sections, **Specification** and **Appendices**, preceded by a foreword and listing of the 'Cooperating Organisations'.

The **Specification** includes: Introduction; Scope; References; Definitions; Materials; Form; Sizes; Information to be provided by the purchaser; Appearance; Classification and compressive strength; Marking; Manufacturer's certificate; Independent tests; Samples for tests.

The **Appendices** are taken up with a detailed description of the tests to which samples of bricks must be put in order to comply with the Standard. These form no part of this text but would be studied under materials science and replicated in the laboratory by students in a later part of their course.

Introduction. This stresses changes made since the previous Standard was written, in particular the necessity of using the strength of bricks taken when they are wet since this is the lower value. Also highlighted is the omission of a water absorption test since for these bricks it bears no relationship to durability nor is it an easy indicator of weather resistance as rain penetration is more likely through the mortar.

Scope emphasises that the Standard covers only calcium silicate bricks laid on mortar and to a coordinating size of $225 \times 112.5 \times 75$.

Definitions important to the student at this stage are similar to BS 3921: frog, solid brick, cavity, cellular brick, frogged brick.

Materials for the manufacture of these bricks include sand of siliceous gravel or crushed siliceous gravel or rock with a lesser proportion of lime. The mixture is pressed into shape and *autoclaved* (subjected to steam and pressure). Suitable pigments may be added.

Form or Type. The bricks may be solid or frogged. *See BS 3921 for a description.*

Sizes. Coordinating and work sizes similar to those for clay bricks, BS 3921. Note that sample size for checking the dimensions is only 10 bricks.

Information to be provided by the purchaser. The purchaser must state whether loadbearing or facing bricks are required. In addition a strength classification from Table 2 may be stated. If not stated then the lowest strength class for the type and kind of brick may be inferred by the supplier.

Appearance. Bricks must be free from visible cracks and noticeable balls of clay, loam or lime. Facing bricks shall be of the colour and texture required by the purchaser. Arrises of bricks should be relatively free of damage.

Classification and compressive strength. There are five classes of bricks given in Table 2, Class 7, 6, 5, 4 and 3, with compressive strengths ranging from 48.5 N/mm^2 down to 20.5 N/mm^2. A further class, Class 2 for Facing brick and common brick, has no minimum required compressive strength listed.

Marking. This refers to the data which must be supplied by the manufacturer with a consignment of bricks and includes:

- Manufacturer's identification
- The BS reference
- The type of brick, i.e. solid or frogged, and the work size
- The strength class as given in Table 2. This can be indicated by colour coding: 7 green, 6 blue, 5 yellow, 4 red, and 3 black. Class 2 has no colour code.

Manufacturer's Certificate. The purchaser can request a certificate of compliance with this Standard.

Independent tests. Again an option for the purchaser to agree with the supplier. Both the tests required and the laboratory carrying out the work must be mutually agreed.

Samples for test. Gives the number of bricks required for testing, and the method by which the sample is taken.

British Standard 493

Airbricks and gratings for wall ventilation

This précis is intended only as an introduction to a particular British Standard to place particular information in the correct context within this text. The précis, therefore, does not include reference to the entire technical content of the Standard. Tables included in the précis are NOT the tables from the British Standard but may follow the same general pattern including additional or excluding superfluous information as is thought appropriate for this text. At the appropriate stage in any course, students will be referred to the full Standard. Other readers should have recourse to their local public library or technical college/university library. Additional comment is given in italics.

The Standard comprises a **Specification** in eight parts including tables and figures, followed by two **annexes**. The eight parts are: Scope; References; Definitions; Materials; Manufacturer's certificate; Marking; Airbricks for external use (Class 1 units); Wall ventilators or gratings for internal use (Class 2 units)

Scope is more concerned with what is *not* covered, i.e. durability of air bricks or gratings (other than colour change in plastic materials), fire properties of thermoplastic materials and compressive strength of thermoplastic units.

References to other British Standards or other publications.

Definitions. An airbrick is defined as a unit which is built *into* a wall, i.e. it actually displaces a brick or block or a part of one of these. A wall ventilator or grating is defined as being either built into or fixed to internal walling, usually in conjunction with an airbrick. The wall hold of the brick or grating is defined as the area in contact with the wall.

Materials. The list of materials includes a note of the test(s) with which the materials must comply, these being from the British Standards for these materials. There is no need to refer to these other standards at this time. The list is:

☐ Concrete
☐ Clay, fireclay or terracotta
☐ Thermoplastics
☐ Copper or copper alloy
☐ Aluminium
☐ Plaster (internal use only)
☐ Steel, carbon steel sheet or plate, which must be galvanised or plastic coated or stainless steel.

Manufacturer's certificate. Must be produced on request by purchaser and to the effect that the product complies with this Standard in all respects.

Marking. Airbricks should be marked with 'BS 493 Class 1' and gratings with 'BS 493 Class 2'.

Airbricks for external use (class 1 units). This part covers the 'design' of the airbrick, i.e. the hole or slot made in the brick for the passage of air can be circular, square, rectangular or of a louvred design. Reference is made to Figures 1, 2 and 3. Ten coordinating and work sizes are given, three of which are designed for use with standard metric bricks (see BS 3921). These three sizes are as listed below.

Coordinating size	Work size	Minimum clear air space*
225 × 75	215 × 65 One brick long and one course high	1290 mm²
225 × 150	215 × 140 One brick long and two courses high	2580 mm²
225 × 225	215 × 215 One brick long and three courses high	4500 mm²

*Minimum clear air space refers to the aggregate of the area of all the holes or slots in the brick. It is not the area used when determining the number of air bricks required in a wall by the Building Regulations.

Other sizes listed can be used in blockwork or with modular brick walls.

Wall ventilators or gratings for internal use (class 2 units). This part repeats the reference to the design of the apertures as for 'airbricks'.

The **Annexes** cover the information to be supplied by a purchaser when ordering, and details of an impact test which airbricks and ventilator gratings must pass in order to comply with this Standard.

British Standard 743

Materials for damp-proof courses

This précis is intended only as an introduction to a particular British Standard to place particular information in the correct context within this text. The précis, therefore, does not include reference to the entire technical content of the Standard. Tables included in the précis are NOT the tables from the British Standard but may follow the same general pattern including additional or excluding superfluous information as is thought appropriate for this text. At the appropriate stage in any course, students will be referred to the full Standard. Other readers should have recourse to their local public library or technical college/university library. Additional comment is given in italics.

Note that this Standard is partly superseded but remains current and it is referred to in current editions of the Building Regulations.

This Standard comprises a brief **Specification** for a number of materials used for the manufacture of damp-proof courses. This specification is preceded by a foreword and a list of cooperating organisations. *Early editions of the standard did have appendices but these have been deleted in the current edition, tests etc. on materials being the subject of individual Standards. Reference to these Standards is not required for this text.*

The sections in the **Specification** are: Scope, Lead, Copper, Bitumen, Mastic Asphalt, Polyethylene, Slates, Bricks, Mortar, Designation for ordering.

Scope simply lists the materials already given above.

Lead *was traditionally described with reference to its mass/unit area but is now described with a reference code. For a damp-proof course, the lead should* weigh not less than 19.5 kg/m^2. This is equivalent to 1.80 mm thick and has the reference 'Code 4'. *Code 4 may seem a strange reference but was the lead industry's way round changing the thickness of their products when the building industry changed to the metric system of measurement Prior to that time lead was described, for example, as '4 pound lead', i.e. weighing 4 pounds per square foot; 4, 5 and 6 pound lead were common 'sizes' used by plumbers for roof work and in DPCs. After metrication, the industry could no longer refer to pound avoirdupois or feet and inches so '4 pound lead' became Code 4, '5 pound lead' became Code 5 and so on.* The reference to BS 1178 is for milled sheet lead for building purposes. There is no need to refer to that Standard at this time.

Copper. Generally an 'annealed' material is required. *When ductile metals are rolled out to form sheet or strip, the metal becomes brittle or 'work hardened'. This is the principle behind continuously bending a piece of wire to get it to break. To soften or 'anneal' the metal it is heated to a cherry red and allowed to cool slowly.* There is no need to refer to BS 2870 at this time.

Bitumen. A short but detailed specification of the technical requirements of a DPC material made by impregnating a 'base' material, e.g. hessian, with bitumen. There is no need to refer to the materials Standards at this time.

Mastic asphalt. Two types of mastic asphalt are given:

☐ Limestone aggregate mastic asphalt
☐ Natural rock asphalt aggregate.

There is no need to refer to the materials Standards at this time. *Sufficient to know that the material is supplied in large hardened lumps or 'cakes' of around 25 kg and melted in a 'pot', usually heated with LPG, and spread while molten. It hardens after a short time but still has a measure of plasticity depending on the ambient temperature. It is possible to get different grades of the material made from artificial bitumens or modified natural bitumens which will remain relatively stable in higher/lower temperatures. For DPC work it is spread in one coat with a thickness of 13 mm.*

Polyethylene. *A plastic material, thin and very tough.* There is no need to refer to the materials Standard at this time.

Slates used as a DPC must be at least 230 mm long and at least 4 mm thick. Two courses must be laid, the full width of the wall and to break bond. The mortar used must be 1:0–$^3/_4$:3, cement, lime, sand. There is no need to refer to the materials Standards at this time

Bricks used as a DPC must comply with BS 3921, clay damp-proof course bricks. Two courses must be laid, the full width of the wall and to break bond. The mortar used must be 1:0–$^3/_4$:3, cement/lime/sand. There is no need to refer to the materials Standards at this time.

Mortar. Reference is made here to the Standards for the materials in the mortar used for slate and brick DPCs.

Designation for ordering. When ordering, the following should be given:

☐ British Standard number
☐ Material to be used
☐ Width, length, weight or number as appropriate.

British Standard 1199 and British Standard 1200

Building sands from natural sources

This précis is intended only as an introduction to a particular British Standard to place particular information in the correct context within this text. The précis, therefore, does not include reference to the entire technical content of the Standard. Tables included in the précis are NOT the tables from the British Standard but may follow the same general pattern including additional or excluding superfluous information as is thought appropriate for this text. At the appropriate stage in any course, students will be referred to the full Standard. Other readers should have recourse to their local public library or technical college/university library. Additional comment is given in italics.

Note that this Standard, although superseded, is remaining current as it is referred to in current editions of the Building Regulations.

This Standard comprises **Specifications** for:

☐ Sands for external renderings and internal plastering with lime and Portland cement – BS 1199

☐ Sands for mortar for plain and reinforced brickwork, block walling and masonry – BS 1200.

The sections in both **Specifications** are: Introduction, Scope, References, Definitions, Sampling and Testing, Quality of sands, Grading, Supplier's Certificate and cost of tests, Additional information to be furnished by supplier, Table 1 – Sands for *as appropriate.*

BS 1199 – Sands for external renderings and internal plastering with lime and Portland cement

Introduction. This emphasises that the grading of sands is based on the process of washing, decantation and dry sieving; the determination of clay and silt content is now part of the grading process; two grading ranges of sand are given. Either grade can be used for rendering; however, where there is a choice, the coarser grade should be selected as this leads to lower water/cement ratios (*fine sands require higher water/cement ratios*) leading in turn to lower rates of shrinkage on drying (*and therefore less cracking/crazing on drying*).

Scope. This part of the Standard refers to the use of mixtures containing naturally occurring sands, crushed stone sands and crushed gravel sands with lime, lime and cement, lime with gypsum, and cement as binders.

References. A list of appropriate Standards is referred to at the end of the Standard relating to testing equipment and related materials.

Definitions. Definitions are given for sand, natural sand, crushed stone sand and crushed gravel sand. Sand is defined as any material which mainly passes a 5.00 mm BS test sieve.

Sampling and testing. BS 812 gives the details of how the sampling and testing of sands is to be carried out. The results are obtained by averaging the individual results from two tests of two different samples. *There is no need to refer to BS 812 at the present time.*

Quality of sands. Basically sands can be from a single source or a mixture of any two or all three. They must be free from deleterious material either in the mix or adhering to the sand particles. Deleterious materials include iron pyrites, salts (*particularly chlorides and sulphates),* coal or other organic impurities, mica, clay silt, shale or other laminated materials or flaky or elongated particles in form or quantity which could affect the strength, durability or appearance of the mortar or of any materials in contact with it.

Very important – The various sizes of particles in the sand must be evenly distributed throughout the mass of sand.

Grading. Two grades, A and B, are defined in Table 1, and sand should fall within the criteria given in the table when tested according to BS 812 – see Sampling and testing above. It is worth noting that the largest sieve size through which *all* particles for both grades must pass is 6.3 mm, and that about 2% is retained on the next size, which is the 5.00 mm sieve. *See Definitions above.*

At this stage in the student's progress there is no need for further detail on the gradings; however, an explanation of what grading is and what its effect is on renders etc. is appropriate. If one could imagine a loose material made from spheres which are all the same size – say 20 mm in diameter – and it was to be bound together with a powder suspended in water where the binder particles were measured in microns, then two possible scenarios arise. The water/binder paste will fill the voids between the large spheres or the paste will merely coat the spheres. In both cases the amount of binder used per unit of final material produced will be very high, much higher than if the material had fewer/smaller spaces between the spheres. Also, lots of spaces left will produce a material which is weak in compression and adhesion. Filling excessively large spaces while being strong in compression will result in a brittle material.

To obtain a reduction in space and an improvement in performance we must therefore use some smaller spheres to fill the spaces up. In fact it would help if the smaller pieces were not spheres at all but little cubes with concave faces which fitted between the spheres exactly – always assuming we could align the little cubes in one particular way every time. None of that is possible. We generally use naturally occurring or randomly produced materials as aggregates or 'filler' in mortars, renders (and concretes), so we must arrange for the reduction in free space by other means. This is done by producing a balance of the proportion of any one particle size in the aggregate used. The sizes and proportions are measured by determining the quantity of material passing through a set of sieves, each with apertures in a progressively smaller fixed sequence. If the quantities retained on each sieve fall within predefined limits, then the sand can be classed as Type A or B or simply 'not compliant'.

Old specifications and bills of quantities may refer to 'Zones', which was the terminology used in previous Standards.

Supplier's certificate and cost of tests. The supplier of the sand is responsible for providing a test certificate to the purchaser if requested to do so and is always responsible for ensuring that the supply conforms at all times with the appropriate parts of the Standard. The costs of tests requested by a purchaser are paid by him if the material does comply with the Standard and by the supplier if it does not comply.

Additional information to be furnished by the supplier. The following must be given by the supplier if requested by the purchaser: source of supply – county, parish, name of quarry or pit and for dredged material the precise location on the river or estuary; group classification according to BS 812; external characteristics – shape and surface texture of the particles according to BS 812; physical properties – relative density, bulk density in kg/m^3, water absorption, grading by sieve analysis according to BS 812.

BS 1200 – Sands for mortars, plain and reinforced brickwork, block walling and masonry

Introduction. See comment for BS 1199 above.

Scope. This part of the Standard refers to the use of mortars containing naturally occurring sands, crushed stone sands and crushed gravel sands with binders for building plain and reinforced brickwork and blockwork of clay or concrete units.

References. See comment for BS 1199 above.

Sampling and testing. See comment for BS 1199 above.

Quality of sands. See comment for BS 1199 above.

Grading. Two grades, S and G, are defined in Table 1 and sand should fall within the criteria given in the table when tested according to BS 812. It is worth noting that the largest sieve size through which *all* particles for both grades must pass is 6.3 mm and that about 2% is retained on the next size, which is the 5.00 mm sieve. *See Definitions above*. It should also be noted that the proportions retained on each sieve are very different from BS 1199, reflecting the differing needs with regard to mortars and renders or plasters. The set of sieves used is identical and the method of test is still BS 812.

Supplier's certificate and cost of tests. See comment for BS 1199 above.

Additional information to be furnished by the supplier. See comment for BS 1199 above.

British Standard 1282

Guide to the choice, use and application of wood preservatives

This précis is intended only as an introduction to a particular British Standard to place particular information in the correct context within this text. The précis, therefore, does not include reference to the entire technical content of the Standard. Tables included in the précis are NOT the tables from the British Standard but may follow the same general pattern including additional or excluding superfluous information as is thought appropriate for this text. At the appropriate stage in any course, students will be referred to the full Standard. Other readers should have recourse to their local public library or technical college/university library. Additional comment is given in italics.

Note that this Standard is no longer current but is included here as it is referred to in current editions of the Building Regulations.

This Standard comprises a **Guide** for a number of materials used for the preservation of timber and timber-based products and components. This Guide is preceded by a foreword and a list of cooperating organisations. *As this Standard is only a Guide and* not *a specification one cannot annotate drawings and details with the injunction that preservative treatment will 'comply' with BS 1282.*

The sections in the **Guide** are: Scope, References, General, Wood-destroying organisms in the United Kingdom, Wood-destroying organisms abroad, Classification and description of wood preservatives, Specification for wood preservatives, Preservatives without published specifications, Hazards to health, Preparation of timber for treatment, Methods of treatment, Selection of appropriate preservative treatments, Codes and Standards for the preservative treatment of specific timber products, Checking the standard of preservative treatment.

The Guide also includes two Tables:
Table 1 – Published preservative specifications
Table 2 – Recommended codes and standards for the preservative treatment of specific timber products.

Scope reiterates that the Standard gives guidance only on selection of appropriate methods of treatment, preservatives to use and how these are affected by conditions in which the timber is used, etc. It contains references to other Standards and codes of practice where additional information can be found on the treatment of materials such as plywood and components such as windows and window frames.

References. The Standard refers to many other Standards and codes of practice and these are listed on the rear cover of this Standard.

General. Destruction of wood takes place naturally through attack by fungi, insects, marine borers and some bacteria – all described in later parts of the Guide. Two ways to deal with such attack:

☐ Select a naturally resistant timber species – not always possible or always economical
☐ Treat a suitable, and generally cheaper, timber to resist the expected attack.

Two basic factors affect the efficacy of the treatment:

☐ the properties and the efficiency of the preservative
☐ the method by which it is applied.

Wood-destroying organisms in the United Kingdom. Fungi – the 'wet' and 'dry' rots require a moisture content in excess of around 22%. Some timbers such as oak, teak and western red cedar are very much more resistant to such attack even at high moisture contents. Others, such as beech and all sapwood of other species, are very susceptible to attack.

Insects – the most common throughout the UK is the common furniture beetle, *Anobium punctatum*, which attacks the sapwood of timbers. Death watch beetle, powder post beetle and the house longhorn beetle also occur but are not so widespread.

Marine borers are not of interest in terms of this text.

Wood-destroying organisms abroad are not of interest in terms of this text.

Classification and description of wood preservatives. There are three main types: the tar oil type, Type TO; the organic solvent type, Type OS; and the water-borne type, Type WB.

These all have advantages and disadvantages and the descriptions of each which follow attempt to highlight the more important features but of necessity are much condensed versions of the Standard, although sufficient for the student at this stage.

Type TO preservatives are obtained by the destructive distillation of coal and are in effect coal tar oils – creosote. Their chemistry is very complex. Some features are:

□ Suitable for exterior work
□ Their odour generally makes them unsuitable for interior use
□ Resistant to leaching
□ Degree of water 'repellency'
□ Non-corrosive to metals
□ Help protect iron and steel
□ Not readily flammable
□ Cannot normally be painted over
□ Can be glued provided the surface is free of creosote
□ Porous material in contact can 'wick' out the creosote
□ Odour can be picked up by other materials and foodstuffs in the vicinity – not necessarily in contact with the timber.
□ Timber which has been pressure treated can 'bleed' in hot weather.

Type OS preservatives come as a basic material and in two further variants. The basic material is a solution of one or more organic fungicides and sometimes an insecticide in an organic solvent, usually a petroleum distillate such as white spirit. The active ingredients are copper naphthenate, zinc naphthenate, pentachlorophenol, pentachlorophenyl laurate, orthophenyl phenol, tributylin oxide, chlorinated naphthalene, dieldrin and gamma-BHC. Some features are:

□ Suitable for exterior and interior use
□ Most are resistant to leaching but some loss may occur due to evaporation. Generally not corrosive to metals but copper compounds should not be used where there will be in contact with aluminium or aluminium alloys, plastics or rubber
□ Treatment leaves the timber with a clean appearance

□ Once the solvent has dried off timber can be glued and painted
□ No raising of the grain or swelling of timber so can be used on accurately machined timbers
□ Once the flammable solvent has evaporated, the timber is no more flammable than untreated timber
□ May taint nearby foodstuffs.

The first variant will additionally contain a wax or a resin as a water repellent, which can make the timber more difficult to glue or paint over. The second variant consists of the preservative with a water repellent and a pigment either to enhance or maintain the appearance of the timber.

Type WB preservatives are a solution of inorganic preservative salts and a 'fixing' agent, usually a dichromate all in water. The most common solution contains copper sulphate, sodium dichromate and arsenic pentoxide. Other mixtures are: copper sulphate and potassium dichromate; sodium fluoride, potassium dichromate, sodium arsenate and dinitrophenol; potassium and ammonium bifluorides; disodium octoborate on its own. They feature:

□ Copper/chrome and copper/chrome/arsenate solutions undergo chemical change in the timber and 'lock' themselves to the timber and so cannot leach out – they are suitable for exposed exterior use
□ Other solutions leach out to some degree or other and so can only be used where protected from moist conditions – painted, under cover, etc.
□ Generally not corrosive to metals but solutions containing copper compound should not be used where there will be contact with aluminium or aluminium alloys, plastics or rubber
□ The preserved timber is usually clean, odourless and can be glued and painted. No staining of adjacent materials
□ Non-inflammable
□ Treatment with water-borne preservatives means the timber has to be dried out and will swell and shrink, grain will be raised and distortion may take place.

Specification for wood preservatives. Reference is made in Table 1 to the relevant British Standards and documentation supplied by the British Wood Preserving Association and government departments.

Preservatives without published specifications. Refers to new formulations being marketed by the industry and the need to specify by 'performance'.

Hazards to health. Warns that many of the compounds used are hazardous to the health of animals and humans, being toxic to fungi, insects and bacteria. Reference to various sources for advice on handling and other precautions to be taken for the safe use and application of the preservatives and the timber and timber products.

Preparation of timber for treatment. All bark to be removed. Timber should be seasoned except where treated by the diffusion process. All cutting and machining, notching, boring, etc. should be done before treatment. Some timbers such as Douglas Fir must be 'incised' (slit-like cuts 20 mm deep in the direction of the grain) to assist penetration of the preservative.

Methods of treatment. These are described in broad detail in Appendix C and so only those given in the Standard are listed here:

☐ Pressure impregnation
☐ Low pressure impregnation
☐ Vacuum impregnation, double vacuum process

☐ Liquefied gas process
☐ Hot and cold open tank process
☐ Immersion, deluging, spraying and brushing
☐ Diffusion process
☐ Boucherie process*
☐ Gewecke processes*
☐ Oscillating pressure method*
☐ Boulton process*

*Rarely used in the UK

Selection of appropriate preservative treatments. Factors affecting the choice are:

☐ The effectiveness of the preservative
☐ Method of application
☐ How the timber absorbs the preservative
☐ The timber's natural durability
☐ Where the timber is to be used.

Codes and Standards for the preservative treatment of specific timber products. See Table 2.

Checking the standard of preservative treatment. The principal point for the ordinary user is the need to use a reputable supplier who is a member of the appropriate trade association, is licensed to provide the treatment of timber and can give a certificate guaranteeing that the treatment has been carried out properly. Larger organisations such as British Telecom can afford to employ inspectors to carry out such work.

British Standard 3921

Clay bricks

This précis is intended only as an introduction to a particular British Standard to place particular information in the correct context within this text. The précis, therefore, does not include reference to the entire technical content of the Standard. Tables included in the précis are NOT the tables from the British Standard but may follow the same general pattern including additional or excluding superfluous information as is thought appropriate for this text. At the appropriate stage in any course, students will be referred to the full Standard. Other readers should have recourse to their local public library or technical college/university library. Additional comment is given in italics.

Note that this Standard has been partly superseded but is included here as it is referred to in current editions of the Building Regulations.

This Standard is divided into two sections, **Specification** and **Appendices**, preceded by a foreword and a listing of the committees responsible for its production.

The **Specification** includes: Introduction; Scope; Definitions; Sizes; Dimensional deviations; Durability; Compressive strength; Water absorption; Sampling for tests; Marking.

The **Appendices** are taken up with a detailed description of the tests to which samples of bricks must be put in order to comply with the Standard. These form no part of this text but would be studied under materials science and replicated in a laboratory by students in a later part of their courses.

Introduction. This stresses the importance of three features of clay bricks and which form part of the testing process for compliance with the Standard. They are water absorption, durability (*frost resistance and soluble salt content*) and tolerance (*the need to have the bricks within a uniform overall size*).

Scope repeats the listing of requirements in the Introduction, adding that the Standard refers to bricks intended to be laid on a bed of mortar and only to a 'format' of 225 × 125 × 75 mm (*commonly referred to as a 'Standard Metric Brick'*).

Definitions are given for compressive strength, water absorption, coordinating size and work size. The first two refer to the testing procedures in the appendix. The reference to sizes is:

- **Coordinating size.** The size of a coordinating space allocated to a brick, including allowances for joints and tolerances.
- **Work size.** The size of a brick specified for its manufacture, to which its actual size should conform within specified permissible deviations.

Sizes. External sizes and void sizes are given; the former are given in Table 1, the coordinating size (the space for a brick and its mortar) being given as 225 × 112.5 × 75. The work size is given as 215 × 102.5 × 65, derived from the coordinating size by deducting a nominal 10 mm thickness of mortar.

Void sizes are given as a percentage of the external volume of the brick and divide bricks up into four '*types*':

- **Solid bricks** have no holes, cavities or depressions. (Cavities are described as hole closed at one end. A depression would be a 'frog')
- **Cellular bricks** shall not have holes but only cavities or frogs exceeding 20% of the external volume of the brick
- **Perforated bricks** have holes, the area of each not more than 10% of the area of the brick. The thickness of material left across the width of the brick must not be less than 30% in total. The volume of the holes shall not total more than 25% of the external volume of the brick
- **Frogged bricks** have only cavities or depressions on one or two beds, which should not exceed 20% of the external volume of the brick.

Dimensional deviations. The 'Limits of Size' are given in Table 2 for a sample of 24 bricks.

On the length, the difference between the maximum and minimum overall sizes is 150 mm. Distributed evenly over all 24 bricks, this approximates to 6 mm per brick. *Such a large difference cannot be taken up simply by varying the thickness of the mortar joints but must be allowed for when setting out the brickwork and the bonding for the structure.*

Durability of brickwork depends on two factors which arise from the use of any particular brick: resistance to frost and the soluble salts content. Frost resistance falls into three classes: Frost resistant (F), Moderately Frost Resistant (M) and Not Frost Resistant (O). Soluble salts content is classed as either Low (L) or Normal (N).

So, one could have a brick which is frost resistant with normal soluble salt content and this would be classified as FN. Similarly, a brick which had no frost resistance and had low soluble salt content would be classed as OL.

Compressive strengths for five *'kinds'* of bricks are listed in Table 4. The *'kinds'* are given as Engineering A, Engineering B, Damp-proof course 1, Damp-proof course 2 and 'All others'.

Water absorption as a percentage by mass is given in Table 4 for each of the five *'kinds'* of brick.

Taking samples of bricks for the test given in the appendix is described in detail. The number required for each test is given in Table 5.

Marking. This refers to the data which must be supplied by the manufacturer with a consignment of bricks and includes:

☐ Manufacturer's identification
☐ The BS reference
☐ The type of brick, i.e. solid, cellular, perforated, frogged
☐ The name of the brick, e.g. sand faced buff.

British Standard 4449

Carbon steel rods and bars for the reinforcement of concrete

This précis is intended only as an introduction to a particular British Standard to place particular information in the correct context within this text. The précis, therefore, does not include reference to the entire technical content of the Standard. Tables included in the précis are NOT the tables from the British Standard but may follow the same general pattern including additional or excluding superfluous information as is thought appropriate for this text. At the appropriate stage in any course, students will be referred to the full Standard. Other readers should have recourse to their local public library or technical college/university library. Additional comment is given in italics.

The standard has two main parts, a **Specification** and **Appendices**, both preceded by a short foreword and a list of the responsible Committees.

Specification. *It is important for the student to realise that the Standard covers bars which can be purely circular in cross-section or which can be 'deformed', i.e. the general cross-section is (1) circular with projections formed on the surface of the bar or (2) square section bars which are twisted and the corners are left plain or chamfered, the purpose of 1 and 2 being to improve the 'grip' in the concrete. The steel used for plain round bar is 250 Grade and for deformed bar is 460 Grade, this being termed high yield steel. These grade numbers refer to the tensile strength of the steel. Grade 460 for example has a tensile strength of 460 N/mm^2. Deformed bars can be produced by either hot rolling or cold working.*

A number of tables are included in the Specification. The most useful of these to the student at this early stage is Table 2, Cross-sectional area and mass of bars of nominal diameter 6, 8, 10, 12, 16, 20, 25, 32, 40 and 50 mm. Note that the range 8 to 40 mm are the 'preferred sizes' and are those readily available from the manufacturers. See also 'Sizes' below.

Also included are sections on Scope; Definitions; Sizes; Cross-sectional area and mass; Length; Steel-making process; Chemical composition; Bond classification of deformed bars; Routine inspection and testing; Mechanical properties; Fatigue properties of deformed bar;

Retests; Verification of characteristic strength; Marking.

Scope. Much of the important information from scope has been explained above.

Sizes. Table 1 gives the preferred sizes for bars of 250 Grade steel in the range 8 to 16 mm and for 460 Grade steel in the range 8 to 40 mm. Tolerances are also given.

Cross-sectional area and mass. Reference has already been made to Table 2 included in this section.

Length. This section sets out tolerances on cut lengths of bar.

Steel-making process. Refining molten iron either in a top-blown basic oxygen converter or by melting in a basic-lined electric arc furnace.

Chemical composition. Well beyond the scope of this text.

Bond classification of deformed bars. Generally square twisted bars are classified as Type 1 and bars with proturbences on the surface as Type 2.

Marking. Bars should carry 'rolled-on' legible marks no further apart than 1.50 m to indicate the origin of manufacture, i.e. a maker's mark.

The remaining sections in the specification are taken up with the provision of a number of tests, and refer to the appendices, where the tests are described in detail.

British Standard 4471

Sizes of sawn and processed softwood

This précis is intended only as an introduction to a particular British Standard to place particular information in the correct context within this text. The précis, therefore, does not include reference to the entire technical content of the Standard. Tables included in the précis are NOT the tables from the British Standard but may follow the same general pattern including additional or excluding superfluous information as is thought appropriate for this text. At the appropriate stage in any course, students will be referred to the full Standard. Other readers should have recourse to their local public library or technical college/university library. Additional comment is given in italics.

Note that this Standard is no longer current but is included here as it is referred to in current editions of the Building Regulations.

This Standard comprises a brief **Specification** and an **Appendix** dealing with 'Surfaced softwoods of North American Origin' (*i.e. from Canada and the USA*). This specification is preceded by a foreword and a list of cooperating organisations.

The **Specification** comprises sections on Scope; Definitions; Moisture Content; Sizes and maximum permitted deviations; Maximum permitted reductions from basic sizes by planing.

Scope. It is important to note that this Standard does not apply to softwood trim, softwood flooring and softwood for structural purposes. *This does not mean that timber complying with this standard cannot be used for trim (See Table C below) or structural purposes, and flooring board could be, but it is very unlikely, machined from it. The softwood trims given in BS 1186 Part III are of limited use. They do cover trims with simple splays and pencil rounds but do not include trims with torus or ogee mouldings common in better class work. These would be machined from softwood covered in this Standard. If the softwood specified here was to be used for structural purposes it is likely that ordinary 'rules of thumb' would be applied, whereas if the structure was the subject of detailed design, then timber complying with BS EN 336 would be appropriate.*

Definitions. Students should familiarise themselves with the definitions for *parcel, planing, moisture content, regularising*. Particular note should be made that moisture content is calculated as the mass of water against *dry* mass of timber.

Moisture content. Measurement is to be carried out using an electric moisture meter. *This instrument consists of a pair of metal probes insulated along their length, leaving only the sharpened tips bare. The probes are driven into the timber and the electrical resistance of the timber between the ends of the probes is measured. The resistance is taken to be a function of the moisture contained in the timber.*

Sizes and maximum permitted deviations. Basic cross-sectional dimensions are given in Table A and basic lengths in Table B, both given below. These tables include annotation on the current economic effect of choosing a particular size and length of softwood.

The maximum allowable deviations depend on which size is being measured: in cross-section sizes not exceeding 100 mm −1 mm, +3mm; and over 100 mm, −2 mm +6 mm. And on length, there is no limit to oversize but undersize is not permitted. *Timber is frequently sawn along its length – 'rip sawing or ripping'– to give smaller cross-sections. This practice is called* **re-sawing**. *Obviously the sawdust removed means a reduction in the sizes of the pieces obtained. The 're-sawing' allowance on each piece obtained is a maximum reduction of 2 mm. For example, a 100 × 25 board is reduced in width by re-sawing. We cannot obtain two pieces 50 × 25 because of the material lost in the saw cut but each piece must not be less than 48 × 25.* Actual sizes of timber vary with moisture content, *particularly the cross-sectional dimensions.* All sizes are taken to be at 20% MC. The allowance to be made at different MCs is given in the Standard.

Maximum permitted reductions from basic sizes by planing. *By planing is meant the* *machining of the opposite faces, generally with rotary knives, to bring the surfaces to a smooth, even finish and the timbers to the same overall dimensions. The surfaces will only require sanding before a decorative finish is applied – paint, varnish, stain, etc.*

This reduction in overall size is controlled within the limits set out in Table C and depends on the end use of the timber and the basic sawn dimensions. The reductions vary from 4 to 13 mm and are inclusive of any 're-sawing'

Table A Common cross-sections with availability and relative cost.

	Width								
Thickness	75	100	125	150	175	200	225	250	300
16	R 40	R 48	R 60	S 78					
19	R 40	R 48	R 60	G 78					
22	R 48	G 54	R 70	G 82	S 101	S 120	S 139		
25	S 48	S 63	R 85	R 99	S 120	S 140	S 160	S 190	S 220
32	S 62	S 87	R 110	R 140	S 170	S 200	S 220	S 260	S 300
36	S 62	S 87	R 110	R 140					
38	S 62	G 87	R 110	R 140	S 170	S 200	S 220	S 260	S 300
44	S 81	S 106	S 132	S 159	S 185	S 212	S 238	S 270	S 340
47	G 81	G 106	G 132	G 159	G 185	G 212	G 238	G 270	S 340
50	S 106	S 125	S 150	S 175	S 200	S 225	S 250	S 300	S 350
63		S 125	S 150	S 175	S 200	S 225	S 250	S 300	S 350
75		G 180	S 230	G 280	S 330	G 380	G 430	S 500	S 580
100		S 250		G 400		G 580	G 690	S 750	S 830
150				S		S			S
200						S			
250								S	
300									S

R – Re-sawn as required
G – Generally held in stock
S – Seldom held in stock, ordered if necessary
123 – Cost per metre in pence at first quarter 2004, local delivery free, no price shown – subject to quotation for best cost at the time

Table B Common lengths showing availability and effect on cost.

1.80	2.10	3.00	4.20	5.10	6.00	7.20
W DL	W DL	G N	G N	G N	G N	G DL
	2.40	3.30	4.50	5.40	6.30	
	W DL	G N	G N	G N	G DL	
	2.70	3.60	4.80	5.70	6.60	
	W DL	G N	G N	G N	G DL	
		3.90			6.90	
		G N			G DL	

W – Sawn from longer lengths at an additional charge
G – Generally held in stock
N – Normal cost
DL – More expensive because of length

Table C Maximum permitted reductions from basic sawn sizes to finished sizes by planing two opposing faces (mm).

Applications	Basic sizes			
	15 up to and including 35	Over 35 and including 100	Over 100 and including 150	Over 150
Matching and interlocking boards	4	4	6	6
Wood trim other than specified in BS 1186 Part 3	5	7	7	9
Joinery and cabinet work	7	9	11	13

allowance described above. The variation in size after planing shall be −0 mm, +1 mm.

Note that this is the overall reduction in the cross-sectional dimension, e.g. a 50 × 50 sawn timber is planed for wood trim and becomes 46 × 46 overall.

Match and interlocking boards generally have some form of joint such as a tongue and groove machined on the edge. The allowances above do not include any reduction in the dimensions caused by the machining of the joint.

British Standard 4483

Welded steel fabric reinforcement

This précis is intended only as an introduction to a particular British Standard to place particular information in the correct context within this text. The précis, therefore, does not include reference to the entire technical content of the Standard. Tables included in the précis are NOT the tables from the British Standard but may follow the same general pattern including additional or excluding superfluous information as is thought appropriate for this text. At the appropriate stage in any course, students will be referred to the full Standard. Other readers should have recourse to their local public library or technical college/university library. Additional comment is given in italics.

This Standard is divided into two sections, **Specification** and **Appendices**, both preceded by a short foreword and list of Committees responsible for the Standard.

The **Specification** is further divided into a number of paragraphs: Scope; Definitions; Information to be supplied by the purchaser; Dimensions; Cross-sectional areas and mass; Process of manufacture; Quality; Fabric classification; Tolerances on mass, dimensions and pitch; Bond classification of fabric; Testing and manufacturer's certificate; Mechanical tests; Retest.

The **Appendices** detail the mechanical tests and the notation and classification of the fabric.

Scope. The Standard covers fabric or mesh made from plain or deformed wire, rod or bar complying with BS 4449, BS 4461 or BS 4482 for the reinforcement of concrete.

Definitions are given for all terms used in describing fabric reinforcement. All terms are used in accordance with their common everyday meaning and so will not be repeated in this précis.

Information to be supplied by the purchaser.

☐ The number of this Standard
☐ The number of the Standard referring to the wire quality (see Scope)
☐ Whether the wire is plain or deformed
☐ The wire sizes and mesh arrangement required
☐ The dimensions of each sheet required
☐ The quantity of each sheet required.

Dimensions. The dimensions of the individual wires, rods or bars shall comply with the appropriate standards listed above in Scope. While it is possible to have made up any combination of wire in any mesh size and configuration, it is more usual for the designer to specify from the preferred range of designated mesh types and sheet size listed in Table 1, from which the following information has been extracted. The stock sheet size is 4800 long × 2400 wide giving an area of 11.52 m^2. The nominal size of the wires ranges from 2.5 mm to 12 mm diameter.

Cross-sectional areas and mass. The mesh pitch range is 100 × 100, 100 × 200, 200 × 200 and 100 × 400. Each combination of mesh pitch and wire size given in Table 1 has a unique reference. Square meshes are designated A followed by a number(s); Structural meshes (100 × 200 pitch) are designated B followed by a number(s); Long meshes (100 × 400) are designated C followed by a number(s); and Wrapping meshes (200 × 200 and 100 × 100 pitch) are designated D followed by a number(s).

The cross-sectional area of the wires in each direction is given as the mass in kilograms per square metre. *Both of these latter pieces of information are used in detail reinforcing calculations and in costing by the builder or quantity surveyor.*

Process of manufacture. The fabric is machine-made under factory controlled conditions, the wires being electrical resistance welded at every crossing.

Quality. Wire, rod or bar must be of Grade 460 steel except for wrapping mesh, which may be of Grade 250. *These grade numbers refer to the tensile strength of the steel. Grade 460 for example has a tensile strength of 46 N/mm^2.*

Wires etc. may be butt welded in their length and broken welds not exceeding 4% of the total in any sheet are permissible; however, these broken welds should not exceed half the number of welds along the length of any one wire.

Fabric classification. If the designer does not wish to use one of the meshes listed in Table 1, this paragraph refers to the correct way to designate the required fabric to the manufacturer. *Appendix B explains how the designer should specify a specially designed fabric and also how to draw up a schedule of fabric reinforcement for concrete members. An understanding of the latter is important later in QS courses when measuring reinforcement for bills of quantities and for costing purposes.*

Tolerances on mass, dimension and pitch refer to the maximum permissible deviation from the mass per square metre of fabric, the maximum permissible deviation on the length of the wires and the maximum permissible deviation in the size of the spacing of the wires.

Bond classification of fabric refers to the weld quality of the intersections of the wires. *Really only of interest to designers of reinforced concrete structures.*

Testing, Mechanical test and Retests all refer to the testing required by a fabric in order to comply with the requirements of the Standard, the tests being fully described in the appendices.

British Standard 6398

Bitumen damp-proof courses for masonry

This précis is intended only as an introduction to a particular British Standard to place particular information in the correct context within this text. The précis, therefore, does not include reference to the entire technical content of the Standard. Tables included in the précis are NOT the tables from the British Standard but may follow the same general pattern including additional or excluding superfluous information as is thought appropriate for this text. At the appropriate stage in any course, students will be referred to the full Standard. Other readers should have recourse to their local public library or technical college/university library. Additional comment is given in italics.

This Standard comprises a **Specification** for damp-proof course material made from woven or felted, fibrous material impregnated with bitumen. This specification is preceded by a foreword and a list of cooperating organisations. Appendix A gives details of tests and testing procedures. Appendix B includes a table listing the recommended uses for various types and classes of DPC; Appendix C covers high bond strength bituminous DPCs, which are beyond the scope of this text. This standard supersedes the reference to bituminous DPCs in BS 743.

The sections in the **Specification** are: Scope, Definitions, Classification, Base materials, Bituminous materials and fillers, Assembly of damp-proof course, Marking and packaging.

Scope gives the all important information that the classification of these DPCs is by 'base material'.

Definitions. Only one is given: nominal mass per unit area, defined as 'A numerical designation of the mass per unit area, which is a convenient round number approximately equal to the actual mass per unit area expressed in kg/m^2'.

Classification. Classification is given in columns 1 and 2 of Table 1. There are six classes, A to F. Classes A to C are hessian-based, fibre-based and asbestos fibre-based respectively. Classes D to F are similar to these but with the inclusion of a layer of lead.

Base materials. The hessian base is to be of a single layer of woven jute fibre. The fibre base is to be of one or more absorbent sheets of mixed, felted animal and vegetable fibre. The asbestos base is to be of an absorbent sheet containing not less than 80% asbestos fibre. *Presumably the remaining fibre can be of animal and/or vegetable origin.*

Bituminous materials and fillers. Reference is made to minimum quantities of bitumen which should be used to saturate the base material, and how this is to be tested and measured.

Two types of bituminous material are used: one for saturating the base and the other for coating both sides of the saturated base. The second type of bitumen is mixed with a very finely divided mineral – talc, mica, etc. Once the coating is applied it must be coated with a layer of mineral dust to prevent the material from sticking to itself when rolled up for shipment. This material need not be so finely divided as the filler material and can be made from sand, mica, slate, etc.

Assembly of damp-proof course. 'The base shall be impregnated completely with saturating material. Any surplus saturant shall be removed, after which the coating material shall be applied. When a lead sheet is included, this shall be laminated with the base and the two sheets shall be covered on both sides with coating material'.

Marking and packaging. The material is packed in rolls of at least 8 m length, each roll marked with the BS number and the classification letter A to F.

Classification of bitumen damp-proof courses. Table 2 of Appendix B gives a table of situations for which bitumen DPCs are both suitable and unsuitable. In the context of this text, all classes are suitable for use where the compressive load

Class	Description	Mass per unit area of assembled DPC (kg/m²)
A	Hessian base	3.8
B	Fibre base	3.3
C	Asbestos base	3.8
D	Hessian base laminated with lead	4.4
E	Fibre base laminated with lead	4.4
F	Asbestos base laminated with lead	4.9

is in the range 0.10 to 0.50 N/mm^2. This means buildings up to four storeys in height. All classes are suitable for water movement in any direction – up, down and horizontally. Where high shear or flexural stresses are involved, these are not suitable and reference should be made to the more stringent requirements for DPC laid down in Appendix C.

British Standard 6515

Polyethylene damp-proof courses for masonry

This précis is intended only as an introduction to a particular British Standard to place particular information in the correct context within this text. The précis, therefore, does not include reference to the entire technical content of the Standard. Tables included in the précis are NOT the tables from the British Standard but may follow the same general pattern including additional or excluding superfluous information as is thought appropriate for this text. At the appropriate stage in any course, students will be referred to the full Standard. Other readers should have recourse to their local public library or technical college/university library. Additional comment is given in italics.

This Standard comprises a brief **Specification** of the material used for the manufacture of polyethylene damp-proof courses. This specification is preceded by a foreword and a list of cooperating organisations. Appendices describe a variety of tests which are beyond the scope of this text. However, Appendix D includes a table of uses for polyethylene DPC and data from that table are summarised at the end of this précis.

The sections in the **Specification** are: Scope, Definitions, Composition, Thickness, Finish and impermeability, Marking and packaging.

Scope lists the above subsections of the specification.

Definitions refers the reader to BS 6100, section 1.0.

Composition gives a technical description of the material forming the DPC. *It is interesting to note that to comply with the Standard, the DPC should contain not less than 2% by mass of carbon black. This is mixed into the plastic as a finely divided powder, and so the DPC is black in colour. This implies that blue or green polyethylene DPCs do not comply* with this standard, although many rolls of such material are sold and used for that purpose.

Thickness. 'Nine specimens . . . shall have a single layer of thickness not less than 0.46 mm.'

Finish and impermeability. The sheet shall be free from air bubbles and with no visible pinholes. The test for the latter is outlined in Appendix C and consists of viewing the sheet against a strong light. *If the sheet is coloured black, pinholes would be readily seen in that test.*

Marking and packaging. The material is packed in rolls of at least 8 m length, each roll marked with the BS number and date.

Recommended uses for polyethylene damp-proof courses. Table 1 of Appendix D gives a table of situations for which these DPCs are both suitable and unsuitable. For the purposes of the present text, they are suitable for use with any compressive load but not if there is any lateral, shear or flexural load or stress. They are suitable for water movement upwards and horizontally, above ground level, but not for water moving downwards such as at parapets and chimneys or in cavity trays.

Index

blocks, 22
    autoclaved aerated concrete, 23
    concrete, 22
      dense and lightweight, 23
    dimensions of standard metric block, 23
    materials, 22
blockwork substructure, 71
bolts, 333
    bulldog timber connectors, 333
    coach bolts, 333
    hexagonal-headed, 333
    self-locking nuts, 332
bonding, of bricks to form walls
    common bond, 9
    english bond, 10
    flemish bond, 10
    garden wall bond, 13
    quetta bond, 13
    rattrap bond, 14
    scotch bond, 13
    sectional bond, 6, 11, 12, 22
    stretcher bond, 9
brick, 2
bricks, by function, 4
brick and blockwork in superstructure, 81
brick materials, 5
brick sizes, 2
    coordinating sizes, 3
    nominal sizing, 3
bricks and blocks standards and dimensions, 2
British Standards, 356–79
    air bricks and gratings for wall ventilation,
      360
    bituminous damp proof courses for masonry,
      377
    calcium silicate bricks, 358
    carbon steel rods and bars for reinforcing
      concrete, 371
    clay bricks, 369
    general, 356
    materials for damp proof courses, 362
    polyethylene damp proof courses for masonry,
      379
    sands for mortars, plasters and renders, 364
    sizes of sawn and processed timber, 372

    welded steel fabric reinforcement, 375
    wood preservatives, 366
building masonry walls from foundation up to
    DPC level, 57

cavity fixings, 337, 338
ceiling finishes, 124
central heating, 252
    emitters, 255
    piping for central heating systems, 253
    pressurised system, 254
    TVR, 255
    underfloor heating, 254
chemical anchors, 339
coach screws, *see* screw nails
cold bridging at wall openings, 132
common and facing brickwork
    bucket handle or grooved jointing or pointing,
      21
    facing brickwork, 18
    flat jointing or pointing, 20
    flush jointing or pointing, 20
    keyed jointing or pointing, 21
    pointing and jointing, 19
    recessed jointing or pointing, 20
    reverse weather jointing or pointing, 21
    tuck pointing, 20
concrete, 316
    general,
      formwork, 316
      in-situ concrete, 316
      moulds, 316
      plain concrete, 316
      precast concrete, 316
      reinforced concrete, 316
      shuttering, 316
    materials and mixes
      aggregate
        coarse, 316
        fine, 316
        grading, 316
      concrete proportions, 317
      designed mixes, 318
      no-fines, 319
      prescribed mixes, ordinary, 318

prescribed mixes, special, 318
   water, 317
   water/cement ratio, 317
   weigh batchers, 320
  reinforcement, 318
   bent ends, 319
   cold rolled, 318
   deformed, 318
   fabric, 318
   grip, 319
   high yield steel, 318
   hooked ends, 319
   hot rolled, 318
   mild steel, 318
   rectangular mesh, 318
   round bar, 318
   square mesh, 318
   square twisted, 318
concrete screws, 336
conservation of energy, 355
contact adhesives, 340
convention on thicknesses of walls, 8
cutting bricks
  bevelled closer, 8
  cutting bricks, 6
  half batt, 7
  king closer, 8
  quarter batt, 7
  queen closer, 8
  three-quarter batt, 8
  whole brick dimensions, 6

damp proof courses and membranes, 344–7
  materials, 345
dimensional stability of walls, 79
dooks, 335
door hanging, 190
door ironmongery, 196
door types,
  15 pane doors, 190
  bound lining doors, 185
  fire resistant doors, 193
  flush panel doors, 187
  glazing, 196
  ledged and braced doors, 185
  panelled doors, 188
  pressed panel doors, 189
  smoke seals, 195
doors and windows,
  functions of doors and windows, 182
drawing symbols and conventions, 353–4
dumpy level, 285
durability of bricks, 3

earthwork support, 325
electrical work, 266
  accessories, 277
  circuits, radial, 271
  circuits, ring, 271
  flexible cord, 277
  fuses, 269
  IEE regulations, 266
  lamps, 277
  miniature circuit breakers, 270
  more on protective devices, 275
  phases, red, yellow and blue, 267
  power generation, 266
  residual current circuit breaker,
   269
  sub-circuits, 270
  sub-mains and consumer control units,
   268
  wiring diagrams, 276
  wiring installation types, 267
  work stages, 272
   earth bonding, 273
   electrician's roughing, 272
   final fix, 275
   testing and certification, 275
excavation, general, 34
excavation, detail, 53
  excavation for and placing concrete
   foundations, 53
  marking out the excavation, 53
  poling boards, 325
  shoring, strutting & waling, 325
  steel sheet piling, 326
  steel trench support, 327
  strutting, 325
  timbering, 325
  timber sheet piling, 326
  walings, 325

fastenings and fixings, 328–40
  see also nails, screw nails and bolts
  cavity fixings, 337, 338
  chemical anchors, 339
  concrete screws, 336
  contact adhesives, 340
  dooks, 335
  frame fixings, 338
  gravity toggle fasteners, 337
  plugs and plugging, 335
  rawlbolts, 336
  rawlnuts, 338
  rawlplugs, 335
  spring toggle fasteners, 337

flat roofs in timber, 162
   insulation and vapour control layers, 164
   options for structure, 162
   voids and ventilation, 164
flitched beams, 108
floor boarding
   man-made board, laying, 313
   man-made board, material, 312
   timber, 312
   tongue and groove forms, 314
floor finishes, 124
foundations, 37
   see also substructures
frame fixings, 338
frog up or frog down, 17
   economics, 17

general principles of bonding, 21
gravity toggle fasteners, 337
ground floor construction, 59
   detail drawings, 59
gypsum wall board, 341–3
   Ames taping, 343
   collated screws, 342
   dry lining screw bits, 342
   dry lining screws, 342
   paper faces, 341
   plasterboard nails, 342
   sheet sizes, 341
   square edged, 341
   taper edged, 341
   taping and jointing, 343

half brick thick walls, 16
honeycomb brickwork, 16
hung floors, 64
   concrete floors, 67
   timber floor alternatives, 66
   timber floors, 64

insulation of external walls, 84
internal partitions, 96
   acoustic partition, 99
   foundations, 97, 98, 99
   stud partition details, 98
intersections of masonry walls, 9

levelling, 285
   booking readings, 288
   calculating levels from readings, 288
   collimation, 287
   dumpy level, 286

   and staff, 287
   use of, 287
   EPDM (electronic position and distance
      measuring) equipment, 290
   setting to a level, 289
   stadia wire, 288
loadbearing and non-loadbearing internal
      partitions, 96

maps and plans, 279
   1:1250, 1:2500 maps, 279
   Ordnance Survey, 279
   plans
      1:500, 279
      details, 284
      house, 282
      scales, 280
      sections, 283
mortar
   additives, 30
   cement, 25
   'fat' mixes, 28
   general rules for selection of mortar,
      29
   joints, 3
   lime, 26
   mixing in additives, 30
   mixing mortar, 31
   sand, 27
   water, 27
   which mortar mix?, 27
   whys and wherefores of mortar, 25

nails
   boat, 329
   collated, 330
   copper tacks, 330
   cut, 328
   Hilti gun, 335
   improved, 330
   masonry, 334
   materials, 328
   shot fired, 335
   star dowels, 334
   wire, 328

openings in upper floors, 113
   for flues, 114
   insulation of flue, 115
   integrity of flue material, 115
   isolation of flue, 115
   for pipes, 113
   for stairs, 116

pipe sleeves, 126
plasterboard, *see* gypsum wall board
plugs and plugging, 335
plumbing, 233
  air locking, 263
  appliances, 255
    baths, 258
    kitchen sinks, 256
    showers, 258
    taps, 258
    WCs, 256
    wash hand basins, 257
  capillary fittings, 236
  compression fittings, 234
  corrosion, 263
  equipment, 247
    cold water cistern, 248
    feed and expansion cistern, 251
    hot water cylinder, 248
  first fixings, 264
  fusion joints, 238
  hot and cold water services, 243
  insulation, 262
  joints to accessories, 238
  joints to appliances, 238
  overflows, 246
  pipe fittings, 234
  pipework, 233
  push-fit fittings, 236
  range of pipe fittings, 239
  services, 243
  soil and ventilation stacks, 246
  solvent weld fittings, 237
  valves and cocks, 241
    schematics, 242
  waste disposal piping
    air admittance valve, 261
    systems, 259
    traps, 260
  water hammer, 263
  water supply from the main, 246

quoins, 9
  alternative definition, 16

rawlbolts, 336
rawlnuts, 338
rawlplugs, 335
reveals, 9
risbond joints, 14, 15

roof tile and slate
  abutment, 178–81
  bituminous shingles, 176
  eaves, 178–81
  interlocking tiles, 176
  materials, 171
  pantiles, 177
  plain tiles, 175
  ridge, 178–81
  slates, 175
  Spanish and Roman tiles, 177
  timber shingles, 176
  verge, 178–81
roof, 148
  Belfast truss, 153
  box beam purlin, 152
  bracing, 160
  classifications, 148
  forms, 149
  gang nail plate, 155
  hammer beam truss, 153
  insulation, 169
  king post truss, 152
  mansard truss, 152
  prefabrication, 149
  purlined roof, 151
  queen post truss, 152
  roof (gable) ladders, 158
  terminology, 149
  TRADA truss, 153, 154
  traditional, 167
  trussed, 150
  trussed rafter, 153
  truss shapes, 157
  trussed purlin, 152
  verges meet eaves, 159

screeds
  bonding agents, 323
  granolithic, 322
  heated screeds, 324
  laitance, 322
  laying
    in bays, 323
    bonded screeds 1, 322
    bonded screeds 2, 323
    joint treatment in bays, 323
    monolithic screeds, 322
    unbonded screeds, 323
  materials
    cement/granite dust mixes, 322
    cement/sand mixes, 322
    cement/whinstone dust mixes, 322

type
    bonded, 322
    monolithic, 322
    unbonded, 322
screw nails
    clearance hole, 331
    coach screws, 332
    collated screws, 331
    counterbores, 332
    countersinking, 331
    drywall screws, 342
    materials and finishes, 331
    pelleting over, 332
    Phillips heads, 330
    pilot hole, 331
    plug cutter, 332
    Pozidriv heads, 330
    slotted head, 330
    traditional wood screws, 330
    twin threaded screws, 331
    wood screws, 330
scuntions, 9
setting out, 49
    equipment required for basic setting out, 49
    procedure, 50
    the site plan, 49
    where do we put the building?, 49
solid concrete floors
    single and double layer concrete floors with hollow masonry wall, 62
spring toggle fasteners, 337
stairs, 221
    balusters, 223
    balustrade, 223
    flight, 221
    handrail, 223
    joining steps to stringer, 225
    kite steps, 227
    landings, 222
    measurements, 224
    newel, 224
    rise and going, 224
    rough carriage, 226
    staircase, 221
    steps, 222
    strings, 221
    winders, 227
subsoils, 36
    general categorisation of subsoils and their loadbearing capacities, 37
substructures
    bearing strata, 48
    critical levels and depths, 46
    depths and levels, 48
    failure of wide, thin, strip foundations, 44
    finished ground level, 47
    foundation width and thickness, 41
    level, 46
    mass of buildings, 39
    mass, load and bearing capacity, 40
    principal considerations, 38
    reinforced concrete foundations, 44
    simple foundation calculations, 39
    step in foundation, 49
    trench fill foundations, 45
    ventilators, 349–51

terminology
    of bricks, 2
    of roofs, 149
    of wall openings, 129
testing of bricks, 5
timber, 291
    batten, 291
    baulk, 291
    board, 291
    conversion, 291
    cross sections available, 292
    deal, 291
    defects
        natural, 295
        seasoning, 295
    four sider machine, 293
    lengths available, 292
    moisture content, 294
    natural fire resistance, 298
    North American timber, 294
    planed, dressed or wrot timber, 292
    planing and thicknessing machine, 293
    plank, 291
    preservation, 296
    preservatives
        fire resistance, 298
        organic solvent types, 296
        tar oil, 296
        treatment methods, 297
        water borne, 296
    quarter sawn, 291
    regularised, 291, 292
    sawn, 291
    scantling, 291
    seasoning, 294
        distortion, 295
        kiln, 295
        natural, 295
    slab sawn, 291

tolerances
   regularised timber, 294
   sawn timber, 292
   wrot timber, 294
  wane, wany edge, 291
timber casement windows, 205
  depth and height of glazing rebates, 206
  draught stripping materials, 206
  hanging the casements, 207
  joining the frame and casement members, 209
  timber for casement windows, 206
timber frame construction, 88
  balloon frame construction, 90
  breather membranes, 95
  cavity ventilation in masonry skin, 93
  DPC over fire stopping, 95
  fire stopping in cavity, 94, 95
  hold down straps to foundations, 93
  masonry skin materials, 93
  metal strapping tying at upper floors, 94, 95
  modem timber frame construction, 91
  platform frame construction, 89
  spread of fire in cavity, 94
  traditional timber frame, 88
  wall ties for masonry skins, 94
timber sash and case windows, 211
  the case, 212
  the sashes and case together, 214
timber vertical sliding sash windows, 214
timber windows
  for ordinary glazing work, 218
  glazing, 218
tipping, 17
topsoil, 35
type of bricks by shape, 4

upper floors, 103
  alternative materials for joisting, 118
  brandering for ceiling finish, 110
  ceiling finishes, 124
  cheek pieces for ceiling finish, 111
  floor finishes, 124
  herring bone strutting, 111, 112
  joist support
   in masonry walls, 105
   in timber frame walls, 106
   on beams, 108
   on flitched beams, 108
  linear and point loadings, 112
  modem sound and fire proofing, 121
  pugging, 120–22
  solid strutting, 111, 112
  sound proofing, 120

steel herring bone strutting, 111, 112
support of masonry walls from floors,
  123
upper floor joist spacing, 110
upper floor joists, 103
U-value, 355

ventilators in substructures,
  calculating number required, 351
  clay and plastic, external, 349
  clay liners, 349
  dead spots, 351
  positioning, 351
  telescopic liners, 350
vertical alignment in masonry, 14
  risbond, 15

wall-floor interfaces
  ground floors, 62
  precautions, 62
walls, environmental control, 75
  air infiltration, 77
  expansion joints, 99
  fire, 79
  heat loss and thermal capacity, 75
  noise control, 79
  resistance to weather, precipitation, 75
walls
  general, 73
   dimensional stability, 79
   insulation of external walls, 84
   requirements, 74
   support of masonry walls from floors,
    123
  mutual, 228
   calculation of surface density, 228
   fire resistance, 231
   transmission of sound, 228
   wall types, 229
  openings, 126
   alternative lintelling, 134
   alternative sills, 136
   cold bridging, 132
   large openings in masonry walls, 127
   for larger pipes and ventilators, 127
   in partitions of masonry, 139
   in rainscreen cladding, 142
   for small pipes and cables, 126
   threshold arrangements, 137
   in timber frame walls, 141
weather for building, 32
weeps, 347
wood screws, *see* screw nails

woodworking, 298
  circular saws, 299
  plunging router, 300
  stress grading, 310
  stress grading machines, 311
  stress grading marks, 312
  timber
    barefaced mortice and tenon joint, 306
    blind mortice and tenon joint with foxtail
        wedges, 307
    butt joint, 304
    cogging, 310
    dovetailing, 303, 310
    draw boring mortice and tenon joints,
        308
    finger joints 1 and 2, 305
    grooving, 301
    half checked or halved joint, 304
    halving, 301
    joining, jointing or housing, 304

morticing, 302
moulding, 303
notching, 301
operations on, 298
plain housing, 305
plain mortice and tenon joint, 306
rebating, 301
scarfed joints, 308
scarfed joint variations 1, 309
scarfed joint variations 2, 309
shouldered housing, 305
shouldered mortice and tenon joint, 306
single and double notchings, 309
splaying, 303
tenoning, 302
toe joint, 308
tonguing, 301
trenching, 301
tusk tenoned joint, 307
using a pressed steel connector, 310